Springer

Berlin
Heidelberg
New York
Hong Kong
London
Milan
Paris
Tokyo

Jian-Gou Ma (Ed.)

Third Generation Communication Systems

Future Developments and Advanced Topics

With 142 Figures and 14 Tables

Editor:

Dr.-Ing. Jian-Guo Ma
Associate Professor & Research Coordinator
Division for Integrated Circuits & Systems
School of Electrical and Electronic Engineering
Nanyang Technological University
Nanyang Avenue
639798 Singapore
Singapore

E-mail: ejgma@ntu.edu.sg

ISBN 3-540-43806-8 Springer-Verlag Berlin Heidelberg New York

Cataloging-in-Publication Data applied for.
Bibliographic information published by Die Deutsche Bibliothek. Die Deutsche Bibliothek lists this publication in the Deutsche Nationalbibliografie; detailed bibliographic data is available in the Internet at <http://dnb.ddb.de>.

Springer-Verlag is a part of Springer Science+Business Media

http://www.springeronline.com

Printed in Germany

Typesetting: Dataconversion by author
Cover-design: medio, Berlin
Printed on acid-free paper 62 / 3020 hu – 5 4 3 2 1 0

Preface

The wireless communication revolution is bringing fundamental changes to both communications and society. Undisputedly, this growth will continue and evolve, particularly because of the rapid growth of personal mobile communications, which is one of the hottest growth areas of the twenty-first century.

At the start of 1990s, there were just over 10 million mobile cellular telephone users around the world. At the end of 2001 this figure reached over 905 million, or has grown nearly one hundred-fold. The mobile phone population shows little sign of slowing as the number of subscribers added grows each year – there were 234 million new mobile subscribers in 2000 alone. Growth has been steady, at an average of 50% per year since 1996. The solidification of the ultimate vision has been realized: inexpensive handsets that can communicate by voice, messages, etc., regardless of location, especially through the introduction of second-generation networks such as GSM.

The market has headed toward the ability to communicate by data, pictures, and videos in addition to voice, and even the ability to access the Internet through a single cellular phone. The technology in support of the new challenges is known as third-generation (3G) cellular telephony. Will third-generation networks, launched in 2002, also push up mobile growth? Can the third-generation mobile communication systems satisfy the requirements for variable speed, latency, and connection? What lies beyond 3G? Are these technologies safe? How will the possible hardware support to the market requirements?

The topics of this book are covered in a relatively basic manner, to address the trends and the advanced developments in wireless technologies. Chapter 1 provides an overview of the wireless technology towards IMT-2000 and its standards and explains the underlying acronyms. Wireless can be divided into the categories fixed wireless, short-range wireless, and mobile wireless. Many issues about third-generation mobile communications such as the over expectations, the financial burden, and its compatibility with the existing mobile networks and so on are covered.

Chapter 2 looks at mobility management in heterogeneous networks and how different protocols could be used to provide a complete solution to both terminal and session mobility. Mobile Internet protocol (IP) and Session initiation protocol (SIP) are used together to provide terminal mobility for both real-time and non-real time applications. This is one view of a fourth-generation (4G) network – with an IP core and multiple access technologies using IP mobility management for vertical handover.

Wireless mobile devices will become the main form of communications in the next decade. Support of this mobility will require rethinking of several aspects of the interworking of all existing and emerging fixed and mobile networks. Enabling mobility in IP-based networks is an important issue for making use of the many devices appearing in the market. Predicting the right time to perform handovers is another important issue in mobility management. Application and link-layer mechanisms could be used to predict handover events and provide seamless vertical handovers. However, unplanned handovers cause session disruption and packet loss. These types of handovers are unpredictable and less frequent. All these issues are presented in Chapter 2.

Chapter 3 describes some issues stemming directly or indirectly from concerns of the public about the safety of radio frequency (RF) energy used by wireless technology. Several issues related to setting exposure limits for RF energy from wireless communications and other technologies are considered, and implications for industries are presented. The wireless industry needs to educate its personnel in ways to communicate honestly and effectively with the public about these very sensitive issues. The potential consequences of public fears in the political and legal arenas are significant. Health concerns, even if unsupported by scientific evidence, can limit the potential usefulness of the technology as surely as channel capacity and propagation losses.

The following parts of the book discuss some “hard” engineering problems of design and implementations of the technologies. Chapter 4 reports an all-planar millimeter-wave integration approach for wireless local area network (WLAN) video transceiver modules. The concept of making millimeter-wave RF front-end modules using planar guiding structures and printed-circuit-board technology is introduced and its potentials is explored. Design examples are also given in the chapter.

Chapter 5 discusses the multiple access (MA) schemes underlying the air interface. Some basic MA schemes are briefly covered. The forthcoming 3G mobile systems apply digital transmission technologies, which allow the use of different hybrid MA schemes. Time-slotted CDMA has been selected as a key technology for the air interfaces of the 3G mobile communication systems by the Third-Generation Partnership Project (3GPP). The basics of time-slotted CDMA, including mathematical descriptions, are presented. Applications of adaptive antenna and advanced joint-detection (JD) algorithms as well the mitigation of intercell multi-access interface (MAI) are investigated in detail.

In recent years, there have been a wide variety of needs for communications. To meet such needs, a complex and flexible network topology is required. The technology of microwave photonics will give us a promising solution concerning this issue. Microwave photonics integrates both the merits of microwaves as a medium propagating in the atmosphere and light waves as a medium transmitting in a flexible optical-fiber cable. Microwaves give us a movable wireless link with very low cost, whereas fiber optics give us a low-loss and very broadband link, free from electrical interference and fading. In Chapter 6 the characteristic features, applications, and fundamental problems and technologies of microwave photonics are discussed.

All the technologies and wireless protocols mentioned above lead to a great demand on RFIC to physically realize the systems. For commercial applications the cost is always the first consideration. As advanced complementary metal-oxide semiconductor (CMOS) technologies are introduced, the unit-gain frequency f_T is increasing, and it is now sufficient for RF applications. Compared to the other parts of a wireless system, RF circuits contain very few transistors. However, they are one of the three key technoligies of a wireless system, namely RFIC, baseband IC, and software. Although RFICs are simpler than traditional IC designs in terms of numbers of transistors, RFIC designs are more difficult than traditional IC designs. RFIC designs require multidisciplinary knowledge of the trade-off among the many contrasting conditions. The design considerations from the system-level point of view are covered in Chapter 7.

In summary, this book provides detailed information on the future developments and the advanced topics of third-generation mobile communication systems from the system protocol to the implementation technologies. This book can be used as a reference for engineers involved in wireless communications to gain an overview of the future developments.

I want to most warmly thank all the authors who have contributed to this book and have spent their valuable time preparing the manuscripts. Many thanks to Dieter Merkle and Heather King of Springer-Verlag, who have been very patient and supportive throughout the development of the book.

Nanyang Meadows, Singapore *Jian-Guo Ma*
Summer 2003

Contents

1 Wireless Technology Towards IMT-2000

1.1 Introduction

At the start of the last decade, there were just over 10 million mobile cellular telephone users around the world. By the end of 2001, this figure had grown by almost 100 times to over 955.5 million, or one mobile phone for every six inhabitants. The mobile phone population shows little sign of slowing as the number of subscribers added grows each year – there were 234 million new mobile subscribers in the year 2000. Growth has been steady at an average of 50% per year since 1996. The introduction of second-generation networks such as global system for mobile communication (GSM) sparked an increase in mobile growth. Will third-generation networks, planned to be launched this year, also push up mobile growth? At current growth rates, the number of mobile subscribers will surpass that of fixed telephones by the middle of this decade.

There are 35 markets – both developed and developing – where this transition has already taken place. The mobile phone is becoming away of life for many, transcending the limitations of fixed telephones. One phenomenon is that fixed household telephone penetration is holding steady or even dropping as users opt for mobiles.

In Finland, the percentage of households with a fixed telephone has dropped from 94 to 83% over the last ten years, while that of mobile phones has increased from 7 to 60%. In developing countries, competition and prepaid cards are proving a potent combination for driving mobile growth. The rise of mobile phones in developing countries is perhaps most powerfully conveyed in the fact that, based on current growth, China will surpass the United States as the world's largest cellular market sometime in 2002. Unlike fixed telephone penetration, which generally peaks at around one telephone for every two inhabitants, mobile penetration has not yet reached an upper limit. The highest mobile penetration is found in Taiwan ROC.

Ten years ago, there were fewer than 100,000 mobile phones in Taiwan; today four out of every five Taiwanese has one. Mobile penetration increased by 25% in 2000 among the top ten most wireless economies. At this rate, most adults will soon have at least one mobile phone.

Mobile telephones used to be an exotic extravagance. Having become a mainstream voice communication medium, they are poised to take on new challenges, transmitting high-speed data, video, and multimedia traffic as well as voice signals to users on the move.

The technology in support of the new challenges is known as third-generation cellular telephony, IMT-2000. The early analog cell phones are labeled the first generation, and similar systems featuring digital radio technologies are labeled the second-generation.

This chapter discusses the wireless technology supporting IMT-2000 and its standards and explain the acronyms underlying. Wireless technology can be divided into "fixed wireless", "short-range wireless" and "mobile wireless".

1.1.1 Fixed Wireless

Fixed wireless refers to a system like local multipoint distribution services (LMDS) dish receiver/transmitter on a roof, where the active part of the network is fixed, but the transmission path is not. The signal in LMDS is sent through the air using radio frequency (RF), not through a cable. Once the receiver receives a signal, it can be sent into homes/offices over wires or by wireless means.

It is often uneconomical to lay fiber to a building, and the other terrestrial options, including digital subscriber lines (DSL) and cable modems, do not deliver equivalent bandwidth and are unavailable in many locations. Under the circumstances, broadband wireless might turn out to be the best solution. It can be cheaper and faster to install than fiber, and it can deliver high bandwidths, as much as 10 Gbps in exceptional cases. Like everything else, though, broadband wireless has its downfalls.

Some of the technologies seem theoretically possible but have to be proved in practice. Some are prone to interruptions in transmission from rain, fog, birds, airplanes, or swaying buildings. In addition, there is a host of other requirements such as needing a line of sight between buildings, licensing requirements, manufacturing challenges and the absence of standards.

Fixed wireless can also be divided into "free space optics", also known as optical wireless, "local multipoint distribution service", "multichannel multipoint distribution service" and unlicensed bands.

Free Space Optics

Using low-powered infrared lasers, vendors have created two types of products carrying data over light. The first are point-to-point products used to provide high-speed connections between two buildings. The second type provides multiple high-speed connections through the air at much shorter distances, either in a point-to-multipoint fashion, or in a meshed architecture.

The point-to-point technology evolves first and has been most often used in large campus local area networks (LANs), but more recently service providers have become interested in the higher capacities that are available. In these scenarios, rooftop lasers are aimed at receivers within their line of site, and optical data connections are established through the air. Under ideal atmospheric conditions, these products can reach somewhere between 2 and 4 km with data rates between 155 and 622 Mbps.

The first issue for any free-space optical system is fog. The frequency at which the signal travels is very high, which means that the wavelength is very short (in nanometers). While rain or snow can also disrupt service, fog is a bigger issue because the moisture particles are so small and dense that they act like millions of tiny prisms, and when a band of light shines through them the signal distorts and dissipates. This is most acute as the light beam fades and spreads out over longer distances. For safety reasons, high-power lasers cannot be used to push through the fog particles. To solve this problem, wireless optical vendors recommend that in foggy cities transmitters be positioned closer together, while in less foggy areas the distances can be increased.

The second issue has to do with the natural movements of buildings. Although we are rarely aware of it, buildings often sway from side to side or even settle farther into the ground. This is a problem, because for free-space optics to work, the lasers and receivers must be lined up exactly. Several of the vendors say they have auto-tracking capabilities that automatically move the lasers back into position if they detect a change in alignment.

The third issue has to do with objects that could obstruct laser beams. The most notable instance is the flying pigeon scenario. For a strict point-to-point or hub-and-spoke configuration that only provides one path for traffic, this can be a problem. Nevertheless, for meshed systems, like that from AirFiber (USA), there is always an alternate path.

Local Multipoint Distribution Service (LMDS)

While LMDS does not provide the same capacity as fiber, it does provide sufficient bandwidth for most business customers. Because it operates at a higher frequency than MMDS or many of the unlicensed bands, it provides higher capacity than these systems. Its distance capabilities are limited, but its minimum 2 km reach in any direction makes it ideal for businesses in metropolitan area networks (MAN).

While point-to-point LMDS products have been available for a while and are used by carriers to create fixed links to aggregate traffic, many of vendors have developed point-to-multipoint LMDS/millimeter radios and receivers. In a typical architecture, a carrier divides the area surrounding the central base station into sectors, each sector getting a particular amount of bandwidth that is shared across its subscribers.

Customers are usually allowed to burst to certain rates. Because it is a fixed connection, there must be line-of-site access between the base station and the customer antenna. Buying a license is expensive, and since the USA Federal Communications Commission (FCC) last auctioned off the LMDS frequencies in 1999, there are few left in the major markets. This greatly limits the number of new entrants, but discussions are underway to open more frequencies above 39 GHz.

Because high-frequency radio components are difficult to manufacture in volume, vendors are not able to meet price points that make deploying the technology attractive. Therefore, there have not been many deployments of LMDS networks, particularly in the USA. The customer-premises equipment like Ethernet interfaces are cheap and easy to make; however, the radio parts for higher frequencies are more exacting and more expensive.

Compounding the problem is the fact that governments doling out frequencies in different parts of the world allocate different spectra. For example, most licenses in the USA center around 28 GHz and 31 GHz, as well as the 24 GHz and the 39 GHz bands. In Europe most licenses center around 26 GHz, while in the UK most center around 10 GHz. Therefore radios made for use in one country cannot be used in another, which means vendors have to develop separate products for different regions.

The reach of LMDS equipment is limited by rain fade, which is distortion of the signal caused by raindrops scattering and absorbing the millimeter waves. Because the waves have high amplitudes, walls, hills and even leafy trees can block, reflect and distort the signal. Decreasing the distance between the transmitter and the customer helps keep the signal strong enough to handle these issues, according to vendors.

Another issue affecting LMDS is the lack of agreement on standards. One of the biggest disputes still pending revolves around frequency-division duplexing (FDD) versus time-division duplexing (TDD). FDD uses two channels, one for receiving data and one for transmitting it. TDD uses one channel for both transmitting and receiving, scheduling time slots for each.

Multichannel Multipoint Distribution Service (MMDS)

Multichannel multipoint distribution service (MMDS) was originally licensed for one-way video transmission to provide an alternative to cable television. After several service providers failed to compete with satellite-based video programming providers, the FCC ruled in 1998 to allow MMDS to be used for bidirectional services. The long reach and low throughput speeds of MMDS have made it a good match for residential and rural applications. Over the last eighteen months, Sprint and MCI/Worldcom have bought several MMDS license-holding providers and are poised to emerge as the leaders in MMDS voice, video and data service.

Because MMDS services operate at such low frequencies, heavy rain and fog, which affect transmission for LMDS and optical wireless services, do not affect MMDS. Instead, the biggest problem for MMDS is maintaining direct line of sight between the transmitter and receiver. Since dense tree coverage, buildings or mountains can interfere with transmission, more transmitters and antennae are needed to overcome obstructions. Multipath distortion, which results from signal reflections off buildings or other structures, can also cause problems.

MMDS advocates say these impediments can be overcome by dispersing a number of transmitters throughout a market in a cellular architecture and by using an advanced modulation solution, like orthogonal frequency-division multiplexing (OFDM) technology. Essentially, OFDM allows more data to be transmitted at high speeds than traditional networks, while also allowing the signal to transmit through some obstacles. Cisco Systems (USA) has developed its own twist on the technology, which it calls vector OFDM. Using this technology, the Cisco MMDS product can support 22 Mbps per downstream 6 MHz channel and up to 18 Mbps upstream.

While LMDS and optical wireless require expensive components such as high-frequency radio parts or lasers, MMDS technology gear is inexpensive to assemble because the price point is lower, it better suited for small businesses and residential use. LMDS is appropriate for larger customers, who are willing and able to pay a premium to build the necessary infrastructure. One of the biggest challenges for any fixed wireless solution is getting the price points low enough to compete against fiber, DSL, cable and other technologies.

1.1.2 Short Range Wireless

The use of devices in certain limited-range communications and control systems is referred to as short-range wireless. The main technology used in short-range wireless is Bluetooth, but it is seriously hobbled by a lack of standardized code, which means that devices of different brands often cannot communicate with each other. This is a large flaw for a technology hailed as the next step in computer interconnectivity.

A new standardized version of Bluetooth has been developed, but the first device using it will not be ready in the near future, and there is no guarantee that existing Bluetooth devices will be compatible with the new version. The technology is seen as ideal for short-range connections compared with infrared beams, which require a direct line of sight between the devices and cannot travel as far as Bluetooth radio waves.

The challenge, however, is making sure all Bluetooth products can communicate with each other. Right now, the standard is defined, but companies are using different specifications. To make matters worse, Bluetooth cards for home computer are too expensive, and it may not be compatible with the devices people want to use.

1.1.3 Mobile Wireless: Today

Mobile wireless is the operation of autonomous, battery-powered wireless devices or systems outside the office, home or vehicle, for example, the mobile phone. The success of mobile wireless lies in the ability to provide instant connectivity anytime and anywhere and the ability to provide high-speed services to the mobile user: first voice, then simple text and now data. Ultimately, the goal is complex data like video streaming, and so on.

The quality and speeds available in the mobile environment must match the fixed networks if the convergence of the mobile wireless and fixed communication networks is to happen. The challenge for the mobile networks lies in providing both a very large footprint of mobile services to make the movement from one network to another as transparent to the user as possible, and reliable high-speed data services along with high-quality voice.

The mobile wireless market has been voice-oriented with low-speed data services slower than a fixed 56 kbps modem. The popularity of mobile voice services has been the deciding factor for the development of mobile networks so far. Data, mainly in the form of short message service (SMS), was simply an add- on, which was supposed to be a business tool but emerged as a "killer application", with many European subscribers spending more on SMS than voice.

Both the voice and data markets continue to grow, and the second-generation networks are evolving to keep up and, in fact, are generating demands for newer services. Unfortunately, until serious network speeds are experienced, applications are evolving faster than networks, creating hype, forced rollouts of technologies and ultimately disappointment, which is potentially damaging to all players in mobile networks.

A typical cellular systems as shown in Fig. 1.1 may be divided conceptually into five parts: radios, switching systems, databases, processing centers and external networks. Often, though, multiple parts will be realized in a single physical entity – a database combined with a switching system.

The mobile connection of a cellular subscriber and a fixed telecommunications network is, of course, realized with radios. The mobile station (MS) is usually a small handset these days. Its corresponding base site radio, or base transceiver station (BTS), works through a base station controller (BSC), which keeps the rest of the fixed network happily in the dark about the radio details of the mobile MS - BTS connection. Together, the BTS and BSC are called a base station (BS).

The first-generation analog cellular systems, such as advanced mobile phone system (AMPS), use narrowband FM radio techniques that need only 10–30 kHz

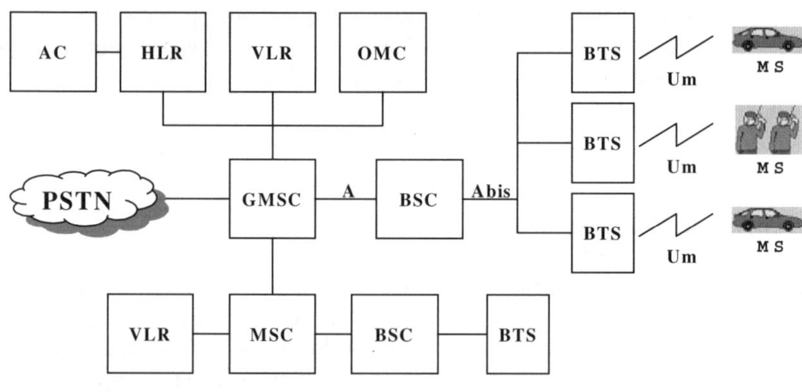

A : MSC + BSC interface
Abis : BSC + BTS interface
AC : Authentication Center
HLR : Home Location Register
OMC : Operation & Maintenance Center
GMSC : Gateway Mobile Switching Center
MSC : Mobile Switching Center
BSC : Base Station Controller
BTS : Base Transceiver System
Um : Air interface
VLR : Visited Location Register
PSTN : Public Switching Telephone Network
MS : Mobile Station

Fig. 1.1 In a typical cellular system, the base transceiver stations (BTS), one of which is at the center of each cell, are connected to base station controllers (BSC), which are in turn connected to mobile switching centers (MSC), and eventually to the public switched network (PSTN). Unique to mobile systems are location registers (HLR and VLR), which keep track of users on the move

of spectrum for each channel. Today, the variety of radio link technologies feeds the controversies in third-generation proposals. Since the radio link defines the base and mobile stations, a great deal is at stake. The arguments tend to mask continuing innovation in the fixed parts of mobile networks.

The switching function is a combination of computing platforms and transmission facilities that route user information and signaling among nodes throughout the mobile network. The functional entity is called the mobile switching center (MSC). The center, its attached base stations and any interworking functions to terrestrial networks or other kinds of networks are collectively called a mobile switching center, and an interworking function, BMI. The BMI is said to communicate with mobile stations over an air interface.

Several types of databases are queried by network entities while providing services to mobile subscribers. Location registers manage mobility and are unique to mobile networks. The home location register (HLR) permanently stores subscriber data relative to network intelligence, while the visitor location register (VLR) maintains temporary working copies of active subscribers in the network.

Peripheral computing platforms enhance the profitability of mobile networks. The authentication center (AC), for instance, performs functions that validate a mobile station's identity. Voice-announcement systems and message centers are other functions.

The Second Generation

GSM, TDMA (IS-136), and CDMA (IS-95) are the leading technologies in the second generation. All these systems have different features and capabilities. Although GSM and TDMA both use time-division multiplexing on the air interfaces, their channel sizes, structures and core networks are different. CDMA has an entirely different air interface.

GSM

GSM has 646.5 million subscribers worldwide (December 2001, EMC World Cellular Database). It uses air interface based on narrowband TDMA technology, where available frequency bands are divided into time slots, with each user having access to one time slot at regular intervals. Narrowband TDMA allows eight simultaneous communications on a single 200-kHz carrier and is designed to support 16 half-rate channels. The fundamental unit of time in this TDMA scheme is called a burst period, which lasts 15/26 ms. Eight burst periods are grouped into a TDMA frame (120/26 ms), which forms the basic unit for the definition of logical channels. One physical channel is one burst period per TDMA frame. A GSM mobile can seamlessly roam nationally and internationally, which requires that registration, authentication, call routing and location updating functions exist and be standardized in GSM networks.

GSM offers a variety of data services. GSM users can send and receive data, at rates up to 9.6 kbps, to users on plain old telephone service (POTS), integrated services digital network (ISDN), packet switched public data networks, and circuit switched public data networks. A unique feature of GSM, not found in older analogue systems, is SMS, a bidirectional service for short alphanumeric (up to 160 bytes) messages.

The European version of GSM (also used in Africa and parts of Asia - Pacific) operates at 900 MHz and 1800 MHz. Since the North American version of GSM operates at the 1900 MHz frequency, the phones are not interoperable. However, the SIMs are, so users can take SIM cards from one area to the other and simply rent phones. Dual-band 900–1800 phones exist and allow international roaming in Europe, Africa and Asia - Pacific. Tri-band 900 – 1800–1900 GSM phones allow international roaming anywhere agreements are in place.

GSM networks consist of mobile stations talking to the base transceiver station, on the Um interface. Many BTS are connected to a base station controller via the Abis interface, and the BSC connects to the mobile services switching center (the core switching network) via the A interface.

Home location register and visitor location register provide customized subscriber services and allow seamless movement from one cell to another. The authentication register and the equipment register provide security and authentication. An operation and measurement center and a cell broadcast center allow configuration of the network and provide the cell broadcast service in the GSM network. The GSM network supports automatic handovers.

Since the mobiles are not transmitting or receiving at all times battery consumption can be conserved. Further using discontinuous transmission and discontinuous reception (DTX and DRX, respectively), the mobile transmits or receives only when there is a voice activity detection, so power can be conserved even more, which is a highly desirable characteristic of any network. Also, since the mobile is not transmitting or receiving at all times, this allows the mobile to listen to control channels and to provide useful information about other channels back to the cell.

Recent Developments and Initiatives

Launched in June 2000 by the GSM Association, the Global Roaming Forum is an association of companies with roaming products and services that utilize GSM, iDEN, CDMA, TETRA and TDMA technologies. The forum's mission is to develop technical requirements for terminals, networking and smart cards, as well as commercial standards for services, billing, financial settlements and fraud management. The majority of European GSM operators plan to implement general packet radio service (GPRS) as their network evolution path to the third generation.

Mobile station application execution environment (MExE) will allow operators to provide customized, user-friendly interfaces to a host of services from GSM, through GPRS and eventually universal mobile telephone system (UMTS). The first implementations of MExE are expected to support wireless application protocol (WAP) and Java applications. MExE can extend the capabilities that currently exist within WAP by enabling a more flexible user interface, more powerful features and security.

Number portability will allow customers to retain their mobile numbers when they change operators or service providers. Location services will standardize the methods for determining a GSM subscriber's physical location.

TDMA (IS-136)

TDMA (IS-136) has 94.7 million subscribers worldwide (December 2001, EMC World Cellular Database). It is so named because frequency bands available to the network are divided into time slots, with each user having access to one time slot at regular intervals. Three users share a 30-kHz bandwidth by splitting a 30-kHz carrier into three time slots. IS-136, an evolved version of IS-54 (the original standard), is the US standard for both the cellular (850 MHz) and PCS (1.9 GHz) spectrums. TDMA (IS-136) utilizes time-division multiplexing for both voice and control channel transmissions.

Digital control channels allow residential and in-building coverage, dramatically increased battery standby time, several messaging applications, over the air activation and expanded data applications. GSM also has the same characteristics. The digital control channel allows for the creation of micro-cell applications, making it suitable for wireless PBX and paging applications. TDMA networks transmit at a higher data rate on a relatively low bandwidth channel, resulting in chances of co-channel interference. As described above for GSM, the time slot structure allows the mobiles to conserve battery power and to collect information about other channels.

TDMA (IS-136) exists in North America at both the 800 MHz and 1900 MHz bands. TDMA (IS-136) normally coexists with analog channels on the same network. One advantage of this dual-mode technology is that users can benefit from the broad coverage of established analog networks while TDMA (IS-136) coverage grows, and at the same time take advantage of the more advanced technology of TDMA (IS-136) where it exists.

IS-136 Revision A has introduced several new features such as adaptive channel allocation (ACA), depending on the instantaneous channel quality determined by the level of interference, the private system identification (PSID), which allows development of large-scale corporate private systems either as multilocation or in-building closed-user groups and two way SMS (256 chars). IS-136 Revision B standard includes all IS-136+ proposals from the UWC-136 RTT proposal for voice and circuit switched features. Notable features are packet data service, mobile assisted handoff, improved SMS and intelligent roaming.

CDMA (IS-95)

CDMA has 112.2 million subscribers worldwide (December 2001, EMC World Cellular Database). It is based on the IS-95 protocol standard first developed by Qualcomm (USA). CDMA differs from the other two technologies by its use of spread-spectrum techniques for transmitting voice or data over the air. Rather than dividing the RF spectrum into separate user channels by frequency slices or time slots, spread-spectrum technology separates users by assigning them digital codes within the same broad spectrum. Advantages of CDMA technology include high user capacity and immunity from interference by other signals. Like TDMA, CDMA operates in the 1900 MHz band as well as the 800 MHz band.

Work on developing the CDMA standard is conducted mainly by the CDMA Development Group (CDG). Whilst work to develop CDMA as a third-generation technology has attracted a great deal of attention over recent years, the CDG has also been working to improve the current performance of CDMA as a second-generation technology. The CDG has formally adopted the cdmaOne name and logo as a technology designator for all IS-95 based CDMA systems. The term represents the end-to-end wireless system and the necessary specifications that govern its operation. The designation cdmaOne incorporates the IS-95 air interface, the ANSI-41 network standard for switch interconnection and many other standards that make up a complete wireless system.

The CDMA (IS-95) maximizes spectrum efficiency and enables more calls to be carried over a single 1.25-MHz channel. In a CDMA system, each digitized voice is assigned a binary sequence that directs the proper response signal to the corresponding user. The receiver demodulates the signal using the appropriate code. The resulting audio signal will contain only the intended conversation, eliminating any background noise. This allows more calls to occupy the same space in the communication channel, thereby increasing capacity.

As a simple example, let us assume a user is talking into a mobile phone on a CDMA network. The transmitted portion of a voice signal has frequency components, approx. 300~3400 Hz. This analog signal is digitally encoded, using quadrature-phase shift keying (QPSK), at 9600 bps. The signal is then spread to approximately 1.23 Mbps using special codes that add redundancy. Some of these codes include a device identification that is unique to the phone, like a serial number. Next, the signal is broadcast over the channel. When broadcast, the signal is added to the signals of the other users in the channel. On the receiving end, the same code is used to decode the incoming signal. The 9600 bps signal is obtained, and the original analog signal is reconstructed. When the same code is used on another user's signal, the redundancy is not removed, and the signal remains at 1.23Mbps.

The problems are the quality of reception and voice squeakiness. To address this major personal communication service (PCS) carriers are using 13 kbps vocoders instead of 10 kbps. This improves quality at the cost of capacity. The technology has been widely adopted by major cellular and PCS carriers in the USA and internationally.

CDMA networks provide operators with reliable digital systems that offer higher capacity, large coverage area and improved voice quality and above all a good third-generation upgrade path, CDMA2000. It also offers simplified system planning - using the same frequency in every sector of every cell. Factors contributing to CDMA's capacity gains are:

- Frequency reuse,
- Soft handoffs,

- Power control,
- Variable rate vocoders.

Some of the benefits of using cdmaOne are:
- Capacity gains of eight to ten times that of AMPS analog systems;
- Improved call quality, with better and more consistent sound as compared to AMPS systems;
- Simplified cell planning through the use of the same frequency in every sector of every cell;
- Enhanced privacy through the spreading of voice signals;
- Improved coverage characteristics, allowing for fewer cell sites;
- Increased talk-time for portables.

The cdmaOne technology improves quality of service by soft handoffs, which greatly reduce the number of call drops and ensure a smooth transition between cells. In soft handoff, a connection is made to the new cell while maintaining the connection with the original cell. This transition between cells is one that is almost undetectable to the subscriber. The cdmaOne technology also takes advantage of multipath fading to enhance communications and voice quality. Using a rake receiver and other improved signal-processing techniques, each mobile station selects the three strongest multipath signals and coherently combines them to produce an enhanced signal.

The cdmaOne data capabilities are based on IS-95A, which can provide data speeds of 14.4 kbps. Standards IS-95B and IS-95C are designed to enhance the data capability of CDMA. IS-95B can provide data speeds of up to 64 kbps by aggregating existing channels. IS 95-B can provide these enhanced data rates through software upgrades only.

IS-95C aims to offer a minimum of 24.4 kbps per channel and aggregated data speeds of more than 115 kbps. It is expected that IS-95C will define CDMA's capability as a third-generation system. CDMA already supports asynchronous data and faxing (IS-99) and has standardized packet data (IS-657).

The major development initiatives being taken by the CDG for the second-generation CDMA systems enhancements include enhanced roaming, which enables transparent roaming across cellular and PCS networks, with selection of networks and location services. Enhanced roaming will provide roaming between CDMA systems similar to that on GSM: registration, authentication and credit checking are automatically carried out between the networks when users simply switch on their mobiles. Roaming agreements will still be needed between operators.

Mobile Wireless Market: Technology Forecasts

SK Telecom (Korea) and NTT DoCoMo (Japan) will roll out the third-generation network in 2001. Both are running test and evaluation programs for the global industry: will customers use it, what will they use it for and what are the applications that can start revenue streams like video, music, gaming and so on.

The third-generation technologies may not have any major impact until 2003, or possibly 2004, and then may coexist with the second-generation technologies for another 2 to 3 years before gaining prominence. The global forecast by technology is shown in Table 1.1.

Table 1.1 Global forecasts by technology (millions)

	2001	2002	2003	2004	2005
Total subscriptions	703.870	940.226	1180.694	1406.284	1616.125
GSM	406.605	566.176	730.449	875.608	991.706
CDMA	86.797	128.155	167.744	202.662	230.671
PDC	48.742	53.973	60.084	63.677	66.141
D-AMPS	81.569	130.691	182.444	228.707	274.182
The third generation		11	2.573	15.519	39.426

(Source: Ovum, Jan. 1. 2001)

Standards

The debate as to which standard is superior revolves around two basic underlying technologies, TDMA and CDMA. Time-division multiplex systems like GSM have the advantage of having very high market penetration, having been in operation for many years. In addition, they are easy and cheap to upgrade to packet-based data services, have almost global roaming, because of recent developments towards roaming between TDMA and GSM networks and the possibility of a common air interface and subsequent evolution towards the third-generation after implementing enhanced data for global evolution (EDGE).

These technologies have evolved and matured successfully over many years of existence. The proponents of TDMA claim that their networks are more rugged compared to the CDMA technology, which, in their view, is not mature and suffers from problems like deteriorating speech quality with increases in traffic load and voice squeakiness. CDMA proponents, on the other hand, claim substantial improvements in capacity, security and speech quality. They claim that many more users can be supported over the same bandwidth in CDMA as compared to TDMA. The number of users is fixed in case of TDMA networks. Another claim from CDMA proponents is the comparatively much clearer, simpler and more well-defined evolutionary path to the third generation. The third-generation networks use CDMA in the air interface, and so CDMA networks like IS-95 have an advantage in evolving towards the third generation. In addition, although the evolutionary path to the second-and-a-half generation may be cheaper for GSM/TDMA, the total costs involved in moving to the third generation will be substantially higher in the case of GSM/TDMA.

The basic third-generation standards were developed largely by the private sector rather than by formal standards organizations. However, the International Telecommunication Union has adopted International Mobile Telecommunications 2000 or IMT-2000 to formally standardize the already developed versions of third-generation wireless, to let service providers offer a consistent set of mobile telephone services throughout the world and to provide a roadmap for upgrades.

Most of Europe and Japan have settled on W-CDMA, an upgrade to the GSM standard widely used in those areas. Korea is still debating on both W-CDMA and CDMA2000. TD-CDMA probably will be used only in China, where the specification was developed. The USA is working with all three major third-generation standards. When a region adopts a single standard, users can roam with the technology throughout the area and there is no splintering of support and usage. However, others say multiple standards let users adopt the technology with which they prefer to work and determine which, if any, should become dominant.

CDMA over TDMA

A CDMA system uses a combination of frequency division and code division to provide multiple-user access. Although the capacity of a CDMA system is not unlimited, its limitations are considerably higher than those of a TDMA system. It can provide 8~10 times more users than traditional FDMA/CDMA. The advantages of a CDMA system are:

- Because a number of transmissions are possible over the same bandwidth, the frequency reuse in CDMA networks can be very high;
- Better signal quality;
- Privacy of coded digital communications;
- Easy addition of more users, however, there exists a “soft” capacity limit since additional users add more noise to the cell.

If CDMA is the superior technology then why do the figures in the market forecast section indicate otherwise? Why are service providers going towards a supposedly inferior technology? For any technology to succeed in the market there are two essential requirements, the capacity of the system and the ease to implement the system. A major concern about CDMA was that it had very little field experience, whereas TDMA systems had been operable for quite some time.

Fast time to market is essential to companies due to the phenomenal pace of today's wireless communications market, and many providers choose to invest in TDMA systems that have already been developed and proven. The very fact that CDMA technology is so new and unproven makes investment very risk for companies when there is the option of going with a time-proven technology such as TDMA. In addition, existing service providers who have already spent heavily on TDMA networks would like to protect their investments. Continuous developments and improvements in the GSM networks, for example, have added value to their investments and have so far been able to keep up with the demands for higher capacities and data applications.

Therefore, the share of these technologies in the second-generation/the second-and-a-half generation market will remain almost the same, while the demand and numbers continue to increase. The CDMA technology will have a much larger influence by means of the third-generation technologies over the next 5-10 years.

1.1.4 Mobile Wireless: Third-Generation and Beyond

The third-generation systems aim to provide enhanced voice, text and data services to user. The main benefit will be substantially enhanced capacity, quality and data rates than are currently available. This will enable the transparent provision of advanced services to the end user (irrespective of the underlying network and technology, by means of seamless roaming between different networks). It will also bridge the gap between the wireless world and the computing/Internet world, making interoperation apparently seamless. The third-generation networks should be in a position to support real-time video, high-speed multimedia and wireless Internet access.

All this should be possible by means of highly evolved air interfaces, packet core networks and increased availability of spectrum. Although the ability to provide high-speed data is one of the key features of the third-generation networks, the real strength of these networks could be providing enhanced capacity for high-quality voice services. The need for landline-quality voice capacity is increasing more rapidly than the current second-generation networks will be able to support. High data capacities will open new revenue sources for the operators and bring the Internet closer to the mobile customer.

Technologies like "general packet radio service (GPRS)", "high-speed circuit-switched data (HSCSD)" and "enhanced data for global evolution (EDGE)" fulfill the requirements for packet data service and increased data rates in the existing GSM/TDMA networks. GPRS is actually an overlay over the existing GSM network, providing packet data services using the same air interface by the addition of two new network elements, the "serving GPRS support node (SGSN)" and "gateway GPRS support node (GGSN), and a software upgrade. Although GPRS was designed for GSM networks, the TDMA will also support GPRS. This follows an agreement to follow the same evolution path towards the third-generation mobile phone networks concluded in early 1999 by the industry associations that support these two network types.

General Packet Radio Service (GPRS)

General packet radio service (GPRS) is a wireless service that is designed to provide a foundation for a number of data services based on packet transmission. Customers will only be charged for the communication resources they use. The operator's most valuable resource, the radio spectrum, can be leveraged over multiple users simultaneously because it can support many more data users. Additionally, more than one time slot can be used by a customer to get higher data rates.

The SGSN is the same hierarchical level as an MSC. The SGSN tracks packet-capable mobile locations, performs security functions and maintains access control. The SGSN is connected to the BSS via frame relay.

The GGSN interfaces with external packet data networks (PDN) to provide the routing destination for data to be delivered to the MS and to send mobile- originated data to its intended destination. The GGSN is designed to provide interworking with external packet-switched networks, and is connected with SGSN via an IP-based GPRS backbone network.

A packet control unit is also required and may be placed at the BTS or at the BSC. A number of new interfaces have been defined between the existing network elements and the new elements, and between the new network elements. Theoretical maximum speeds of up to 171.2 kbps are achievable with GPRS using all eight timeslots at the same time. This is about 3 times as fast as the data transmission speeds possible over today's fixed telecommunications networks and 10 times as fast as current circuit-switched data services on GSM networks. Actually, we may not see speeds greater than 64 kbps, however, they will be much higher than the speeds possible in any second-generation network. In addition, another advantage is that the users are always connected and are charged only for data transferred and not for the time connected to the network.

Packet switching means that GPRS radio resources are used only when users are actually sending or receiving data. Rather than dedicating a radio channel to a mobile data user for a fixed period, the available radio resource can be concurrently shared between several users. This efficient use of scarce radio resources means that large numbers of GPRS users can potentially share the same bandwidth and be served from a single cell. The actual number of users supported depends on the application being used and how much data is being transferred. Because of the spectrum efficiency of GPRS, there is less need to build in idle capacity that is only used in peak hours.

High-Speed Circuit-Switched Data (HSCSD)

High-speed circuit-switched data (HSCSD) is the evolution of circuit-switched data within the GSM environment. The use of HSCSD will enable the transmission of data over a GSM link at speeds of up to 57.6 kbps. This is achieved by adding together consecutive GSM timeslots, each of which is capable of supporting 14.4 kbps. Up to four GSM timeslots are needed for the transmission of HSCSD, which allows theoretical speeds of up to 57.6 kbps. This is broadly equivalent to providing the same transmission rate as that available over one ISDN B-channel.

HSCSD is part of the planned evolution of the GSM specification and is included in the development of GSM Phase 2. In using HSCSD, a permanent connection is established between the called and calling parties for the exchange of data. As it is circuit switched, HSCSD is more suited to applications such as video conferencing and multimedia than "bursty" type applications such as e-mail, which is more suited to packet-switched data. In networks where HSCSD is deployed, GPRS may only be assigned third priority, after voice (first priority) and HSCSD (second priority).

In theory, HSCSD can be preempted by voice calls, such that HSCSD calls can be reduced to one channel if voice calls are seeking to occupy these channels. HSCSD does not disrupt voice service availability, but it does affect GPRS. Even given preemption, it is difficult to see how HSCSD can be deployed in busy networks and still confer an agreeable user experience, i.e. continuously high data rates. HSCSD is therefore more likely to be deployed in start-up networks or those with plenty of spare capacity, since it is relatively inexpensive to deploy and can turn some spare channels into revenue streams.

An advantage for HSCSD could be that while GPRS is complementary for communicating with other packet-based networks such as the Internet, HSCSD could be the best way of communicating with other circuit-switched communications media such as the PSTN and ISDN. However, one potential technical difficulty with HSCSD arises because in a multitimeslot environment, dynamic call transfer between different cells on a mobile network (handover) is complicated unless the same slots are available end-to-end throughout the duration of the circuit-switched data call.

Because of the these technologies, evolution, the market need for HSCSD and the market response to GPRS, the mobile infrastructure vendors are not as committed to HSCSD as they are to GPRS. So, we may only see HSCSD in isolated networks around the world. HSCSD may be used by operators with enough capacity to offer it at lower prices.

Enhanced Data for Global Evolution (EDGE)

Enhanced data for global evolution (EDGE) is a high-speed mobile data standard to enable GSM and TDMA networks to transmit data at up to 384 kbps. It is the technology for network operators who failed to win spectrum auctions for the third-generation networks to allow high-speed data transmission. EDGE provides speed enhancements by changing the type of modulation and making better use of the carrier, which is 200 kHz in GSM systems. EDGE also provides an evolutionary path to the third-generation, IMT-2000 compliant systems, such as universal mobile telephone systems (UMTS), by implementing some of the changes expected in the later implementation in the third-generation systems.

EDGE builds upon enhancements provided by GPRS and HSCSD technologies that are currently being tested and deployed. It enables a greater data transmission speed to be achieved in good conditions, especially near the base stations, by implementing eight-phase-shift keying (8PSK) modulation instead of Gaussian minimum-shift keying (GMSK). GPRS is based on modulation by GMSK.

The GMSK modulation technique does not allow as high a bit rate across the air interfaces as 8PSK modulation, which was introduced in EDGE systems. 8PSK modulation automatically adapts to local radio conditions, offering the fastest transfer rates near the base stations in good conditions. It offers up to 48 kbps per channel, compared to 14 kbps per channel with GPRS and 9.6 kbps per channel for GSM. By also allowing simultaneous use of multiple channels, the technology allows rates of 384 kbps using all eight GSM channels.

For EDGE to be effective it needs to be installed along with the packet-switching upgrades used for GPRS. This entails the addition of two types of nodes to the network: GGSN and SGSN. The GGSN connects to packet-switched networks such as internet protocol (IPl) and X.25, along with other GPRS networks, while the SGSN provides the packet-switched link to mobile stations.

The additional implementation of EDGE systems requires only one EDGE transceiver unit to be added to each cell, with the base stations receiving remote software upgrades. EDGE can coexist with the existing GSM traffic, switching to EDGE mode automatically.

Because the basic infrastructure interfaces with the existing GPRS, GSM or TDMA infrastructure, the major vendors are the incumbent GPRS and GSM suppliers such as Ericsson, Nokia, Motorola and Alcatel. By providing an upgrade route for GSM/GPRS and TDMA networks, EDGE forms part of the evolution to IMT-2000 systems. Since GPRS is already being deployed, and IMT-2000 is only expected by 2002, there is a definite window of opportunity for EDGE systems to fill in as a stopgap measure.

Standardization of the third-generation mobile systems is based on ITU recommendations for IMT-2000. IMT-2000 specifies a set of requirements that must be achieved fully for a network to be called the third generation. By providing multimedia capacities and higher data rates, these systems will enhance the range and quality of services provided by the second-generation systems.

The main contenders for the third-generation systems are wideband CDMA (W-CDMA) and CDMA2000. The ETSI/GSM players, including infrastructure vendors such as Nokia and Ericsson, backed W-CDMA. The North American CDMA community, led by the CDMA Development Group, backed CDMA2000. Universal mobile telephone system (UMTS) is the widely used European name for the third generation.

The proposed IMT-2000 standard for global third-generation mobile networks is a CDMA-based standard that encompasses three optional modes of operation, each of which should be able to work over both GSM and CDMA network architectures. The three modes are shown in Table 1.2.

Table 1.2 The three modes in the third-generation standards

Mode	Title	Based on	Supporters
1	Direct-sequence frequency- division duplex	The first operational mode of ETSI's UMTS terrestrial radio access RTT proposal	Japan's ARIB and GSM network operators and vendors
2	Multi-carrier frequency-division duplex	The CDMA2000 RTT proposal from the US Telecommunications Industry Association	cdmaOne operators and members of the CDMA Development Group
3	Time-division duplex	The second operational mode of ETSI's UMTS terrestrial radio access RTT proposal. Unpaired band solutions to better facilitate indoor cordless communications	Harmonized with China's TD-SCDMA RTT proposal

The UMTS frequency bands selected by the ITU are 1885–2025MHz (Tx) and 2110 MHz–2220 MHz (Rx). Higher frequency bands could be added in the future, if necessary, for stationary data. There is still some confusion about all the frequency options, as the US FCC has not yet given clear indications.

The Third-Generation Market

Wireless Internet access is high on the priority lists of major wireless carriers. NTT DoCoMo's i-mode service in Japan already has over 20 million users. It is important to understand that wireline data technologies are advancing very quickly and will support very high data rates at very low costs, which would be prohibitive with foreseeable wireless technology.

The Third-Generation Timeframes

The actual deployment of the third generation will not be a homogeneous occurrence. Japan will lead with the service in May 2001, followed by Western Europe in late 2002. The USA is expected to wait for some time at the second-and-a-half generation before rolling out the third-generation, and is concentrating more on fixed wireless access than mobile access.

The Third-Generation Architecture

The third-generation networks will have a layered architecture, which will enable the efficient delivery of voice and data services. A layered network architecture, coupled with standardized open interfaces, will make it possible for the network operators to introduce and roll out new services quickly. These networks will have a connectivity layer at the bottom, providing support for high quality voice and data delivery. Using IP or asynchronous transfer mode (ATM) or a combination of both, this layer will handle all data and voice information. The layer consists of the core network equipment like routers, ATM switches and transmission equipment. Other equipment provides support for the core bit stream of voice or data, providing quality of service.

Note that in the third-generation networks, voice and data will not be treated separately which could lead to a reduction in the operational costs. The application layer on top will provide open application service interfaces, enabling flexible service creation. This user application layer will contain services for which the end user will be willing to pay. These services will include electronic commerce, global positioning system (GPS) and other differentiating services. In between the application layer and the connectivity layer will run the control layer with MSC servers, support servers, HLR and so on. These servers are needed to provide any service to a subscriber.

Migration Strategies

The migration to the third generation is not just based on evolving core networks and the radio interface to IMT-2000 compliant systems. Migration towards the third generation also is based on the following steps/technologies:

- Network upgrades in the form of EDGE, GPRS and HSCSD, to provide support for high-speed packet data;

- The development of mobile Web portals;
- Development of microbrowsers and operating systems;
- Analysis of the NTT DoCoMo experience in 2000, what works, what doesn't work and why.

GSM and TDMA to the Third-Generation

GSM and TDMA systems have more or less the same set of options for migrating to the third generation. The path to the third-generation is not as simple in the case of GSM/TDMA as it is in the case of CDMA. The main evolutionary standards are GPRS, EDGE and, finally, W-CDMA. Vendors are positioning each of these standards as a step to the next standard, but operators are not as certain. Operators moving from GSM to GPRS to EDGE and then to W-CDMA will have to make three separate investments, would become expensive. At this time, there seem to be four basic options that GSM and TDMA operators are considering:

- Install GPRS, then move straight to W-CDMA;
- Install EDGE, then move straight to W-CDMA;
- Install GPRS, then move to EDGE, then to W-CDMA; or
- Install EDGE, skip the move to W-CDMA, and wait for the next generation, or the fourth generation.

CDMA to the Third Generation

While GSM and TDMA operators have multiple choices ahead for progressing to the next-generation networks, CDMA operators have a single path that truly builds upon itself. Currently, all North American CDMA networks are based on cdmaOne (IS-95), which can be set up to provide data rates up to 14.4 kbps. The next step is to upgrade software from IS-95A to IS-95B, which provides additional voice efficiencies, giving additional capacity, and allows for up to 84 kbps packet data. While this migration does not require any additional hardware, most operators may decide not to move to IS-95B for two reasons:

1. IS-95A in itself is relatively new and carriers have just launched their IS-95A data services.
2. By the time IS-95B becomes available, 1XRTT will be ready.

The Costs

In the shorter term, TDMA and GSM have a much more cost-effective upgrade option by means of moving to GPRS to be in a position to provide data services. As mentioned in the sections 1.1.4.8 and 1.1.4.9, an upgrade to GPRS does not require substantial investments and existing GSM/TDMA service providers can upgrade to GPRS at around 28% cost of their initial second-generation investments. The IS-95 upgrade path to CDMA20001xRTT is comparatively costly at around 40% of the investment in the existing second-generation networks.

It should also be noted that IS-95A has also not been in existence for long. However, in the final run to truly third-generation networks, GSM/TDMA operators may have to incur much higher investments as the cost equations for TDMA or GSM may vary, depending on the exact path taken (EDGE, or no EDGE, or only EDGE). CDMA has the unique advantage of having the same air interface in the second generation as in the third generation (from the same underlying technology).

The Fourth Generation

Several new standards have been proposed that do not fit into the second generation, the second-and-a-half generation or the third generation. These standards either provide only data services and/or provide much higher data rates than those specified by the third-generation systems. They are 1Xplus and 1XTREME. Since they use a single CDMA carrier they may be called the second- and-a-half generation but they provide much higher data rates than the third generation.

The 1XTREME standard will not require additional antennas as HDR will, and it will also keep data on the same spectrum as the voice services, meaning carriers will not have to devote any spectrum specifically to data services. The 1XTREME standard is proposed to deliver the same voice capacity increases as standard 1X, and provide data rates approaching 1.4 Mbps.

The second iteration, expected to be in trials by the first quarter of 2001, is expected to deliver data rates as high as 5.2 Mbps. Motorola expects 1XTREME to be market-ready in the same time frame as HDR: by the end of 2001 to the first half of 2002. Another interesting thing is that these so-called the fourth-generation technologies may start appearing almost at the same time as the third generation. It is not very clear as to how these developments will influence an already very complex set of equations.

1.2 The Destiny of the Third Generation

Recently, one of the hottest topics in telecommunications has been wireless technology. During this time, the technology has attracted many users and has undergone numerous changes, including Internet connectivity. However, even these profound changes may pale in comparison to what may happen during the next few years.

Therefore, it appears that wireless technology has reached a turning point, as vendors and researchers prepare to take it to the next level. Most industry observers agree that next-level wireless technology will offer more bandwidth, security, and reliability, making it more suitable for multimedia, e-commerce, videoconferencing and other advanced applications. Those applications could include video on demand; mobile e-commerce; wireless Web surfing; location-sensitive services, such as programs that find nearby movies or restaurants and customized personal information services, available anytime, anywhere.

A key issue for wireless is what form the technology's next generation will take. Many vendors, service providers, market analysts and other industry observers contend the next level will be the much-discussed third-generation wireless approach, which is actually a set of digital, packet-based, broadband technologies. Vendors are just starting to implement the third generation, but some experts in the field are already questioning its functionality and usefulness.

The third-generation technology itself is not good enough. But the second-and-a-half generation, a variation on today's second-generation approaches, will meet users needs for quite a while and will eliminate the need to adopt the radically different third-generation technology. Still others maintain that wireless LAN (WLAN) or radio-router technology would be better suited than third-generation technology for many advanced applications.

Vendors and researchers are carefully considering these and other issues as they decide the direction of wireless technology. Billions of dollars in revenue are at stake for service providers as well as device vendors and networking-infrastructure companies. For consumer and business users, at stake is the type of technology they will have to use on their wireless devices, including their smart phones, personal digital assistants (PDA) and pagers with Internet access. Therefore, the issue is critical on many fronts.

1.2.1 Driving Forces

Figuratively speaking, the road towards the third generation is bumpy, and some of the bumps are bigger than the car. Wireless is being pushed by technology and pulled by the market. Technology can now take wireless to the next generation, while users and service providers want the applications that new technologies could enable.

The demand exists, in part, because there has been an explosion of diverse devices with smart capabilities. Vendors and users want better wireless-networking technology to take advantage of the devices" online functionality. Meanwhile, providers have seen potential sales if they could provide advanced wireless services. The key markets for next-generation wireless systems will be sales people and other frequent travelers who need access to corporate data while on the road, as well as consumers of video and other data-intensive entertainment applications.

Network providers and device vendors are beginning to roll out the third-generation services and products. The USA trails Western Europe and Japan in adopting them, as has typically been the case with mobile technology. For example, the US Federal Communications Commission (FCC) auctioned second-generation mobile licenses and frequency spectra in January 2001 for US$17 billion to such companies as AT&T and Verizon. However, this was years after many European nations finished auctioning off their second-generation spectra and were already selling the third-generation spectra. To identify available the spectra and study other issues, the FCC wants to postpone by 24 months its original plan to auction off the third-generation licenses by 30 September 2001.

1.2.2 Focus on Technology

In the late 1970s and early 1980s, consumer wireless communications began to grow. The early mobile phones used first-generation technology, which was analog, circuit-based, narrowband and suitable only for voice communications.

For the past few years, commercial wireless devices have used second-generation technology, which is digital, circuit-based, narrowband and suitable for voice and limited data communications. The key question now is: what comes next? Many vendors and industry observers have assumed that the next important wireless approach will be the third-generation technology, suitable for voice and advanced data applications, including online multimedia and mobile e-commerce.

While the third generation is generally associated with mobile wireless, it could also be used with fixed wireless, such as local multipoint distribution services (LMDS) and multichannel multipoint distribution service (MMDS). The third generation promises transmission speeds of up to 2.05 Mbps in stationary applications, 384 kbps for slow-moving users and 128 kbps for users in vehicles. The third generation is thus considerably faster than the second and the second- and-a-half generation technology, as shown in Fig. 1.2.

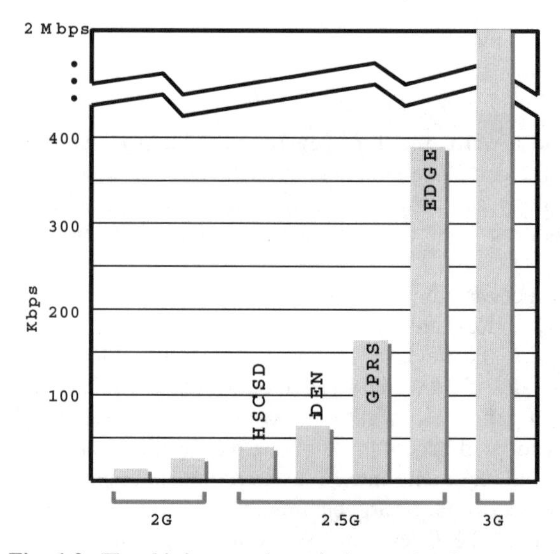

Fig. 1.2 The third-generation wireless technology promises maximum data rates of 2.05 Mbps, which is much more than the throughput that the second and a half-generation and the second-generation approaches offer

The third-generation technology comprises three primary standards: W-CDMA (wideband code-division multiple access), CDMA2000 and time-division CDMA (TD-CDMA). Each standard is based on and provides an upgrade path for at least one of today's primary wireless interfaces: time-division multiple-access (TDMA), global system for mobile communications (GSM), and code-division multiple-access (CDMA).

There are some commonalities in the third-generation standards, but are not fully compatible with each other at the air-interface (radio-transmission) level. The modulation is done differently. Any of the browser technologies will work with the third-generation. Users will eventually determine which browser technologies they want to use with the third generation.

The third generation will work with IP version 6, or IPv6, the new version of the Internet protocol, whose main feature is a larger address space that permits many more Internet addresses than IP version 4, or IPv4. This will be critical as the number of devices with Internet addresses skyrockets.

1.2.3 Dilemmas

The first carriers trying to implement the third-generation have experienced numerous problems and delays, not all directly related to the wireless technology itself. In the spring of 2001, British Telecommunications postponed the rollout by subsidiary Manx Telecom of the world's first commercial third-generation phone network on the Isle of Man, an island of about 75,000 residents located between northern England and Ireland. In December, Manx finally switched on its full third-generation network, which was the first in Europe in operation.

Also in the spring of 2001, NTT DoCoMo (Japan) planned to introduce its next-generation mobile system, based on W-CDMA. However, the company decided to delay commercial activation of the system until October 2001. Meanwhile, NTT DoCoMo had to recall a number of the third-generation videophones because of a software malfunction. And Vodafone, the world's largest wireless carrier, may not launch its third-generation services until 2003, largely because there has been a shortage of handsets. These delays have caused the third-generation concept to lose steam in the marketplace recently.

These problems have brought a healthy dose of realism to the market. This is new technology being aggressively deployed, and there will be bugs. The net result is that timetables, uptake projections and application deployment are now more reasonable.

1.2.4 Threats to Third Generation

A variety of factors will decide whether the third-generation will succeed as the next-generation wireless technology. The frequent determinants of a technology's success are 50% government regulation, 40% economics, and the rest, technology. In other words, the best technology doesn't always win, and technological factors don't always decide the outcome.

With this in mind, a number of factors appear key to the battle over next-generation wireless technology. For example, the US government has been having trouble finding available third-generation spectrum to sell to service providers. The US Defense Department, government agencies, schools, and health facilities are using some of the spectra identified as potentially suitable for the third-generation, as are other types of wireless communications. However, after the

September 11 terrorist attacks, the US military is not likely to give up its share of potential third-generation spectrum for videoconferencing on mobile phones.

1.2.5 Cost Overrun

Governments in North America, Europe and parts of Asia have auctioned off licenses to companies that want to use part of the limited third-generation spectrum to provide wireless services. In most cases, companies are suffering cost overruns due to the huge sums of money paid for the licenses.

Large companies in Europe, such as Deutsche Telekom, France Télécom, Telefónica (Spain) and Vodafone (UK) spent an estimated US$125 billion to US$150 billion on the third-generation licenses. Germany raised about US$44 billion in its third-generation auctions, while the UK earned an estimated US$32 billion. Vodafone, which hopes to begin commercial third-generation service later in 2002, bid about US$6 billion for a single UK license. On the other hand, LG Telecom (Korea) agreed to pay the Korean government about US$900 million for its third-generation license, while SK Telecom (Korea) and KTF (Korea) agreed to pay US$1 billion each.

Industry observers say the desire to participate in the third-generation market, whatever the cost, probably drove telecommunications companies to pay very high prices in the German and UK auctions. However, after the excitement from the auctions died down, financial firms began releasing sobering assessments of the third generation's likely financial returns. Soon, credit ratings began falling for some companies that won the third-generation bids, and this endangered their ability to get loans to pay for the purchases. To recover their costs, some experts say, German and British telecommunications companies may charge consumers high fees for the third-generation services, which could discourage demand.

Market analysts say service providers will probably spend almost as much as they paid for their licenses to buy the necessary equipment and then build their third-generation network infrastructure. One key infrastructure build-out factor is that the third generation uses smaller cells than the second generation and thus needs more base stations and transmission towers. The third generation needs smaller cells because the range of radio transmission decreases with the higher frequencies. Also, the third-generation systems use modulation and power-management techniques that require short distances. To save money, where the second-generation systems already exist it is probable that some third-generation base stations will use the same towers. Nonetheless, it seems likely that third-generation systems will be mainly restricted to urban areas for some years at least, because it will not be economical to install them in large rural areas.

Meanwhile, service providers will face another significant challenges in recovering their large license and build-out investments. The global economic decline, particularly in the technology sector, could scare already-nervous investors and stall widespread investment by potential corporate and consumer customers in new wireless technology.

The investment community is blamed, in part, for making rosy forecasts and statements without being aware of the immaturity of the third generation. The third generation was also precipitated by regulators greed, operator paranoia and by vendors who wanted to sell more equipment, therefore the industry went through a frenzy.

1.2.6 Data Service Capability

A key to the success of the third generation will be how well it works with data. The third generation will have to handle intensive data sets, such as those used in multimedia, because one of its principal purposes will be to take cellular phones beyond voice communications or simple data transfers. A problem is that third-generation technology experiences a performance penalty that will affect ultimate user throughput. Typical third-generation users will get performance up to only 56 kbps, the maximum speed of a PC dial-up modem.

As users move farther from a base station, interference from other cells will weaken the signal and cause channel errors. Also, a system that works with both voice and data will not get the maximum throughput. Users demand that wireless voice communications offer the same quality as wireline phone technology. Therefore, wireless systems must devote resources to voice communications equality of service under all circumstances, which reduces maximum data performance. CDMA2000 has tried to address this issue by putting voice and data traffic on different frequencies. Eventually, third-generation providers in general will run voice and data on separate channels.

1.2.7 Killer Applications

A key to the success of any future wireless technology will be whether users can access interesting or important content with it. In fact, systems could succeed or fail not because of the technology but because of the lack of desirable content and "killer applications". The issue is whether content providers will be able to offer compelling material and whether users will want to bother accessing it over wireless, rather than traditional wireline, networks.

The third generation should have a killer application that is, one that people cannot do without, but none has yet emerged. Service providers and the needs of the end user will drive the emergence of killer applications. The best answer is that there is unlikely to be a single killer application for the third generation. The advantage of the third generation will be its ability to support a wide variety of different applications, and that will be the "killer characteristic" of third generation technology.

Meanwhile, as shown in Fig. 1.3 it is predicted that early third-generation PDA will be more expensive than current-generation PDA, which could limit the new technology's early adoption (source: PA Consulting). One factor will be the higher cost of the application-specific integrated circuits used in third-generation devices.

Security and privacy are concerns for the third generation, as they are with any networking technology, especially one that is untested. Other potential problems are assuring quality of service and congestion control throughout a network that consists of many vendors" wireless networks.

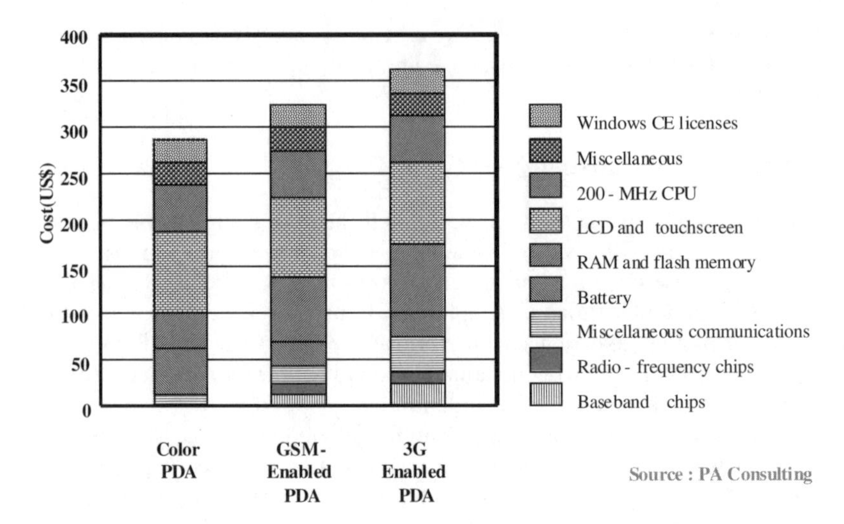

Fig. 1.3 Third-generation personal digital assistants (PDA) will cost more than either their second-generation counterparts or PDA without Internet connectivity. The primary contributors to the higher costs of third-generation PDA are RAM, flash memory and miscellaneous communications-related functionality, as well as radio-frequency and baseband chips

Finally, with high build-out costs, carriers will need time to bring service to a large number of areas. However, many users may not adopt the third generation until they can use it widely. For example, some industry observers say that the third generation will not become widely used in the USA until it is available in every major urban area.

1.2.8 Alternative Technologies

A key threat to the third-generation wireless is the possibility that users will decide to get wireless services from other types of technology, like the second-and-a-half generation. Companies and standards organizations have developed several so-called second and a half-generation systems as upgrades to existing second-generation technologies. Numerous wireless service providers offer the second and a half-generation services in Australia, China, Europe, Korea, Japan, New Zealand and North America.

Like the third generation, the second-and-a-half generation is always on, provides simultaneous voice and data and delivers more speed than today's circuit-switched data connections. The second-and-a-half generation offers more bandwidth than the second generation, but less than the third generation. Service providers can implement the second-and-a-half generation much less expensively than the third generation because the former uses existing second-generation spectra and doesn't require a new network infrastructure, although some system upgrades are necessary.

Some providers see the second-and-a-half generation as a stepping stone or introduction to the third generation. For others, it is an adequate alternative that may let them skip the third generation altogether. One problem is that the second-and-a-half generation technologies were not designed to optimally handle voice and data communications simultaneously and can also experience latency problems that is, problems with service life, display size and quality, contents variety, etc. Also, because the third generation is capable of offering greater bandwidth than the second-and-a-half generation, the quality of video will be considerably greater. This could attract potential second-and-a-half generation users.

The most prominent second-and-a-half generation technologies are integrated digital enhanced network (iDEN), general packet radio service (GPRS), and enhanced data GSM environment or enhanced data rate for global evolution (EDGE). GPRS systems are under development in many areas. However, some carriers have experienced the build-out delays and cost increases that have plagued the third-generation rollouts. In addition, GPRS does not work with most existing GSM phones because the handsets do not have the necessary capabilities yet, although this is starting to change. This change could be problematic, though, because existing GSM phones do not have the packet-data capabilities necessary to work with GPRS, and these features have been difficult to develop.

1.2.9 Wireless LAN

Unlike the third-generation wireless, wireless LAN (WLAN) technologies such as ETSI's high-performance radio LAN (HiperLAN), the HomeRF Working Group's HomeRF, and market-leader IEEE 802.11b (also called Wi-Fi) provide network services via corporate-type networks. Therefore, the third generation is not designed to compete directly against WLAN. LAN and mobile networks are complementary services, targeted to different device types and different environments.

However, some industry observers say that many potential third-generation users who have access to WLAN may prefer to obtain their wireless services via the latter. Therefore, although the third generation is not designed to compete with WLAN, it will have to do so. Many WLAN are already deployed and are growing in popularity and thus have a head start defining high-performance wireless data services in the marketplace. Users will become accustomed to WLAN services and will not want to give them up. Also, users will want to receive wireless services from a technology that, like WLAN will "look" like Ethernet and "act" like a corporate network.

Although WLAN are currently private networks, they eventually could become de facto public networks, which would generate more competition for the third-generation technology. In a number of locations, including San Francisco, individuals with WLAN are establishing access points that nearby residents and businesses can use to tie into the networks.

The principal advantage of LAN over the third generation is that the cost of deployment is low. However, WLAN technology is not suited for wide-area coverage and is better suited for indoor, rather than outdoor, environments.

In addition, WLAN use an unlicensed spectrum. If the networks become popular, they will share a very limited frequency range, which could lead to signal interference. However, other spectra may be available for WLAN in the future. Meanwhile, WLAN may have trouble attracting potential third-generation users unless vendors develop more mobility-oriented applications, such as programs that locate the nearest restaurants, rather than just wireless versions of desktop software.

1.2.10 Radio-Router Technology

Radio-router uses a radio-transmission framework for packet-based, broadband, IP wireless communications. Radio-router technology is designed to make the links in an IP network mobile. Proponents hope that since IP network technology is already well developed and inexpensive, radio-router systems will be relatively easy, quick and economical to implement. A radio-router network can be built atop the existing IP infrastructure, rather than from the ground up like a third-generation network.

The technology uses orthogonal frequency-division multiplexing (OFDM), in which a single channel is divided into several subchannels, each at a different frequency. This boosts bandwidth by letting a system carry several transmissions at the same time. Radio-router systems offer a maximum throughput of 1.5 Mbps, about the same as a T1 line. OFDM, unlike traditional frequency-division multiplexing (FDM), uses signal modulation and demodulation techniques as well as the orthogonal placement of adjacent channels to minimize interference. Therefore there is less emphasis on the quality of individual channels.

Radio-router technology is data-focused that is, it is designed from a data perspective. However, it does support voice packet-switched voice, not circuit-switched voice. Radio routers, IP routers with radio adjuncts, would handle packet traffic and serve as the equivalent of cellular base stations. Consumers could work with radio-router technology via PCMCIA cards in their laptops and via flash-memory cards in handheld devices.

Despite the marketplace potential of the technology, radio routers face an uphill fight against the entrenched cellular businesses. Cellular providers are much larger and better established companies, and cellular service appears to be a safer investment to many managers. Radio-router technology, on the other hand, might seem exotic and thus might not attract big infrastructure investments.

1.3 Conclution: Gloomy Days to Overcome

The next generation of wireless services, besides improving the overall capacity, will create new demand and usage patterns. This will, in turn, drive the development and continuous evolution of services and infrastructure. While development of the third-generation networks will continue and grow in the near future, the second-generation networks will keep evolving in terms of continuous enhancements and towards convergence of existing second-generation standards.

The initial third-generation solutions should coexist with the second-generation networks while slowly evolving to all third-generation networks. While the third generation in its true sense should have transparent roaming across all networks throughout the world, given the penetration and the investments in the second-generation, true roaming and consistent service availability, both across networks and independent of networks, looks to be a very distant proposition.

However, the reality of the third generation is too far from its ideal at the moment. In spite of the huge amount of money that operators invested in obtaining licenses, the degree of confidence in the technology and the potential market demand look depressed and opaque. Now that the first systems are being deployed and the third-generation services and applications are being offered to the public, demand for wireless multimedia development is likely to be flat curve for the next decade.

Considering that it took ten years for GSM to reach true mass-market dimensions, the third-generation may take many more years to reach the level of GSM penetration today, unless killer applications or disruptive technologies appear soon. Implementation of these specifications into systems, equipment and services is a highly complex and risky task, so it should not be surprising if there are some difficulties along the way.

Korea and Japan probably will lead in deployment of the third generation but they do not presently have nationwide coverage. Europe probably will not even begin to significantly adopt the technology until mid-2003.

In the US, W-CDMA deployment will require new spectrum allocations. Therefore, it is unlikely W-CDMA will begin nationwide deployment before 2005. The W-CDMA technology is at least two years behind CDMA2000; it will be more expensive to implement, and for now, will do well only in areas in which regulators favor W-CDMA over CDMA2000, such as Japan and Western Europe. CDMA2000 service providers can build their systems atop CDMA networks, reusing much of the existing infrastructure and cell sites. This is not the case with W-CDMA, so service providers will have to spend more time and money to build the networks.

Meanwhile, some users have complained that W-CDMA is not sufficiently reliable and that its handsets are bigger and heavier than their CDMA counterparts. By the end of 2002 or 2003, will there be enough subscribers enjoying the third-generation services worldwide? About what percentage will be using CDMA2000?

We may have to look for better technology than the current third-generation. We need something that is faster to market with a better price/performance ratio. The third generation will not provide sufficient data service to computer users. The alternative could involve a combination of the third generation and WLAN technologies, with users able to use the latter if they are close to an access point. This could be the best of both worlds.

The third generation suffers from doubt, delay and debt. Nonetheless, vendors are likely to continue moving ahead with the technology. It is predicted that the short-term prospects for the third generation are bleak because of high license fees, delays in obtaining handsets and slow customer adoption. In the long run, however, we may expect brighter prospects for the third generation and predict that the third generation will have about 50% of the worldwide wireless market by 2010.

The financial impact of megaspending for spectrum followed by seemingly unreal financing arrangements has taken its toll. Entry into the third-generation marketplace is not for those with weak hearts or shallow pockets. Nonetheless, service providers will have to make the third-generation succeed because they have already invested too much in the new technology to let it fail. We have gloomy days ahead to overcome.

List of Major Information Sources

1. Cann-Evans (2001). Wireless networks operations – “today to tomorrow”. http://www.europemedia.net/shownews.asp?ArticleID=2290.
2. Garber, L. and Paulson, LD (2001). Will 3G Really Be the Next Big Wireless Technology? http://computer.org/computer/homepage/0102/tn/print.htm.
3. Oliphant, Malcolm W. The Mobile Phone Meets the Internet. IEEE Spectrum, August 1999, pp. 20-28.

To Probe Further

Useful World Wide Web sites include: the Cellular Telecommunications Industry Association at http://www.wow-com.com; the Universal Wireless Communications Consortium at www.uwcc.org; and the CDMA Development Group (CDG) at www.cdg.org for information about cdmaOne and cdma2000.

Standards-setting organizations have sites: the Telecommunications Industry Association (TIA) at www.tiaonline.org; the European Telecommunications Standards Institute (ETSI) at www.etsi.fr; the GSM Association at www.gsmworld.com; and the International Telecommunication Union (ITU) at www.itu.int/home/imt.html.

2 Mobility Management for an Integrated Network Environment

The future of the next generation of wireless communications rests in the palms of our hands. Whether through wireless phones, pocket pagers, wireless-enabled personal computers or personal digital assistants, soon almost everyone will have wireless connectivity to the Internet. The development of wireless network access technology is expanding, and several new alternatives will be offered in the next few years, from those providing broadband wireless hotspots to those providing global coverage UMTS (universal mobile telecommunications system) as well as fixed access such as DSL (digital subscriber line). The networks have very different capabilities. Speed, latency and connection type all vary. At the same time, user demand for continuous connectivity is increasing irrespective of the type of interface or network they are using. With the growth of the wireless Internet and the mobile computing marketplace, we expect to see mobility without any geographical or network boundaries. This is one view of a fourth-generation (4G) network – with an IP core and multiple access technologies using IP mobility management for vertical hand-over. Wireless mobile devices will become the main form of communications in the next decade, and to support mobility, will require re-thinking of several aspects of interworking of all existing and emerging fixed and mobile networks. Enabling mobility in IP-based networks is an important issue for making use of the many devices appearing in the market. Mobile devices have lately become small enough to carry, and portable computers have become by far the fastest growing segment of the computer industry. It is possible that within the decade a vast proportion of these devices will be personal, wireless and mobile workstations. Providing continuous connectivity in different types of networks with fixed and wireless segments is in itself a challenging task. One particular outstanding issue is that of mobility management. In principle, a mobility management protocol should provide a means of terminal, session, service and personal mobility.

2.1 Mobility Management in Heterogeneous Networks

2.1.1 Application Adaptability

As mentioned in Fig. 2.1, providing continuous connectivity in different types of networks with fixed and wireless segments is a challenging task. One particular outstanding issue is that of quality of service (QoS) and in a related issue, of applications being able to adapt seamlessly to the change of QoS. The application should be able to adapt to the new environment in terms of equipment used and available network resources. QoS can change abruptly when a handover occurs or during a vertical handover – defined as a handover between base stations that are using different network technologies, e.g. between a wireless local area network (WLAN) and General Packet Radio Services (GPRS).

As a result, the quality of service perceived by the user (that which is apparent at the boundary between the application and the network) will be disturbed.

Application-aware adaptation is an essential capability of mobile nodes, i.e. the requirement to adjust parameters when there is a change in QoS resulting from a change in network type [1]. However, this needs to be effected in a dynamic manner, since the actual quality of the network will change over time because of its variable load or even because of a change in the network policy (Fig. 2.1). The actual quality of an IP based network (e.g. bandwidth) changes during an ongoing session between two hosts. As an example, consider a two-way multimedia conference audio and video call using the real-time protocol (RTP). In this case, the challenge is to keep the video part of the data stream uninterrupted, even when the available bandwidth is less than that required by the encoded video stream, as could occur with a possible vertical handover of one of the hosts to a lower bandwidth network. There are several ways in which this application adaptation may

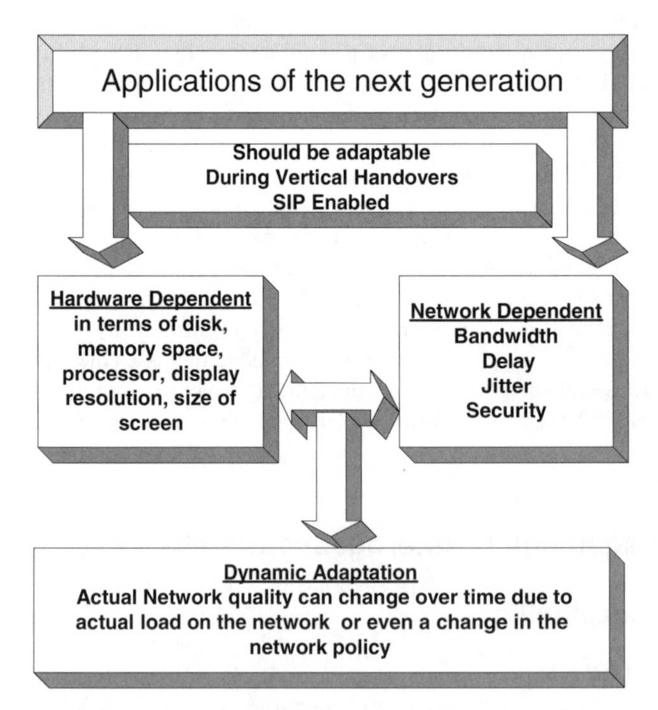

Fig. 2.1 Next-generation adaptable applications

be achieved. The user should at all times be able to make decisions on how the actual quality should be used in an optimal way. This is very important when time-based information is used - different networks might be available to the user at any given time with different bandwidths and costs. These choices should be presented to the user. We have assumed that the difficult problem is a receiver handover. It is assumed that a data source can identify the type of handover that has occurred and provide suitable adaptation. There are several different ways of providing application adaptation.

2.1.2 Receiver-Based Adaptation

One solution is to have the receiver adapt according to its current network connection. For example, if one of the hosts moves to a network of a lower bandwidth, then the application itself will intercept, process and filter all the packets according to the current application profile set by the user. For the example of a move to a GPRS network, the receiver application would remove the video data stream and compress the audio stream. Unfortunately, this approach has one disadvantage. Transferring data, in this case the video stream or a high bit-rate audio stream, in a format that is beyond the capability of the receiving end (the host that moved to the GPRS network), is a waste of bandwidth. This is shown in Fig. 2.2.

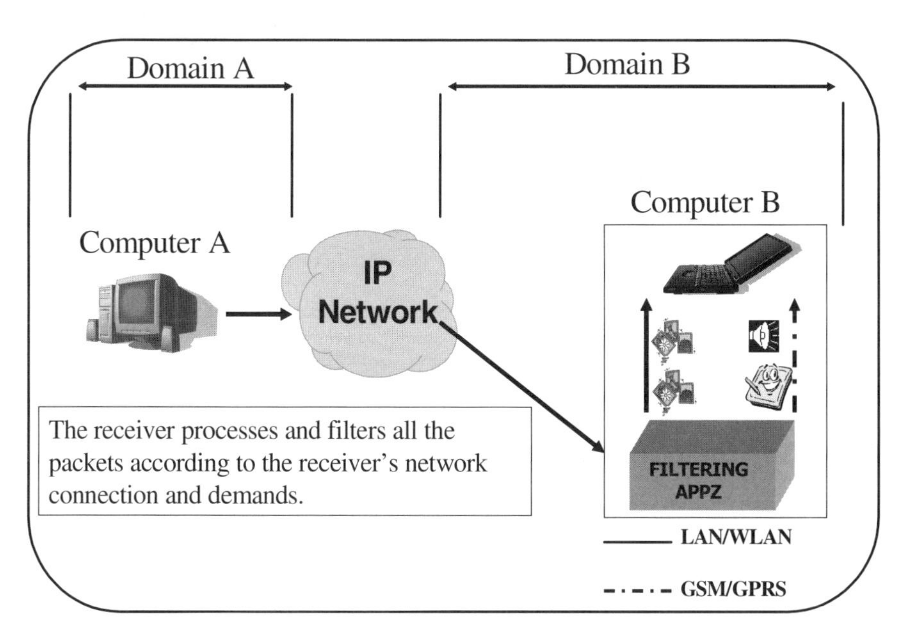

Fig. 2.2 Receiver-based adaptation

2.1.3 Proxy-Based Adaptation

Alternately, a proxy-based solution could be adapted such as the M3A approach [2]. An extended version of mobile IP provides the network layer protocol for M3A, with smart routing and proxy filtering decisions being taken at the home agent. The home agent has an updated profile of its registered hosts and provides filtering according to the host's network connection. Once a host hands over to a different type of network, the routes are updated dynamically by mobile IP. The IP packets are then intercepted, processed and filtered transparently to the user, and sent from the home agent to the mobile host. One disadvantage of using a proxy-based approach is the possibility of overloading the proxy server, resulting in packets being delayed, or even lost. This approach is wasteful of network resources, as all data must be transmitted to the home agent before filtering can take place, thus optimized routing cannot be used (Fig. 2.3).

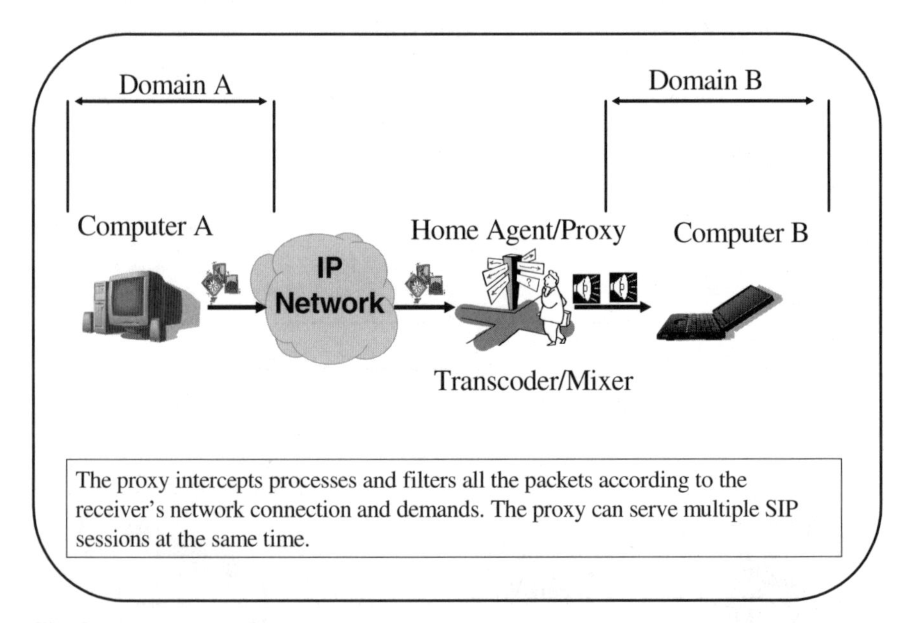

Fig. 2.3 Proxy based adaptation

2.1.4 End-to-end-Based Adaptation

This approach provides an end-to-end solution, where the applications themselves adapt to the changing QoS with the smallest session disruption possible. The approach is based on the session initiation protocol (SIP). SIP "is an application-layer control (signaling) protocol for creating, modifying and terminating sessions with one or more participants" for both unicast and multicast sessions (RFC-2543) [3]. SIP invitations carry session descriptions, which allow participants to agree on a set of compatible media types (specified in the session description protocol, RFC-2327) in order to participate in a multimedia session [4]. SIP could be used with or without mobile IP to provide an end-to-end solution for application adaptation in a mobile environment. End-to-end-based application-aware adaptation is a smart, and bandwidth and processing-efficient way for mobile nodes to adjust when there is a change in QoS (Fig. 2.4).

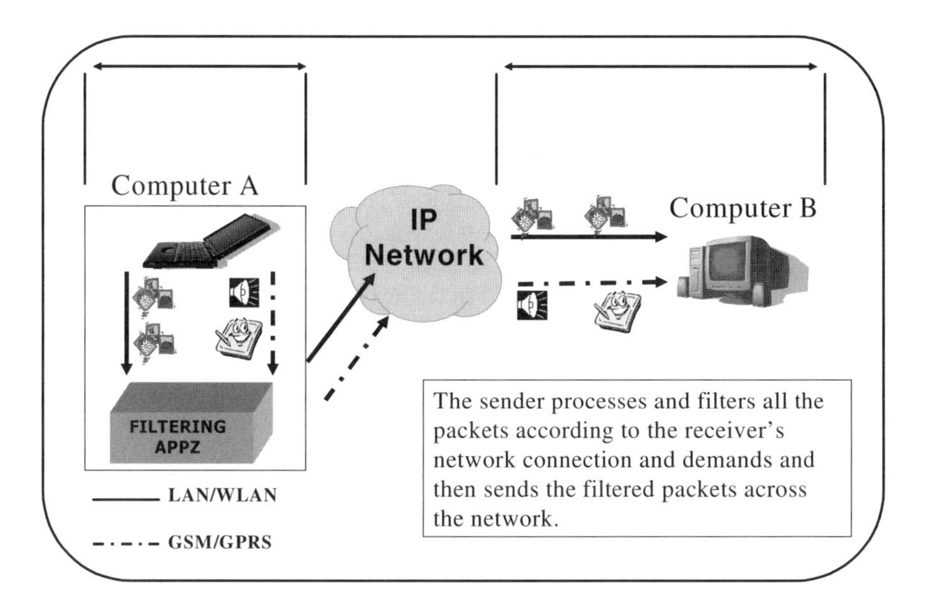

Fig. 2.4 End-to-end-based adaptation

2.2 Terminal Mobility

In this section, we look at terminal mobility and describe different ways in which handovers could be initiated. The main challenge is to enable a typical user to use a real-time multimedia application across many networks with the minimum disruption to the application. As the user moves from networks of high bandwidth to areas of low connectivity, the only visible change to the application should be the limitations of the specific interfaces. For example, a lower bandwidth network may only support low-quality audio, while a high bandwidth network will support both audio and video. In this case, if the user is moving to the lower bandwidth network, the video part of the application has to be terminated. Another challenge is to obtain a low handover latency. One solution is to keep all available terminal interfaces on all the time, and thus minimize the time taken to switch between networks. However, this will increase the power drain of the terminal because all the multiple interfaces are active at the same time. In order to provide low handover latency and at the same time handle power management efficiently, we need to discover the right time to perform the handover [5].

The main reason for introducing mobility management into IP-based networks is to support handover where a change of IP address might mean a large loss of packets. Terminal mobility allows a device to move between IP subnets, while continuing ongoing communication and maintaining sessions across radio cells within the same subnet, or subnets within the same or different administrative domains. Support for both real-time and non real-time services is essential. There are currently three levels of logical/virtual handoff [6]. A cell handoff is one that allows a terminal to move from one cell to another in a subnet within an administrative domain. A subnet handoff allows a terminal to move from a cell within a subnet to an adjacent cell within another subnet that belongs to the same administrative domain. Last, a domain handoff allows a terminal to move from one subnet within an administrative domain to another in a different administrative domain. Over the last few years there have been many twists and turns on the road to deployment. New protocols, such as Media Gateway Control Protocol (MGCP) and the Session Initiation Protocol (SIP), have entered the arena and caused changes in directions.

2.2.1 SIP

The session initiation protocol (SIP), specified in IETF RFC-2543, is a powerful tool for call control and signaling that is gaining support among service providers and vendors. SIP has been chosen as the signaling and call control protocol for the Universal Mobile Telecommunications Service (UMTS) R5 and has direct relevance for both UMTS and DSL/cable networks where SIP signaling is also being adopted. It turns out to be an ideal protocol for providing truly convergent applications, primarily because it borrows so heavily from other protocols, and in particular, the hypertext transfer protocol (HTTP) and the simple mail transfer protocol (SMTP). The main function of SIP is to establish real-time calls and conferences over internet protocol networks. Each session may include different types of data, such as audio and video, although currently most of the SIP extensions address audio communication.

As a traditional text-based Internet protocol, it resembles HTTP and SMTP. SIP uses the session description protocol (SDP) for media description. SIP is independent of the packet layer. The protocol is an open standard and is scalable. It has been designed to be a general-purpose protocol. However, extensions to SIP are needed to make the protocol truly functional in terms of interoperability. Among SIP's basic features, the protocol also enables personal mobility by providing the capability to reach a called party at a single location-independent address. Figure 2.5 shows the basic operation of SIP.

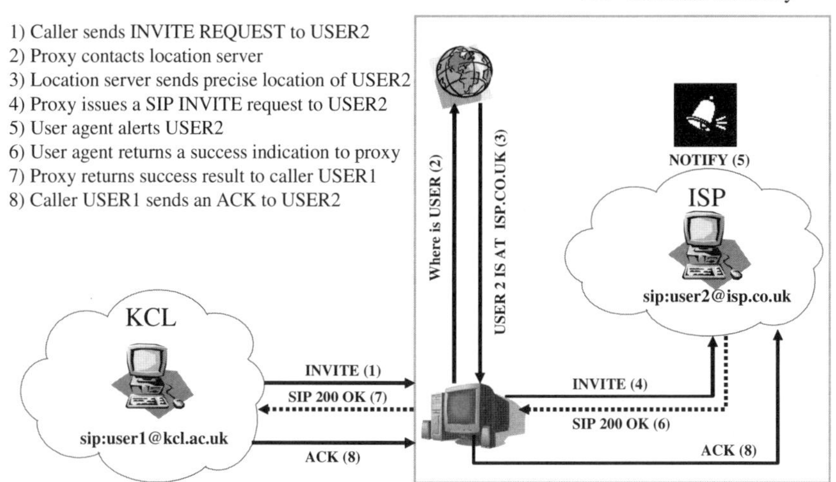

Fig 2.5 SIP operation in proxy mode

2.2.2 Mobile IP

Mobile IP has been proposed as a solution for mobility support and provides users the freedom to roam beyond their home subnet while consistently maintaining their home IP address. This enables transparent routing of IP datagrams to mobile users during their movement, so that data sessions can be initiated to them while they roam. It also enables sessions to be maintained in spite of physical movement between points of attachment to the Internet or other networks. Mobile IP is most useful in environments where mobility is desired and the traditional land line dial-in model or DHCP do not provide adequate solutions for the needs of the users. If it is necessary or desirable for users to maintain a single address while moving between networks and network media, mobile IP can provide them with this ability.

Generally, mobile IP is most useful in environments where a wireless technology is being utilized. This includes cellular environments as well as wireless LAN situations that may require roaming. Mobile IP can go hand in hand with many different cellular technologies like code-division multiple access (CDMA), time division multiple access (TDMA), global system for mobile communication (GSM), and advanced mobile phone service (AMPS), as well as other proprietary solutions to provide a mobile system that will scale for many users. Each mobile node is always identified by its home address, no matter what its current point of attachment to the Internet, allowing for transparent mobility with respect to the network and all other devices. The only devices that need to be aware of the movement of this node are the mobile device and a router serving the user's topologically correct subnet. Figure 2.6 shows the basic operation of mobile IP. In a basic mobile IP operation, packets sent by the correspondent host to the mobile host are always sent to the mobile host's home network first, and then forwarded

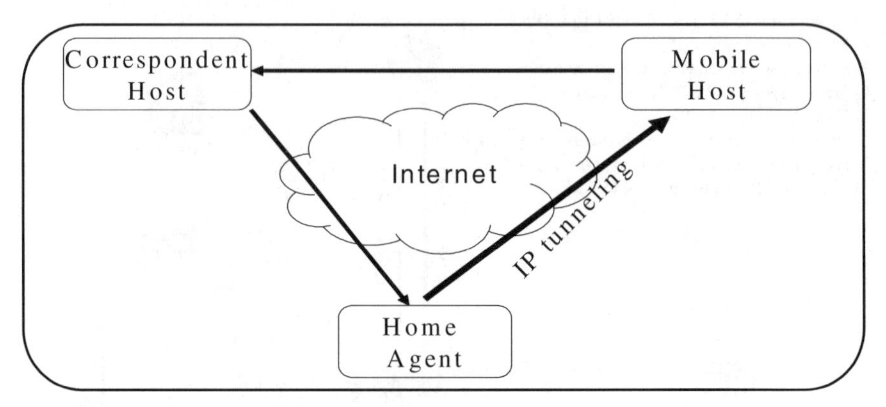

Fig. 2.6 Basic operation of mobile IP

by the home agent to the mobile host's current care-of address. Packets originating from the mobile host are sent directly to the correspondent host, thus forming a triangular route. These packets use the mobile host's home address as their source address to preserve their home identity. The packets forwarded by the home agent to the care-of address are encapsulated in another IP packet.

2.3 Handover Monitoring Mechanisms

There are several ways in which a vertical handover could be initialized [6]. Ideally, a user should stay connected to its existing network until one of the following occurs:

- The quality of the current link is degraded because the user moved outside the network coverage, or because the network is being congested by other users.
- The current network is no longer available.
- The user needs to change the existing session, and the existing network does not support the change.
- The end-to-end quality of the session is degrading and a handover is required in order to keep the required QoS.

The first scenario is the most common one. As the user moves outside the coverage of a wireless overlay network, the signal power received from the base station will be reduced. As a result, the available bandwidth will also be reduced, which has a direct impact on the application itself and the ongoing multimedia session of the user. By monitoring the signal level, a handover can be initiated when the signal strength drops below a threshold value, set by the user.

This information could then be passed up to the higher layers and to the application itself. Furthermore, a similar type of handover would be initiated if the link bandwidth is reduced because more users log in to the network.

The second scenario occurs less frequently. In this case it usually takes more time to perform the handover which results in packet loss. This kind of handover is called unplanned and no signaling is done prior to the handover. Once an unplanned handover is initiated the terminal has to detect other available interfaces, initialize and configure the most appropriate one. Once this is done, the user would perform all the signaling through the new overlay network. We look at these type of handovers in more detail in Sect. 2.4.

The third scenario looks at user-enabled handovers. The user will initiate a handover to an overlay network when its current network cannot support the new required session. For example, while the user is part of an audio conference, he decides to switch to video.

Unfortunately, the user is currently connected to a general packet radio service (GPRS) network with a limited bandwidth. In these circumstances two options are available. The user can either switch to another available network to get the required QoS, or transfer the session to another available device. The first option gives the user the choice of selecting another network while still using the same terminal. This is desirable if the user is on the move and has no other devices available. However, in some cases the user might decide to transfer the session to another terminal.

Finally, in the last scenario the end-to-end quality of the session is monitored, and a handover is initiated if a certain QoS is not met. As an example, consider a two-way multimedia conference audio and video call using the real-time protocol (RTP). By getting feedback from the real time control protocol (RTCP) reports of the session a handover can be initiated if the delay, jitter or packet loss drops below an acceptable level.

The Real Time Protocol

The real-time transport protocol (RTP) provides end-to-end delivery services for data with real-time characteristics, such as interactive audio and video.

Those services include payload type identification, sequence numbering, timestamping and delivery monitoring. RTP also supports data transfer to multiple destinations using multicast distribution if provided by the underlying network. RTP itself does not provide any mechanism to ensure timely delivery or provide other quality-of-service guarantees, but relies on lower-layer services to do so.

It does not guarantee delivery or prevent out-of-order delivery, nor does it assume that the underlying network is reliable and delivers packets in sequence.

The sequence numbers included in RTP allow the receiver to reconstruct the sender's packet sequence, but sequence numbers might also be used to determine the proper location of a packet, for example, in video decoding, without necessarily decoding packets in sequence. RTP does not use the IP address to keep track of associations between end systems. However, it uses the synchronization source identifier (SSRC).

This field identifies the synchronization source. The value is chosen randomly, with the intent that no two synchronization sources within the same RTP session will have the same SSRC. The transport protocol RTP is augmented by RTCP.

Its primary function is to provide feedback on the quality of the multimedia session. Applications may use this feedback to adapt to different network conditions, e.g. initiating a vertical handover. Feedback about transmission quality is also useful to locate problems and diagnose faults. The primary function is to provide feedback on the quality of the data distribution. This is an integral part of the RTP's role as a transport protocol and is related to the flow and congestion control functions of other transport protocols. The feedback may be directly useful for control of adaptivity.

2.4 Unplanned Vertical Handovers

In the following section, we'll look at the session initiation protocol being used with mobile IP to provide real-time applications with the information they need in order to adapt in the best possible way to the changing QoS during an unplanned vertical handover.

2.4.1 Mobile IP and SIP inter-working

In this scenario the primary trigger for a vertical handover is that the current active overlay network is no longer reachable because the mobile host has moved out of coverage of that overlay. Figure 2.7 shows a typical vertical handover.

An upward handover is initiated when several beacons from the current overlay network are not received. The mobile host decides that the current network is unreachable and hands over to the next available overlay network. Even though the mobile host cannot directly hear the old overlay network, it must still instruct the correspondent host of its movement to the new overlay. The basic idea is shown in Fig. 2.8 and explained as follows.

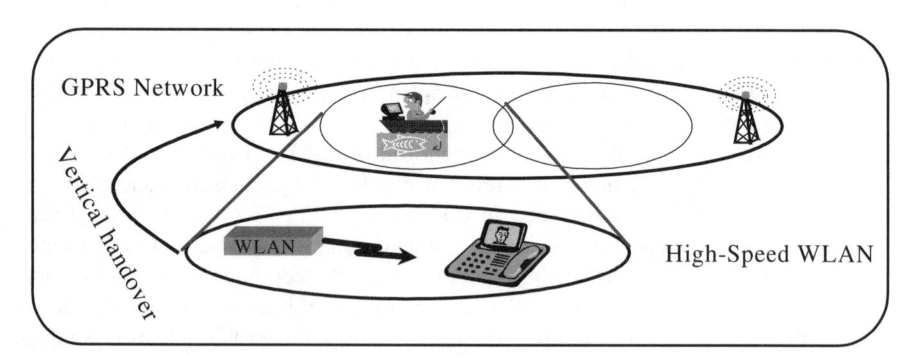

Fig. 2.7 Vertical handover from WLAN to GPRS

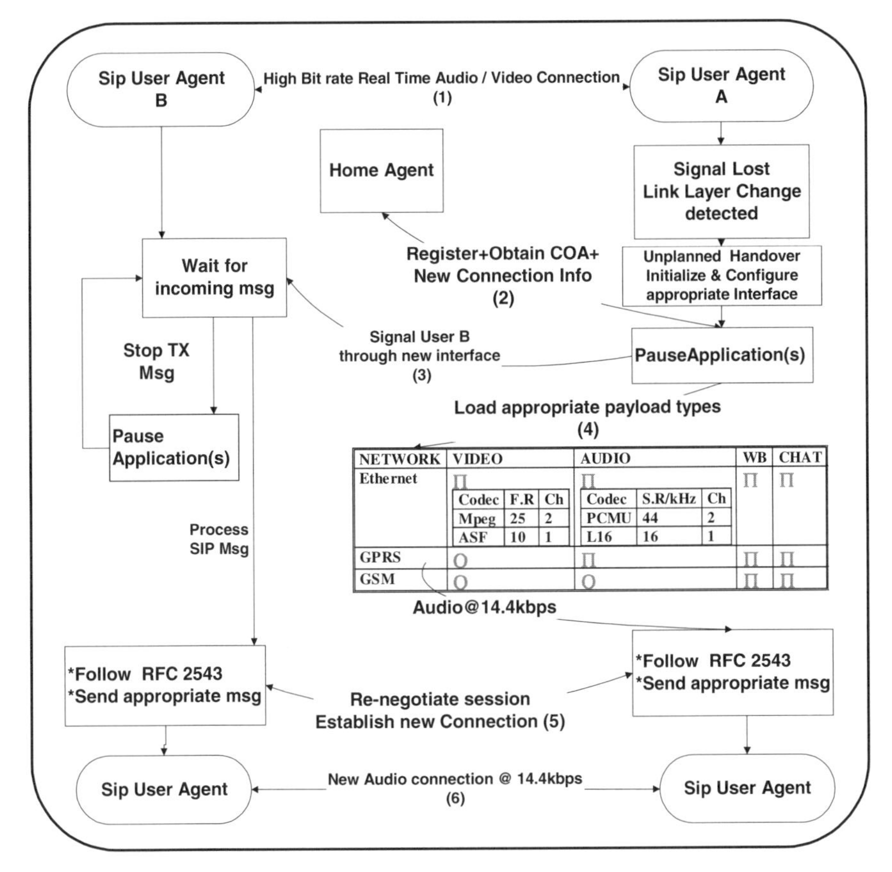

Fig. 2.8 SIP enabled end-to-end session renegotiation

User A is initially connected to its home network and is in a two-way real-time audio communication with user B at a high bit rate. Then at some point, user A loses signal with the current overlay network. This change is detected by the local link layer of the terminal. The sip user agent (SUA) is then informed about the current network status of the terminal and initiates a process by which it will detect other interfaces available on the terminal and, according to user preference characteristics (i.e. bandwidth, cost and services), will select an appropriate overlay to connect to. Once the connection is set up with the new overlay, mobile IP will handle terminal mobility. User A is assigned a collocated care-of address (COA), and a registration request is sent to the home agent. After successful registration, the session initiation protocol is notified and given the appropriate parameters in order to renegotiate the session.

This is done by using the same call-ID to describe the new desired session, taking into account the capabilities of the new active overlay network. Both users then renegotiate the session, load the appropriate application(s) and resume transmitting using the newly agreed coders-decoders (codecs). If the new overlay network has no limitations in terms of supporting the old session then the user can keep the session alive and just transfer it to the new overlay. However, lets us assume that the new overlay network has limited bandwidth and the user's old session needs to change and adapt according to the capabilities of the new overlay. In this case, parts of the ongoing session might need to be terminated (for example the video stream) and only an audio stream will stay available if the user is moving to an overlay of limited bandwidth. This is shown more clearly in Fig. 2.9.

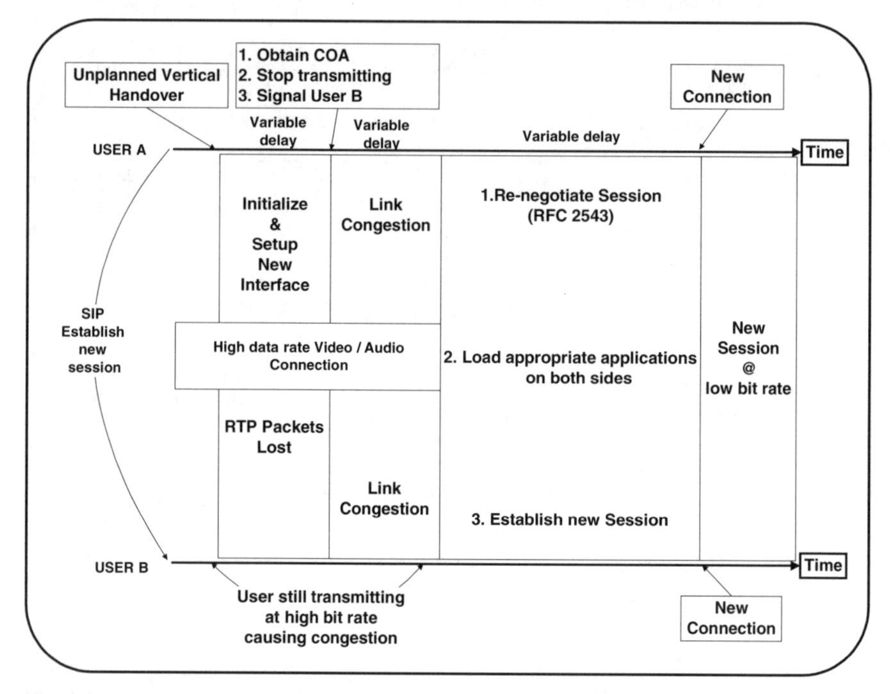

Fig. 2.9 Timing diagram during an unplanned vertical handover

Figure 2.9 is a timing diagram showing the breakdown of an unplanned vertical handover. The handover is split into three main areas: the initialization and configuration of the network interface, the link congestion/signaling and the renegotiation of the existing multimedia session. Let us look at these procedures in more detail.

Initialization and Configuration

An unplanned vertical handover is initiated when several beacons from the current overlay network are not received. The mobile host decides that the current network is unreachable and initiates an unplanned handover to the next available overlay network, taking into account the user preferences. This allows the user to specify specific constrains about which networks to use according to parameters like bandwidth and cost. Once the overlay network is selected, the appropriate network interface has to be initialised and configured. The time taken to initialize and configure the interface is variable and depends on the type of interface being configured. For example, if a dial-up connection is required (i.e. connecting to a GSM overlay network) the delay will vary from (30–35 s). This is the time taken to initialize the modem, dial and establish a point-to-point connection with the remote dial-up server. However, if a GPRS connection is required, the time taken to set up a GPRS interface is reduced to approx. 5 s. One way of reducing this time would be to have the network interfaces turned on and available if needed. However, power management in a multimode terminal is very important. Idle interfaces consume a lot of power and would significantly reduce the battery life of portable wireless device. Once the new interface has been set up, it registers with the home agent and obtains a care of address. During this initialization and configuration interval RTP packets send by the correspondent host are lost.

Link Congestion

Providing continuous connectivity in different types of networks with fixed and wireless segments is a challenging task. This challenge becomes more explicit when an upward vertical handover is considered. Such a handover would be to an overlay with a larger cell size (lower bandwidth per unit area) e.g. WLAN (10 Mbps) to GPRS (56–114 kbps). In this case, the change in QoS is significant and nontransparent to the user.

Let's assume that user A is performing such a handover. Once user A hands over to the new overlay network, it has to reregister with the home agent and obtain a care-of address in order to maintain its home IP address. However, user B does not yet know anything about this unplanned vertical handover of user A and keeps on transferring data at a bit rate that is beyond the new bandwidth capability of the receiving end, in this case the lower bandwidth overlay network, causing link congestion. If care is not taken, this link congestion, will eventually overload user A, causing them to drop their drop their current network connection. Therefore, user B has to be notified about the handover of user A as soon as possible. There are currently two mechanisms by which the correspondent host can be notified about the handover of user A and avoid congesting the link.

End-to-End Signaling

One way in which the correspondent host can be notified is through the host performing the handover. User A can no longer hear the old overlay network, and therefore no signaling can be sent using that network. The earliest time that user A can send any signaling information is when the initialization and configuration of the new network interface is complete and a successful registration is made with the home agent. Only then can user A notify user B about the handover. However, once the registration with the home agent is complete, all the packets send by user

B are now transparently forwarded to user A, causing link congestion. This link congestion prevents user A from making a prompt connection to user B in order to inform user B of the handover. This delay is variable and depends mainly on the amount of congestion on the link i.e. on the initial data rate between the two users. Having an initial high-bit rate session results in a huge amount of congestion, which makes it almost impossible for user A to establish a connection to user B. It is vital that both users stop transmitting data during the unplanned handover until the new multimedia session is negotiated.

In unplanned handovers the new session description is not available to the correspondent host until the mobile host has obtained a new connection through which to inform B of the new session description. Figure 2.10 shows an example of such a handover.

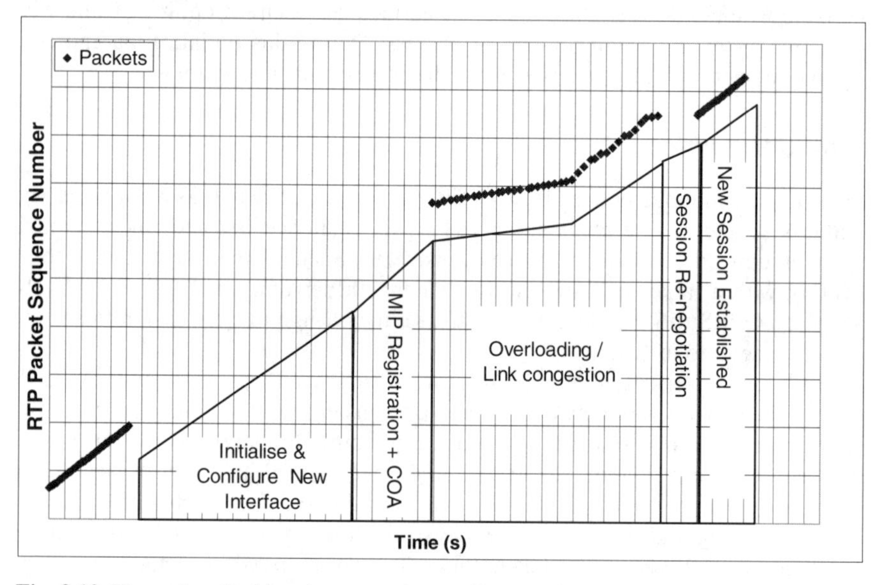

Fig. 2.10 Upward vertical handover – end-to-end approach

Proxy-based approach

The proxy-based approach uses the home agent to take care of all the signaling messages during a handover in an attempt to solve the link congestion problem completely, even at high initial data rates. In this approach, the home agent monitors and keeps track of all active SIP sessions using a SIP cache table. Figure 2.11 shows that user A is currently connected to its home network and is about to set up a two-way real-time audio call with user B. Once the call has been established, user A signals its home agent to update the SIP cache table by adding the IP address of its correspondent host (user B). During the handover, a request is send to

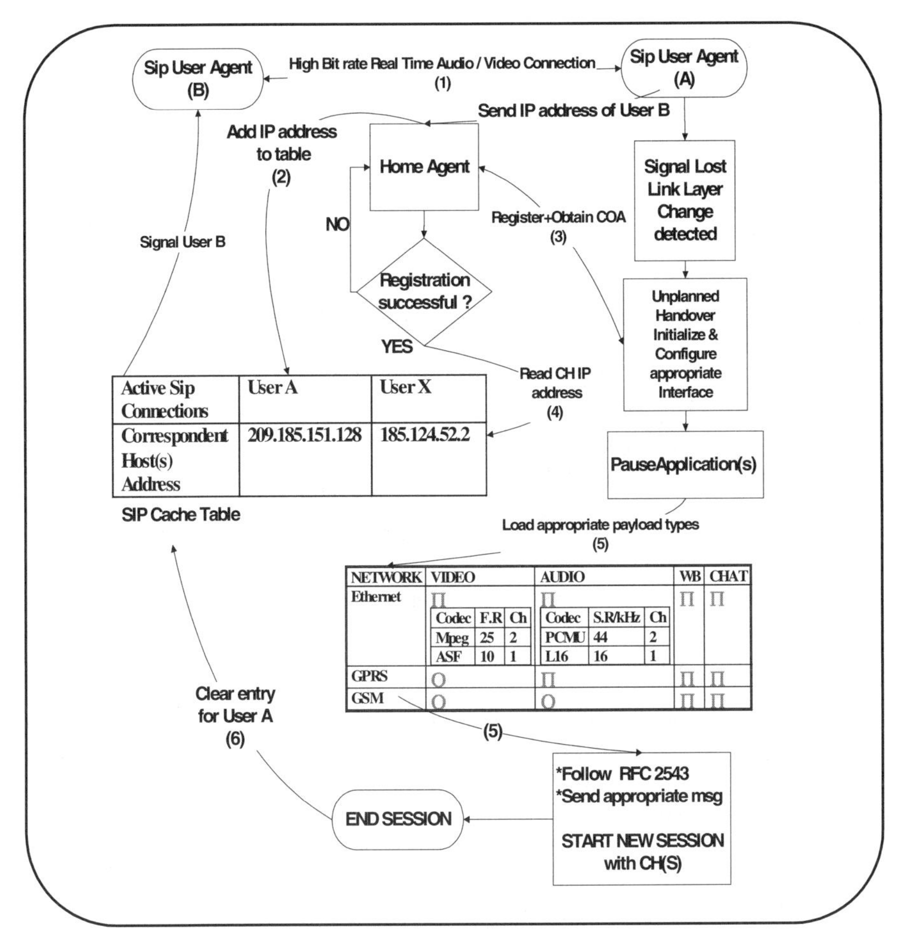

Fig. 2.11 Home agent used as a signaling proxy.

the home agent for a care-of address. After a successful registration, the SIP cache table is consulted and the IP address of user B is obtained.

The home agent then signals user B to stop sending data until they receive the new session description from user A. Meanwhile the SIP user agent of user A receives up-to-date information about the new network connection and generates a reinvite message to describe the new session. A new session is then established, and both users start transmitting using the agreed session parameters. Once the session is finished, user A signals the home agent to clear its entry from the SIP cache table. In this approach the signaling messages are always sent on time, irrespective of the initial bit rate of the users. As a result, user B always receives the signaling messages on time and promptly stops transmitting, avoiding congestion altogether, as shown in Fig 2.12.

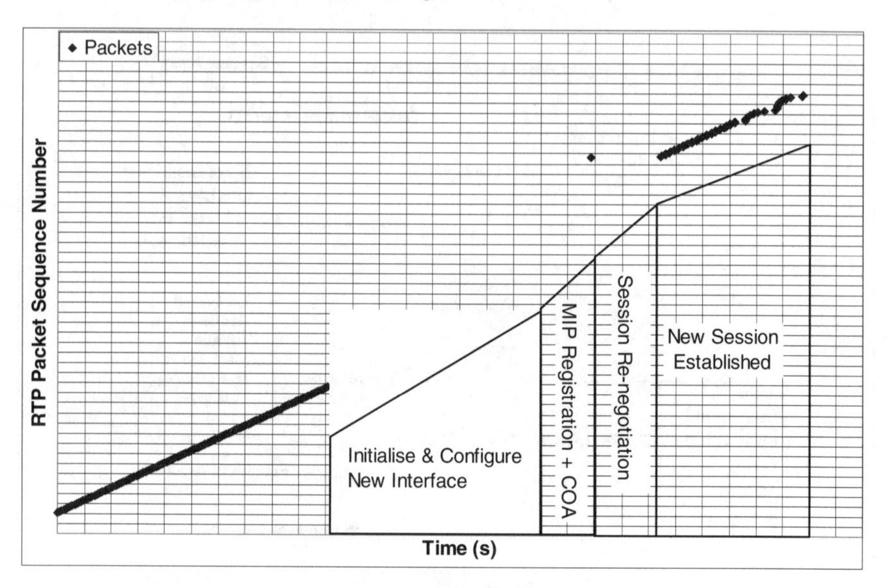

Fig. 2.12 Upward vertical handover – home agent as a signaling proxy

Re-Negotiation Delay

Once the signaling takes place between the users, the multimedia session needs to be modified. The re-negotiation delay is the time taken for both users to pause their applications and re-negotiate the new session. During the handover, user B stops transmitting and waits for a SIP invite message from user A. Having received the SIP message, which contains the new session description, both users renegotiate the session and load the appropriate applications (RFC-2543). The time elapsed is expected to vary from one session to the next and from one user to the other. The terminal itself takes some time to modify/adjust the required applications, depending on the processing power available. A slower terminal takes longer to process and re-negotiate the session, resulting in a higher re-negotiation delay.

2.4.2 Conclusions: Mobile IP and SIP Interworking

In this section we have looked at application adaptation during unplanned vertical handovers. Mobile IP was used for terminal mobility, while SIP was used for signaling and renegotiation of existing sessions. During unplanned handovers, link congestion occurs, preventing any type of signaling taking place between the two users. This link congestion is variable and slows the session renegotiation. We have looked at two different ways of improving the link congestion delay and re-establishing the session as quickly as possible. The first method implements an end-to-end approach where the mobile node informs its correspondent host about the vertical handover. This approach proves to be inefficient, causing link congestion at high data rates. The second approach uses the home agent for the signaling and solves the link congestion problem completely by making sure the signaling arrives on time.

2.5 Planned Vertical Handovers

Planned handovers are predictable, and so some time elapses before the connection degrades or is lost. This gives the terminal enough time to signal and renegotiate the session using the current overlay network before the handover occurs to the new overlay. Planned handovers can be user or terminal initiated. A typical example of a user-initiated handover would be to request a high bandwidth video conference while connected to a GPRS network. In this case, the user that is on the GPRS network has to hand over to an overlay that can support the video stream. On the other hand, terminal handovers are initiated when the signal strength of the wireless base station starts decreasing, indicating that the user is moving away from and eventually outside of the overlay coverage. Sometimes the quality of service (QoS) drops because of a bottleneck somewhere in the path between the two users. In this case the handover will be initiated by monitoring the end-to-end link quality. In the case of the real-time protocol (RTP), this can be done by monitoring the RTCP reports. The primary function of these reports is to provide feedback on the quality of the data distribution. This is an integral part of the RTP's role as a transport protocol and is related to the flow- and congestion-control functions of other transport protocols. Sending reception feedback reports allows the terminal that is experiencing problems to evaluate whether those problems are local or global and accordingly initiate a suitable handover. Figure 2.13

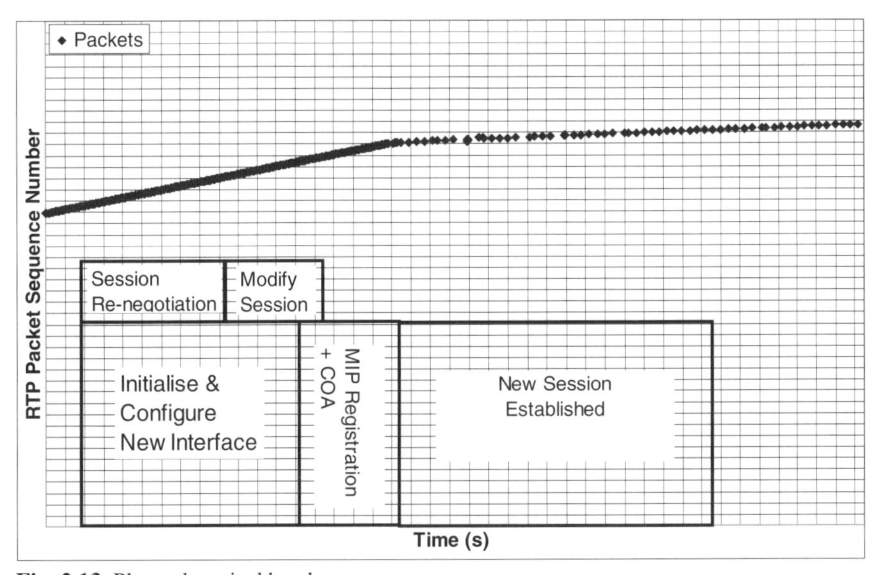

Fig. 2.13 Planned vertical handover

shows a typical planned vertical handover. The mobile host decides to hand over for one or more of the reasons mentioned above. Once a handover is initiated and an appropriate overlay is selected, SIP is used to renegotiate the multimedia session according to user preference characteristics (i.e. bandwidth, cost and services). Once the new session is renegotiated, SIP signals the mobile IP agent to initiate the handover. During a planned handover there is no packet loss or link congestion since the session type is agreed and renegotiated in advance. Figure 2.13 shows a typical planned vertical handover, while Fig. 2.14 shows the timing diagram of the session renegotiation.

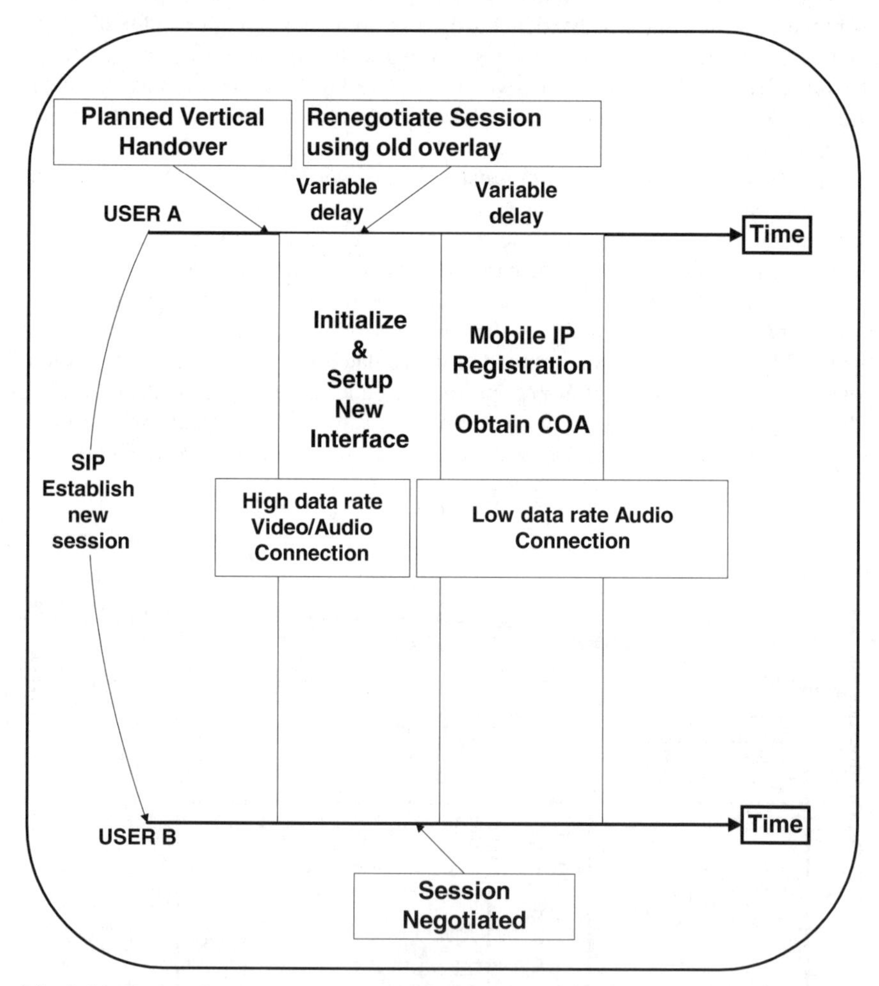

Fig. 2.14 Timing diagram of session renegotiation during planned vertical handover

Limitations of Mobile IP Providing Terminal Mobility

We have already looked at how mobile IP and SIP can work together to handle terminal mobility and to adapt the multimedia session during planned and unplanned vertical handovers. This approach uses a network layer solution, mobile IP, for providing continuous media support when the terminals move around by providing users with fixed home addresses and tunneling packets from the home agent to the mobile user. However, providing terminal mobility using MIP has some serious shortcomings. It does not fully support real time handovers because the mobile host must signal back to the home agent before any packets can be redirected to the new location of the mobile terminal. It uses tunnels in both the forward and reverse directions, and so the headers become invisible, making QoS difficult to implement. Another well-known problem with mobile IP is the triangular routing which is solved by route optimization. However, route optimization also has, many drawbacks. It requires changes in the IP stack of the mobile terminals to enable encapsulation [7].

2.6 Using SIP for Terminal Mobility

SIP-based mobility seems to overcome some of the limitations of mobile IP, nevertheless it is less suitable for TCP-based applications. As mentioned before, in the basic form mobile IP suffers from triangle registration and routing, encapsulation overhead and the need for a permanent home address. In particular, even with route optimization, binding updates must still be tunneled through the home agent, adding handoff delays. Furthermore, they require changes in the operating system of the correspondent host, including authentication mechanisms. On the other hand, SIP does not require any additional changes in the operating system nor the installation of home agents. For multimedia applications, which are typically RTP/UDP (User Datagram Protocol) based, delay and loss are of primary concern; hence we need to decrease the latency as much as possible. So it is advisable to avoid triangular routing and any kind of encapsulation mechanisms. The major advantage of using SIP is that it is primarily designed to offer personal mobility. Personal mobility refers to the ability of the end users to originate and receive calls on any terminal in any location in a transparent manner, and the ability of the network to identify end users as they move across different administrative domains. The SIP Uniform Resource Identifier (URI) scheme and registration mechanisms are some of the main components that are used to provide personal mobility. In order to provide SIP-enabled terminal mobility, the following goals have to be met:

- A mobile device should be able to move between different overlay networks and be able to preserve an ongoing session.
- A mobile device should have the ability to provide the same service, irrespective of the network to which the device is moving.

Terminal mobility can be split into two different stages: precall and midcall mobility. In precall mobility the user acquires a new address prior to receiving or making a call, while for midcall mobility the moving mobile host has to keep the ongoing session alive without having to restart the applications when moving to a new location. Providing continuous connectivity in different types of networks with fixed and wireless segments is in itself a challenging task. Quality of Service (QoS) can change abruptly when a handover occurs.

Precall mobility is the simplest form of terminal mobility. Once the mobile node decides to handover to another overlay network, the new interface is set up and a new IP address is obtained. Once the handover occurs it will be reflected to the application layer which in turn will cause SIP to simply re-register with its home proxy server. This is done every time the terminal obtains a new IP address.

In order to support midcall mobility, we need to add the ability to move while a session is active. By introducing SIP mobility support, we will avoid many of the problems with the limitations of mobile IP. Nonetheless, SIP mobility cannot provide support for TCP connections. In this section, we will look at how SIP could be used to provide mobility for real-time communication. Providing mobility at the application layer means that it can be installed easily and provide mobility to common multimedia applications without the need for mobile IP deployment. Furthermore, using SIP for mobility is possible without making any changes to the IP stack of the terminal.

2.6.1 Vertical Handover using SIP

In order to provide full support for SIP terminal mobility, a common procedure is required to detect handover events and to differentiate between planned and unplanned handovers. During a handover, the user profile preferences should be taken into account in terms of network, cost and bandwidth, and a suitable network selected. Figures 2.15 and 2.16 show how SIP could be used to provide terminal mobility for both planned and unplanned handovers.

An unplanned handover is initiated when several beacons from the current overlay network are not received. The mobile host decides that the current network is unreachable and initiates an unplanned handover. SIP will load the latest user profile and will select an appropriate network to, handover to taking into account the media types of the current multimedia session. Once the new network interface is activated and configured, the mobile host will have a new IP address. Until the session is renegotiated all packets from the correspondent host are lost. SIP will signal the correspondent host, providing the new media description of the required session according to the capabilities of the new overlay network. The session is then renegotiated and modified. Using SIP for terminal mobility during unplanned handovers eliminates the problem experienced with mobile IP. Link congestion does not occur since all packets are not routed to the new IP address of the mobile node until the session is renegotiated. This speeds up the handover time without the need of any for extra signaling which would have been required if link congestion had occurred.

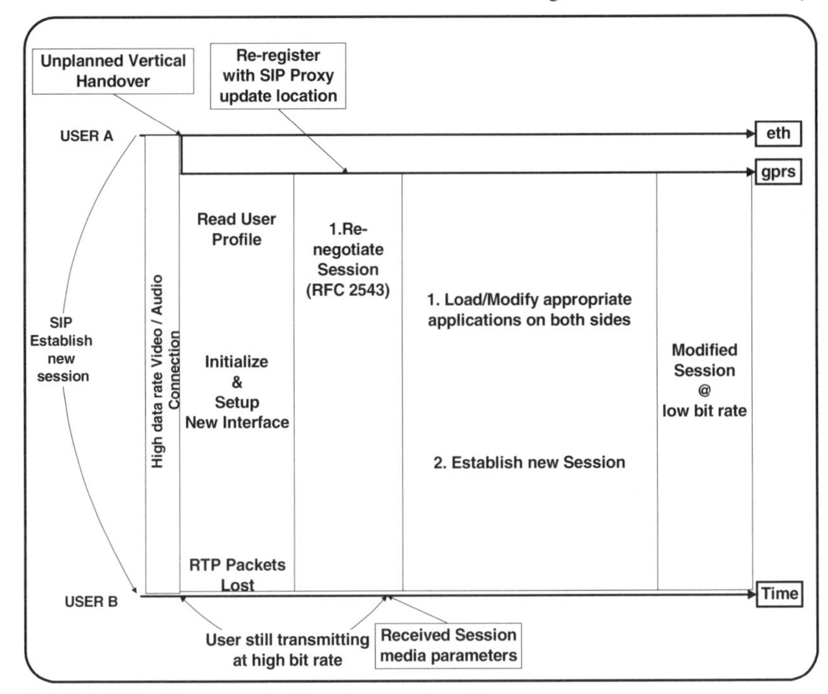

Fig. 2.15 Unplanned vertical handovers using SIP

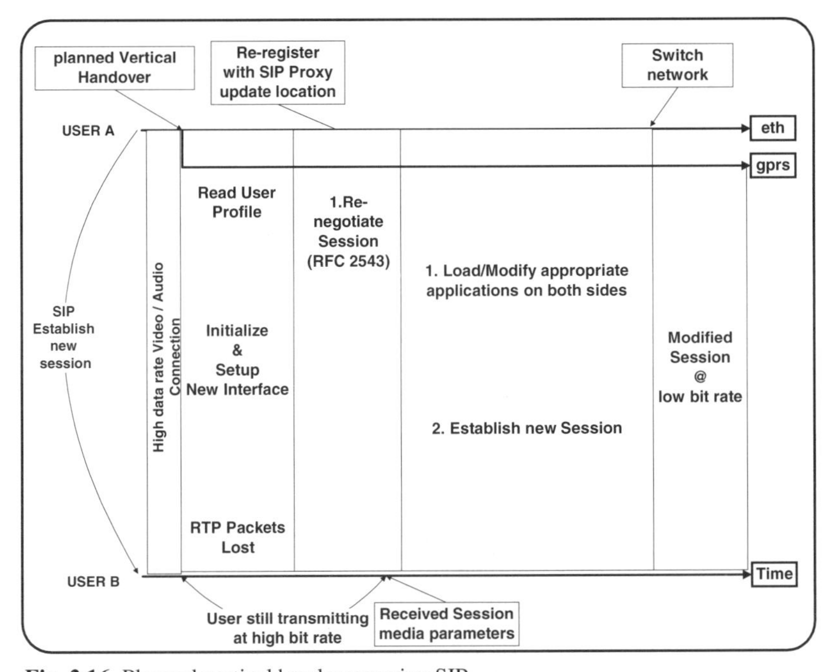

Fig. 2.16 Planned vertical handovers using SIP

Planned handovers are simpler. The main difference, as shown in Fig. 2.16, is that all signaling is done using the first overlay network without experiencing any packet loss. If the new session being negotiated is of a similar nature to the old one, then the handover will be seamless and transparent to the user.

2.6.2 Issues of SIP for Terminal Mobility

One of the issues of using SIP for terminal mobility is the need to change the third-party application to accommodate the IP address change after the SIP reinvite is received with the new contact address. SIP terminal mobility would be very useful if there were a way to avoid changing the application to accommodate the change in IP address when the mobile host performs a vertical handover. One way to achieve this is by using an RTP translator.

Another way is to perform packet filtering within the corresponding node, which looks for UDP packets that have the old IP address of the mobile as their destination address, intercepts such packets, and reroutes them to the new address. The first approach uses a proxy-based solution where the RTP translator (proxy server) intercepts the media packets and directs them to the current location of the mobile host. It also has the ability to buffer media packets and to transmit them to the new location after handovers. Furthermore, such a translator may also be useful for transcoding media to a lower bandwidth or for adding forward error correction. The second approach does not use any proxy servers, but it works on an end-to-end basis. When the mobile host moves to a new overlay network it acquires a new IP address. This information is then passed to the correspondent host using SIP, and as a result, any packets destined for the old address of the mobile node are now locally intercepted and rerouted to the new IP address of the mobile node. This technique does not require any changes in the application itself in order to support the handover. In addition, more functionality could be added to the system to enable it to support end-to-end transcoding. This approach would perform better than the proxy-based one, since the end-to-end delay will be lower if packets are sent directly to the mobile host without being routed via the proxy RTP server.

The extra latency introduced by the RTP proxy would be proportional to the distance between the proxy server and the correspondent host. In both approaches no extra functionality is needed in the SIP proxies or redirect servers.

2.7 Session Mobility

Session mobility allows a user to maintain a media session while moving between different devices, both wireless and fixed. Mobile IP does not support such session mobility. However, this can be supported using SIP. There are not a lot of differences between terminal mobility and session mobility with regard to SIP. In both cases SIP is able to support both the change in IP address as well as the different capabilities of the terminals. The following scenarios will be investigated. In all this we assume that the user has more than one device that is registered with the SIP proxy and available to receive incoming calls.

Caller decides to transfer session to another device. In this scenario the transfer of the session is initiated by the user. The user might want to continue a session begun on a mobile device on the desktop PC while entering his office. One way to perform session transfer would be to use the REFER mechanism [8], as are shown in Fig 2.17. In the basic form user A provides user B with the new terminal location. User B then attempts to establish a session using that contact and reports back results of the attempt to userA@fixed. Once the new connection is established, the initial session is terminated. The diagram shows the sending of the messages. All messages have the same Call-ID.

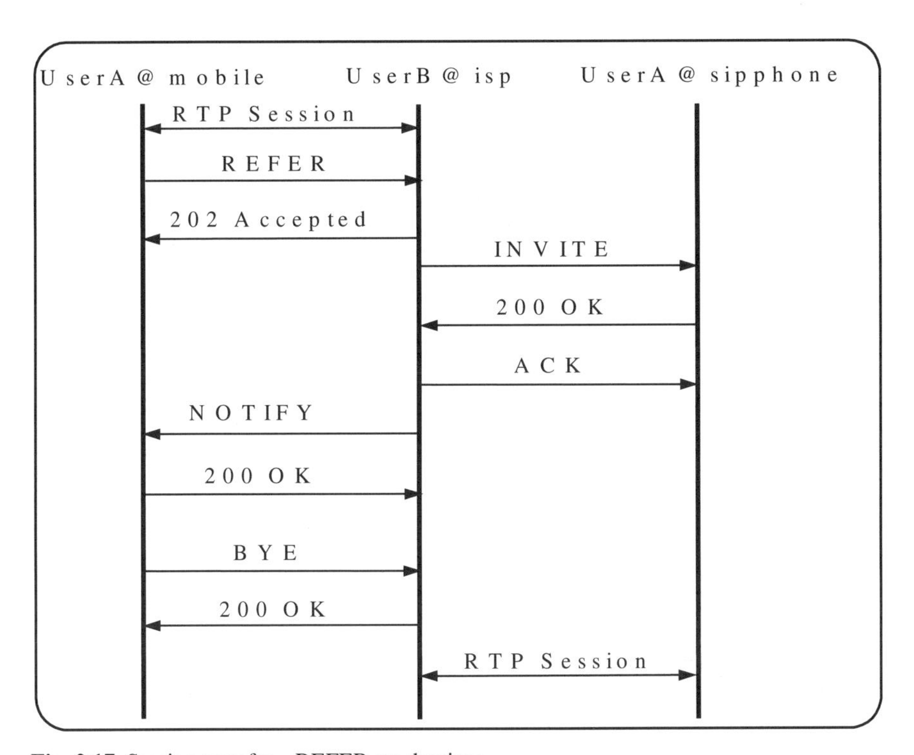

Fig. 2.17 Session transfer – REFER mechanism

Session split up and forwarded to other terminals. On the other hand, the user might decide to split the session and distribute it to different types of devices with different capabilities (Fig. 2.18). For example, the user might be in an audio/video session and would like to transfer the video part of the stream to an overhead pro-

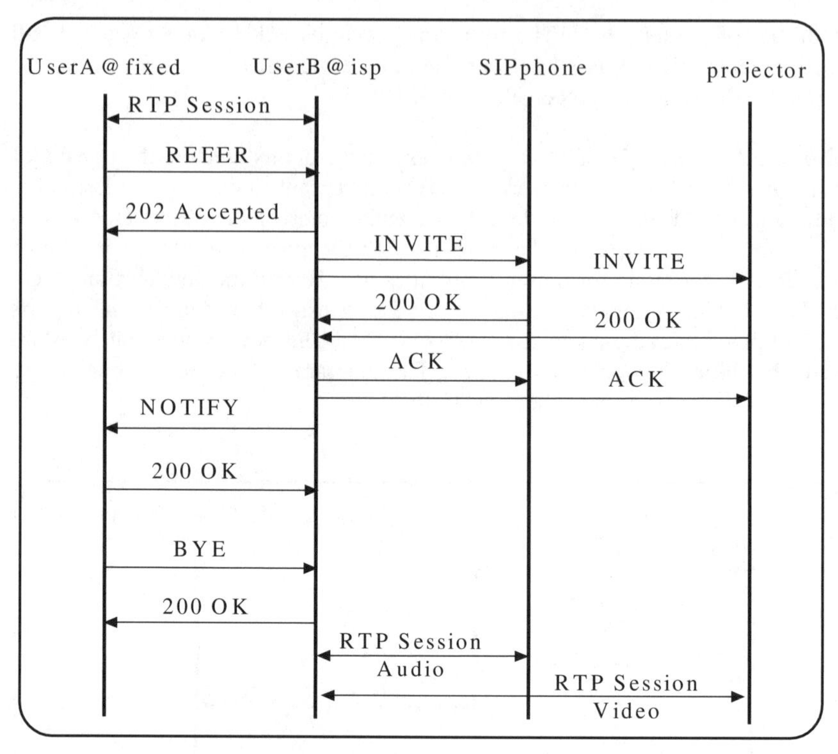

Fig. 2.18 Session split-up

jector. Another example is when the correspondent host would like to change an existing audio session and add a video stream. If the device cannot support this change then the user might like to transfer the video stream to another terminal that supports the new requested session.

Handover detection mechanisms would transfer session to other terminals. As mentioned in Sect. 2.3, the end-to-end quality of service can be monitored – in terms of bandwidth, delay packet loss and jitter, and as a result, a session handover can be initiated. These types of handovers are initiated and executed by the terminal, which also takes into account the user preferences. A session handover may also be triggered if the signal of the current link is degraded as the user moves outside the network coverage or if the network is congested, reducing the available bandwidth. The terminal will either transfer the entire session or parts of it to another device. This gives the user more flexibility.

2.8 Conclusions

In this chapter we have looked at mobility management in heterogeneous networks and how different protocols could be used to provide a complete solution to both terminal and session mobility. Mobile IP and SIP were used together to provide terminal mobility for both real-time and non-real time applications. However, because of some of the shortcomings of mobile IP, we also looked at other ways to replace mobile IP. Application layer mobility can partially replace mobile IP, providing a scalable solution to terminal mobility.

SIP can only provide terminal mobility for real-time UDP traffic, but it does not support TCP-based applications. Predicting the right time to perform handovers is another important issue in mobility management. Application and link layer mechanisms could be used to predict handover events and to provide seamless vertical handovers. However, unplanned handovers cause session disruption and packet loss. These types of handovers are unpredictable and less frequent. Terminal mobility by itself is not enough to integrate the wireless/fixed technologies into one seamless internetwork. The user needs to have the flexibility of maintaining its session when terminal mobility cannot help, i.e. when the users terminal does not have the capabilities to support a specific session. For a complete solution, session mobility is required. Terminal and session mobility working together to provide an ideal environment for mobility management in heterogeneous networks.

References

[1] Alvestrand H, Børseth H, Lovett H, Ølnes J (1997), Final report for IMiS feasibility project. http://www.nr.no/imis/imis-f/Final-report/

[2] http://www.labs.bt.com/projects/m3a/

[3] M. Handley, H. Schulzrinne, E. Schooler, J. Rosenberg Session Initiation Protocol. In: RFC2543 Internet Engineering Task Force (March 1999),

[4] M. Handley, V. Jacobson, Session Description Protocol. In: RFC2327 Internet Engineering Task Force (April 1998)

[5] M. Stemm, "Vertical Handoffs in Wireless Overlay Networks", master's thesis, UC Berkeley, May 1996. Also, UC Berkeley Computer Science Division, Report No. UCB/CSD 96/903, May 1996

[6] F. Vakil, A. Dutta "Mobility Management in a SIP Environment Requirements, Functions and Issues". In: RFC2976 Internet Engineering Task Force (October 2000)

[7] E.Wedlund, H. Schulzrinne, "Application-Layer Mobility Using SIP", ACM SIGMOBILE, Volume 4 , Issue 3 (July 2000)

[8] R. Sparks, "The SIP Refer Method". In: draft-sparks-sip-refer-3265disc-00, Internet EngineeringTask Force (June 2002).

3 Limiting Technology: Problems in Setting Exposure Guidelines for Radiofrequency Energy

3.1 Introduction

As the other chapters in this volume make clear, wireless communications is a highly developed and rapidly growing technology. The hundreds of millions of users of mobile telephones around the world are clear proof of the usefulness of the technology.

Nearly all of the papers in this volume concern "hard" engineering problems such as communications protocols, signal processing, and other technical issues. But wireless communications technology faces other challenges that are partly scientific, and partly social in nature.

Some of these issues stem directly or indirectly from concerns of the public about the safety of radiofrequency (RF) energy used by the technology. I do not pretend to review the enormous scientific literature on the subject. Rather, I consider several issues related to setting exposure limits for RF energy from wireless communications and other technologies. These issues have both scientific and technological components, and great potential impact on the wireless communications industry. Larger questions, for example, the potential effects of wireless communications on the work environment, the political impacts of wireless communications, and privacy issues, are important and interesting, but outside the scope of this chapter.

3.2 Health Effects of Radiofrequency Energy – an Ongoing Controversy

For many years, controversies have occurred over a shifting range of issues related to possible health effects of radiofrequency (RF) energy [1]. It is only natural that this controversy has spilled over to wireless communications technologies as well. Present generations of wireless communications systems operate in or just below the microwave band (which is conventionally defined as 1 to 300 GHz). For example, in the United States, mobile telephones operate in one of two main frequency ranges, near 850 MHz (the older cellular telephone systems) and near 1900 MHz (the newer personal communications services, or PCS). Systems in Europe, Asia, and elsewhere operate at similar frequencies, usually near 900 and 1800 MHz.

While wireless technology, in the form of mass-market consumer appliances, is new, the use of this part of the spectrum is not. Many well-established communications, industrial heating, and other applications operate in nearby frequency bands, sometimes at quite high power levels. But for the first time in history, hundreds of millions of people are spending large parts of their days with low-powered radio transmitters next to their heads. Many of these people have concerns about the safety of the RF energy exposures that they receive. It is surely appropriate to inquire into possible health effects of such exposure.

Also new is the need for many wireless base stations (reportedly 100,000 in the United States alone) to service all of the subscribers. In my city alone (Philadelphia), five wireless telephone providers are building out their systems, and new wireless services are coming across the horizon. Many citizens object to wireless base stations in their neighborhoods (even as they use mobile telephones), in part because of esthetic and in part because of health concerns.

As a scientific issue, the interaction of RF energy with biological systems is hardly new. Scientists and physicians have studied the subject, in one aspect or another, ever since the technology was invented to produce RF energy early in the Twentieth Century. Radiofrequency energy is nonionizing: the energy of a quantum of radiation at 1 GHz is smaller, by a factor of more than 6000, than that of the weakest chemical bond. This is very different from ionizing radiation such as X-rays, where a single photon can disrupt chemical bonds and produce free radicals, a primary mechanism of damage from such fields. RF fields nevertheless interact with biological systems, by exerting forces on charged particles in tissue. However, at realistic exposure levels these forces are many orders of magnitude too small to overcome the effects of random thermal agitation and it seems highly unlikely – some investigators claim "impossible" – that they could produce significant biological effects [2]. By contrast, the heat generated by RF energy as it is absorbed in tissue is a potent mechanism for producing biological effects.

3.2.1 Established Hazards

Sustained research on possible hazards of RF energy began shortly after World War II, when the U.S, military, concerned about possible hazards to their personnel from military transmitters, began funding an extensive research program in government and university laboratories.

By now the literature related to possible bioeffects of RF fields is vast, with thousands of papers in the peer reviewed scientific literature. For an extensive review of this older literature, the 675-page monograph by Michaelson and Lin is highly recommended [3]. A computer database includes more than 30,000 abstracts related to biological effects and medical applications of electromagnetic fields from DC through the microwave range [4].

Certainly RF energy is hazardous, at sufficient levels of exposure: putting a rat in a microwave oven will surely be hazardous to its health. Cataracts can result from exposure to RF energy, at levels similar to those required to produce burns in other tissues. The medical literature has a scattering of reports of burns produced by exposure to intense radiofrequency fields, typically in occupational settings or during medical treatment with RF energy. Tragically, there are two reports of infants being placed in microwave ovens (both survived, but with serious burns) [5].

Because exposure to RF energy at thermally damaging levels is very painful, the exposed individual usually withdraws involuntarily before significant damage occurs. A similar situation occurs when a cook reaches into a broiler to turn over a steak. The cook's hand will be exposed to infrared energy at levels far above government safety limits, but he will withdraw from the oven before sustaining injury.

Consequently, there are few reported injuries from RF energy, and these generally involved intense fields or circumstances in which the victim could not withdraw from the exposure. For example, a few workers have been injured while on towers with high-powered broadcast transmitters (one such worker was suspended in front of a television antenna when somebody turned on the transmitter). The medical literature has a scattering of reports of burns to patients from microwave surgery or diathermy equipment [6].

Other forms of thermal damage can occur, even in the absence of burns. Temporary sterility can be produced in men by exposing their testicles to microwave energy to produce modest heating – a method that was investigated at one time in China for birth control [7]. Whole body exposures at sufficient levels can impose an intolerable heat load on the body, leading to physiological stress and ultimately death. Exposing pregnant animals to RF energy at near-lethal levels can lead to deformations in their offspring.

At lower exposure levels (but still far above those produced by wireless communications systems), exposure to RF energy leads to a range of effects in animals that can be interpreted as normal thermoregulatory responses to manage the added heat load. For example, animals trained to carry out a task (such as pressing a lever for food) while exposed to RF energy will, at some exposure level, stop performing the task and switch to another behavior (in rats, spreading saliva on the tail, a thermoregulatory behavior in that species). This effect, called behavioral disruption, occurs at whole body exposure levels well above the natural rate of heat generation in the animals. This can be interpreted as a normal response of the animal to an uncomfortably warm environment. These effects are hardly mysterious, even as they are technically difficult and expensive to investigate.

3.2.2 Exposure Limits

Major exposure standards in place in the United States, Europe, and elsewhere are designed to exclude all identified hazards from RF energy with a large margin of safety. The oldest major limit (going back to the early 1960s), which in its present version is designated IEEE[1] C95.1-1999 [8], has been widely influential around the world and long has formed the basis for U.S. government regulations. The RF exposure limits developed by the International Commission on Nonionizing Radiation Protection (ICNIRP) [9] have been adopted by most countries in Western

---

[1] Institute of Electrical and Electronics Engineers, a professional organization.

Europe and in many other places around the world. These standards, while differing in many details, are similar in their scientific basis and in their numerical limits.

U.S. and most western European limits were set by committees of experts that evaluated the bioeffects literature, identified potential hazards, and set exposure limits to exclude these hazards with an appropriate margin of safety. The IEEE and ICNIRP committees identified behavioral disruption as the adverse effect occurring at the lowest whole-body exposure level that has been demonstrated reliably in animals. (In reality behavioral disruption is a normal response to excessive heating, and ceases once the exposure has terminated, one might question whether it should be considered "adverse"). The IEEE and ICNIRP limits were designed to provide a safety factor of 10 (measured in terms of incident power density or absorbed energy) for occupational exposures and 50 for the general public. As discussed below, limits in Russia, Switzerland, and a few other places are based on very different assumptions.

In RF bioeffects research, the accepted measure of exposure is the specific absorption rate (SAR), defined as the power dissipated in watts per kilogram of tissue. For an electric field E_i in tissue of conductivity σ, the SAR is

$$\mathrm{SAR} = \frac{\sigma E_i^2}{\rho}$$

where ρ is the tissue density (kg/m^3). In the absence of any heat loss, typical soft tissue exposed to RF energy at an SAR of 1 W/kg will increase its temperature at the rate of approximately $1^{\circ}C$ per hour of exposure. For comparison, the basal metabolic rate in humans is approximately 1 watt per kg of body mass. Heavy exercise, for example playing soccer or football, may cause a doubling of the metabolic rate.

The IEEE and ICNIRP standards are designed to limit whole-body SAR to 0.08 W/kg for general public exposures, i.e. to levels less than one-tenth the rate at which the body generates heat during resting conditions. For exposures in the far field of antennas, it is sufficient to measure the incident power density (W/m^2) or field strength (V/m), which determine the hard-to-measure SAR. For near-field exposures (for example, to the user of a mobile phone) the SAR must be determined directly.

Thus, the exposure guidelines of IEEE, ICNIRP, and other agencies are commonly expressed in terms of incident field intensity (in watts per square meter), with separate limits on SAR for near-field exposure. The standards, which have evolved over the years into complex documents, also include provisions for pulsed fields, contact current, partial body exposure, and other special cases.

Critics often complain that IEEE C95.1-1999 and the ICNIRP standards are "thermal" and that they disregard "nonthermal" hazards. More precisely, the committees examined a large body of evidence for hazards, including reports of "nonthermal" effects. The only persuasive evidence for hazards they identified, however, involved high exposure levels and obviously thermal phenomena.

Both committees were very explicit about the lack of reliable evidence for possible hazards from low-level exposures or "athermal" effects:

IEEE C95.1-1999 "No verified reports exist of injury to human beings or of adverse effects on the health of human beings who have been exposed to electromagnetic fields within the limits of frequency and SAR specified by previous ANSI standards, including ANSI C95.1-1982.... Research on the effects of chronic exposure and speculations on the biological significance of nonthermal interactions have not yet resulted in any meaningful basis for alteration of the standard."

ICNIRP After a long review of the scientific data, ICNIRP concludes: "Overall, the literature on athermal effects of AM [amplitude modulated] electromagnetic fields is so complex, the validity of reported effects so poorly established, and the relevance of the effects to human health is so uncertain, that it is impossible to use this body of information as a basis for setting limits on human exposure to these fields." (The discussion in ICNIRP suggests that this comment would apply to RF energy with other forms of modulation as well.)

3.2.3 Controversies

In some quarters, the possibility of hazard from RF energy at levels below IEEE and ICNIRP guidelines has been intensely controversial. While most of the controversy has occurred outside of the scientific arena, there is evidence of this controversy in the scientific literature as well:

Irradiation of American Embassy in Moscow Readers of a certain age may recall news reports of the discovery by the US government that from the 1950s through the 1970s, its embassy in Moscow had been irradiated, seemingly deliberately, with low level microwave energy. The levels, which ranged up about 0.1 W/m^2, were far below US exposure limits, and also below the much more restrictive Soviet limits. Controversy about possible health effects was raised in the public arena when a former ambassador died of cancer. However, in a 1978 study commissioned by the U.S. government, Abraham M. Lilienfeld (a prominent epidemiologist from Johns Hopkins University) concluded that embassy personnel suffered no ill effects from the microwaves beamed at the chancery [10]. Nevertheless, there are a few articles in the scientific literature disputing this finding [11,12]. The issue is widely discussed on activist sites on the Internet, sometimes accompanied by bizarre speculation about mind control using RF energy and other such topics.

Cataracts from low-level exposure to RF energy Concerns that microwave energy might cause cataracts in people who are occupationally exposed to RF energy at undetermined levels (but almost certainly far below government exposure limits) arose in the 1970s when a scattering of reports appeared in the medical literature. For example, in 1977 Milton Zaret, an American ophthalmologist, published a letter in *The Lancet* (a prestigious medical journal) reporting cases of "chronic hertzian radiation cataracts" in workers such as radar technicians, air traffic controllers and airline pilots, which he thought were due to chronic exposures to RF energy [13]. Another letter to *The Lancet* (1984) reported an apparent increase in cataracts in radio linemen working on radio and television broadcast towers [14].

As a result of concerns raised by these reports, the U.S. government sponsored a number of studies, which were mostly carried out in the 1970s and 1980s. Several high-quality studies showed that microwaves can produce cataracts in experimental animals, but only at high exposure levels that would produce acutely painful heating and probably tissue burns as well. Most investigators believe that the cataracts induced by microwaves in such experiments are thermal in origin, although some investigators argue that nonthermal mechanisms also play a role [15].

The evidence for cataracts in humans from long-term, low-level exposures to RF energy in the workplace remains (in the words of a 1988 review [15]), "equivocal and controversial". One difficulty in establishing such claims has been the absence of adequate (or any) exposure assessment for the individuals who supposedly developed RF cataracts. Another is the difficulty of distinguishing minor defects in the lens supposedly caused by exposure to RF energy, from commonplace defects in the lens or naturally occurring cataracts.

Nevertheless, a few physicians (notably, Zaret) have claimed to be able to diagnose microwave-induced cataracts. A number of workers have filed legal claims for compensation for cataracts presumed to have been caused by occupational exposures to RF energy, generally without success. Joyner describes one suit by a radar worker, which was dismissed by an administrative court that found the biological and epidemiological evidence for such injuries to be inconclusive [16].

Cancer and occupational exposure to RF energy Well before the controversy arose about cancer and mobile phones, a scattering of reports had appeared in the scientific literature claiming an increase in cancer risk in workers whose job titles suggested the possibility of exposure to RF energy. For a variety of reasons, these reports remained unpersuasive. For example, Szmigielski reported an increase in cancer mortality in Polish soldiers whose job classifications suggest exposure to RF radiation, compared to other soldiers used as controls [17]. This study has been widely cited by activists as evidence for hazards from chronic exposure to low-level RF energy. However, the Stewart committee, a blue-ribbon panel of experts in the UK that was formed to evaluate possible health risks of mobile phones, in its final report dismissed the study as "unsatisfactory, [and] can be given little, if any, weight" due to egregious weaknesses in study design [18].

Many biological effects of RF energy The studies mentioned above concern specific health problems that may (or may not) be linked with exposure to RF energy at low levels.

The scientific literature is awash with reports of biological effects of RF and microwave energy, some at exposure levels below current Western limits, and some that the investigators believed were "nonthermal". Few if any of these studies are standard toxicology assays, and their relevance for human health is uncertain. Taken together, they convey the impression that RF fields are biologically active, even at low exposure levels, but they also provide little useful information about possible health problems.

The scientific literature about such effects is rife with controversy. For example, in 1995 Lai and Singh reported that exposure to RF radiation at an average whole-body exposure of about 1 W/kg caused single and double strand breaks in the DNA in the brain cells of rats [19]. (A similar SAR may occur in localized regions in the head of a mobile phone user). Recent studies cast doubts on this finding. A group led by Roti Roti has reported its inability to find any evidence for DNA strand breaks in studies that were specifically designed to confirm Lai's findings. Moreover, the investigators identified a potential artifact in Lai's original studies [20]. Lai continues to defend his studies.

In short, the literature on biological effects and health hazards of RF energy is complex and frequently inconsistent and controversial even among scientists. Such problems are commonplace in environmental toxicology studies, which are exceedingly difficult to do well and are usually open to varying interpretations. Health agencies deal with this problem by taking a "weight of evidence" approach – by evaluating all relevant evidence, and giving little weight to studies that have obvious technical flaws or are apparently contradicted by later followup studies. Thus, the studies cited above have not been persuasive to health agencies, whose expert reviews generally support the major international limits.

However, by picking and choosing among this diverse literature, and giving worst-case interpretations to studies reporting effects, one can paint a frightening picture of the hazards of RF energy. This approach is evident in Web sites of activist groups, which list diverse collections of studies reporting effects of RF energy, while ignoring other studies that may be more relevant to human health risks and had negative results.

This approach was taken in 1977 by Paul Brodeur in his tendentious book, The Zapping of America [21]. Brodeur created a national sensation at the time, with his argument that RF fields, even at low exposure levels, pose deadly hazards that are deliberately concealed by a military/industrial/scientific conspiracy. Conspiracy theories are easy to propose, very attractive to many individuals, and virtually impossible to refute. But hundreds of scientists from academia, industry, and government, in numerous countries, have been involved in assessing risks of RF fields, and it would seem unlikely that any serious hazard could be covered up for any length of time.

3.2.4 Health effects from mobile phones and base stations

Brain cancer and mobile phones Given this background, it is hardly surprising that wireless communications has also engendered controversy on health grounds. A particular controversy developed in 1993, when David Reynard appeared on an American television show and reported that his wife had died of brain cancer incurred through her use of a cell phone, and announced that he was suing the industry for damages. The story was picked up by the international media and created an immediate sensation. The Reynard suit was dismissed in 1995 by a U.S. federal court for lack of valid scientific evidence for its claims.

Reynard's claims prompted governments and industry around the world to support studies on a possible link between mobile phones and brain cancer, which were soon extended to include other possible health problems as well [22]. By late 2002, more than a half dozen epidemiology studies and nearly a dozen animal studies had been reported, bearing on the issue of RF energy from mobile phones and brain cancer. Their results have been overwhelmingly negative [23]. A review in 1999 found that the relevant biological and epidemiological evidence does not suggest that RF fields cause or promote cancer [24]. Nothing has emerged since then to challenge seriously this conclusion.

As many commentators have pointed out, the whole story is not yet in. The epidemiology studies (as with all such studies) had limited statistical power and thus could not have detected small increases in risk or risks that might affect small subgroups in the population. Moreover, brain cancer takes years to develop, and the studies published so far have limited value in identifying health effects that may only become apparent after many years of use of the telephones. A major epidemiology study on mobile phones and brain cancer is underway by the International Agency for Research on Cancer (IARC, a component of the World Health Organization). When completed, this study will provide the most sensitive and reliable test so far for any possible link between use of mobile phones and brain cancer – but it will certainly not be the last word on the subject.

Noncancer health effects to users of mobile phones or residents near base stations A few scientists have published papers claiming that RF energy from mobile phones or base stations can cause health problems other than cancer. The most recent contribution to this literature is a speculative article by two Israeli scientists who argue that the "head serves as an antenna and brain tissue as a radio receiver", capable of demodulating the microwave energy from mobile phones and producing health effects [25]. Apart from lack of direct biophysical evidence for such effects, the slow response time of cell membranes would seem to preclude such effects.

In the first epidemiological study of residents near cellular base stations, published in July 2002, Roger Santini and colleagues reported an increase in nonspecific health symptoms (such as fatigue, irritability, headaches) in people living in the vicinity of cellular base stations [26]. The study had obvious and very serious technical flaws: the investigators had simply mailed questionnaires to many citizens asking them about any health problems, and also how close they lived to base stations, without independently verifying their responses or ensuring a high level of response. Surely individuals who felt that they had health problems connected with base stations would be more inclined than others to respond. Other experts have questioned the feasibility of performing any valid epidemiology studies with residents near base stations, in part because of the difficulty of exposure assessment [27]. Santini, in public statements and in his paper, has recommended that the minimum distance between people and base stations be not less than 300 m.

Given the complexity of the issue, health agencies have been reluctant to state that RF fields within exposure limits are safe, even as they do not claim that they are dangerous. As a result, they send mixed messages to the public. For example, in mid-2000 the Stewart Committee, an expert group in the UK, published an extensive review of the scientific literature that concluded "the balance of evidence to date suggests that exposures to RF radiation below [recommended limits] do not cause adverse health effects to the general population" But "[it] is not possible at present to say that exposure to RF radiation, even at levels below national guidelines, is totally without potential adverse health effects..." [18].

Are mobile phones and base stations safe? Whatever future scientific developments may be, no health agency is likely to declare mobile phones (or any other technology) to be "totally without potential adverse health effects". Such statements are impossible to justify scientifically, because of the impossibility of proving the negative (the absence of effects). Moreover, the causes of cancer are multiple and largely unknown, and agencies are loath to pronounce agents as noncarginogens. For example, IARC has conducted detailed cancer risk assessments on nearly one thousand substances and exposure conditions. It proclaimed only one substance (caprolactam, a feedstock in the production of nylon) to be a "probable noncarcinogen."

Certainly, many people *feel* threatened by RF fields. Particularly in Europe, a substantial fraction of the population feels that they are "electrically sensitive" and suffer adverse effects when exposed to weak electromagnetic fields, including RF fields from wireless base stations and mobile telephones. Organizations such as The Swedish Association for the ElectroSensitive have been demanding protections from "electrosmog". Gro Harlem Brundtland, the Director-General of the World Health Organization, has been quoted in the media as saying that she gets headaches when she uses a mobile telephone. Independent scientific studies, however, consistently fail to find a connection between the symptoms experienced by electrically sensitive people and exposure to electromagnetic fields [28].

3.3 Harmonizing standards

3.3.1 Variation in Exposure Limits Around the World

To everybody concerned with the issue of health effects of RF energy, one vexing problem for many years has been the huge disparity in exposure limits for RF fields that are in place throughout the world. For many years, the differences were most apparent between limits in Russia and most of Eastern Europe (which originated with the Soviet Union and its Warsaw Pact allies) and in the United States and Western Europe. This situation has become even more complicated with the recent adoption of "precautionary" limits by Switzerland, Italy, and a few other countries. Behind these differences is an interesting story of differences in perception of science and health protection.

Table 3.1 compares five different exposure limits for RF energy at 2000 MHz (similar to that used by many cellular telephones throughout the world). The limits are for long-term exposure to the general population. They apply for whole-body exposures in the far field of antennas; limits for partial body exposure in the near field of antennas are given in Table 3.2.

Table 3.1 Exposure limits to RF energy in several countries.

Country	Limit for general public exposure to RF fields (2000 MHz) for extended periods of exposure (W/m^2) [a]	Basis
ICNIRP (adopted in numerous countries worldwide)	10	Science-based
U.S. Federal Communications Commission, Bulletin 65, "Evaluating Compliance with FCC Guidelines for Human Exposure to Radiofrequency Electromagnetic Fields", Washington, DC, 1997	10	Science-based
China , UDC 614.898.5 GB 9175 –88	0.1	Science-based
Russia , Sanitary Norms and Regulations 2.2.4/2.1.8.055-96	0.1	Science-based
Switzerland, Ordinance on Protection from Non-ionising Radiation (NISV) of 23 December 1999	0.1	Precautionary
Typical maximum exposure from cellular base station mounted on 50-m tower (assuming a total effective radiated power of 2500 W in each sector, sum over all channels)	0.01	

[a] Applies to far-field exposure, extended duration.

Limits in the United States, most Western European countries, and many countries in other parts of the world follow IEEE C95.1-1999 or the quite similar ICNIRP limits. Those in Russia together with most of its former Warsaw Pact allies, Switzerland, and a few other countries are as much as one hundred times lower. In Table 3.2, these limits are designated "science-based" and "precautionary" for reasons that will become apparent below.

Science-based limits Since the earliest development of RF exposure limits, the Soviet Union and its Warsaw Pact allies have taken an approach that is completely different from those taken by the United States and Western Europe. This difference has persisted over the years: the latest Russian exposure standards (1996, 1997) are essentially identical to previous Soviet limits (1976, 1978, 1984). While the limits in the United States and Western Europe have evolved over the years, the changes have largely been a result of more refined engineering calculations and a desire to provide a higher level of protection to nonoccupational groups, and not as a result of changes in the scientific understanding of the hazards involved.

As described earlier in this chapter, Western limits (IEEE C95.1-1999 and ICNIRP) were based on studies that identified potentially hazardous effects, largely based on studies involving short-term exposures at high levels to animals.

The rationale for these limits has been spelled out very clearly in the extensive documentation accompanying the standards.

The rationale for the Russian limits is much less clear, and is not described in the standard itself.[2] There is, however, an extensive body of commentary about the Russian standards by scientists who have been professionally involved in RF health and safety studies, including some by Russian and East European scientists [e.g. 29, 30, 31, 32, 33]. Senior Russian scientists have reviewed the biological effects of RF energy from their own perspectives (which differ greatly from those of most Western scientists who are active in this field) [34].

Clearly, the Russian and Chinese limits are not principally designed to protect against thermal hazards; their exposure limits are far below any thermally significant levels. And the limits have other features, for example provisions limiting cumulative exposure over time, that are inexplicable if the goal were to protect against thermal hazards.

These limits reflect the conviction that long-term (hours or more) exposures at levels far below Western limits result in adverse health effects – which is just what activists in the West have been claiming all along. As one Russian commentator put it, the goal of the limits is to ensure that "EMF [electromagnetic field] exposure should not provoke in the person even temporary disturbance of homeostasis… neither in the nearest nor in the remote period of time"[33]. This implies that exposures at the higher levels allowed by Western limits can provoke such disturbances.

Indeed, the Russian and Eastern European medical literature contains many reports of health effects from low-level exposure to RF energy. These include, for example, nonspecific problems (such as headaches, fatigue, irritability, sleep disorders, and dizziness) in workers in radio factories who have presumed but exposure to RF energy at unknown levels [35, 36]. The Chinese literature contains similar reports [37]. The Russian literature contains reference to a "microwave disease" characterized by "asthenic, asthenovegetatic, and hypothalamic syndromes" [38]. This disease is not recognized in Western medicine, and its diagnostic criteria would be unsatisfactorily vague to many Western physicians – a complaint that some Eastern European physicians have raised as well [39, 40].

Many of these reports employ concepts that are unusual in Western occupational health studies. For example, Vasilevskii et al. conclude, on the basis of EEG (electroencephalogram) and other tests, that workers exhaust the "functional reserves" of their central nervous system after 14 years of work with microwaves and other electromagnetic fields [35]. The document that promulgates the Russian standard lists asthenic, astheno-vegetative, and hypothalamic syndromes as "clinical disorders resulting from EMR [electromagnetic radiation] RF exposure" [41].

Evaluating this body of scientific literature poses great problems for Western scientists. Characteristically, the research is only briefly described, lacking basic information such as the frequency and intensity of exposure. Moreover, many of the studies appear to suffer from serious flaws. For example, many of the occupa-

---

[2] I thank Dr. A.G. Pakhomov for providing an English translation of the present Russian standard (SanPin 2.2.4/2.1.8.055-96).

tional health studies are little more than case reports, and not controlled studies. Many occupational health studies appear to rely on post hoc data analysis. In such studies, the investigators applied large batteries of tests to their subjects, and assumed that any variation in the results between the "exposed" and "control" individuals was a direct effect of RF exposure. That may or not may be the case, depending on a host of considerations. In short, this literature contains many claims of health effects of RF energy, but few if any are adequately demonstrated by Western scientific standards.

Precautionary limits. Recently, Italy, Switzerland, and a few other countries instituted exposure limits that are based on a totally different approach, the precautionary principle. The Swiss limits, in particular, were developed in response to public opposition to broadcast transmitters and wireless base stations. Their accompanying documentation cites a variety of reported biological effects of RF energy but does not conclude that levels of public exposure to RF energy is actually hazardous.

Unlike science-based standards such as ICNIRP which seek to avoid identified hazards, the Swiss limits were, in the words of the explanatory document accompanying the limits, "specifically intended to minimise the yet unknown risks" of RF and power-frequency electromagnetic fields. The exposure guidelines were set at the lowest levels that were felt to be technically and economically feasible. In practice, that meant reducing the ICNIRP limits by a factor of 10 in field strength or 100 in power density.

The Swiss are not about to give up their mobile telephones, and it is surely no coincidence that the "precautionary" limits are somewhat higher than RF exposure levels resulting from typical wireless base stations antennas mounted at conventional heights on towers. They do, however, make it difficult to locate base station antennas on apartment buildings and other low structures-and one might suspect that the limits were intended to discourage locating antennas in such places. The Swiss limits do not apply to wireless handsets, which expose the user to far higher levels of RF energy than do wireless base stations, or to medical or industrial exposures.

Such precautionary limits for RF energy have had significant costs, in addition to increased operating expenses for wireless providers. In Italy (as in Switzerland) RF exposure limits had been set at low levels, in large part to address citizens' concerns about health risks from cellular base stations. But reducing the exposure limits served to *increase* levels of public concern. As Italian scientist Vecchia described it [42], the public interprets exposure limits as thresholds for real hazard. Reducing exposure limits for RF energy necessarily means that real-world exposures are closer to the limits and hence are likely to be perceived by the public as more dangerous. In Italy this led to a cascade effect, with different regions in the country (and sometimes individual cities) imposing their own increasingly restrictive limits. A national political crisis occurred when it was discovered that a Vatican radio transmitter outside of Rome produced RF fields that were above the Italian limits. In Italy, overexposing the population to RF energy is a criminal offense, which should give pause to executives in the Italian telecommunications industry.

3.3.2 Can Standards be Harmonized?

"Harmonizing" RF exposure limits has long been a perceived need among standards setting committees, and is one major goal of the World Health Organization's EMF Project. Such harmonization would help meet a variety of needs. From an ethical perspective, it addresses the desire of the World Health Organization (WHO) to provide a consistent level of health protection to different people around the world. From an industry perspective, the large variations in exposure limits create practical problems in serving a worldwide market. From a political perspective, bringing exposure limits around the world into line would help reduce some of the public controversy and fears connected with RF fields.

But harmonization will not be easy to accomplish. "Harmonizing" Russian and Western limits such as those of ICNIRP will never be possible as long as the studies that are the basis for the Russian limits are inaccessible to Western scientists because of language or failure to meet Western criteria for scientific quality. Russian scientists, for their own part, are likely to consider many Western studies, that involve high RF exposure levels and obviously thermal phenomena, to be painfully obvious and besides the point.

Larger differences in philosophy are involved as well. Western standards setting bodies evaluate adverse effects that are rather narrowly defined. The Russian exposure limits for RF energy are based on criteria that have little relation to those usually considered by Western agencies in setting limits. For example, most Western health agencies would undoubtedly consider a "temporary disturbance in the homeostasis" of an individual to be vague and unquantifiable. To a Western scientist, the criteria that Russian scientists use to diagnose it (e.g. subtle changes in heart rate variability or EEG) to be nonspecific with no clear significance for health. But such concepts are familiar ones in Russian medicine, and Russian exposure guidelines for RF energy are based on such concepts. Such differences stem from very different views about health and medicine. They will not be resolved any time soon.

One useful first step would be for different standards groups to sit down and decide on a uniform set of criteria for accepting scientific reports (publication in peer reviewed journals, appropriate exposure assessment, blinded study design, etc.), and then evaluate specific reports for inclusion or exclusion from consideration. A second useful step would be to identify Russian, Chinese or other studies that Russian scientists accept as demonstrating health effects of RF energy. These studies, which may not meet Western criteria for acceptance, should then be followed up by stronger studies with appropriate design and standards of reporting. It would certainly benefit Western societies to find out if there *are* health hazards from RF energy that have been overlooked by Western standards setting committees.

Even more difficult will be reconciling precautionary and science-based limits. The precautionary principle appears in numerous international treaties and enjoys widespread political support, particularly in Europe. However, it remains elusive in meaning, and subject to a variety of definitions [43]. Except among environmental activists, it seems to have garnered little support in the United States, which continues to rely on traditional risk assessment in setting exposure limits and environmental protection.

In the end, harmonization may result more from political and economic pressures than from scientific data. Recently, the Czech Republic revised its limits upwards to those of ICNIRP, as part of integration into the European Union [44]. Other Eastern European nations, which also want to join the European Union, will seriously consider similar changes

There is a strong message in this story to the wireless industry. To a large extent, the precautionary limits adopted by Switzerland, Italy, and other countries resulted from public opposition to wireless base stations. Insofar as industry can mitigate these concerns, for example through more effective communication with the public or more sensitive location of base stations, it may save itself a lot of trouble and cost in the future.

3.4 Limiting RF Exposure from Handsets

A user of a mobile telephone faces a very different exposure situation than a person living near a wireless base station. The exposure occurs in the near field of the antenna, with maximum levels far higher than produced in the body of a person by a nearby base station.

This near-field exposure poses difficult technical problems. The exposure is both difficult to measure, and may approach or exceed regulatory limits. These limits, as they apply to a handset, are expressed in terms of the SAR of RF energy in the body of the user. Table 3.2 compares the SAR limits of two major standards, in effect in the U.S. and most of Western Europe.

Table 3.2 Comparison of IEEE and ICNIRP exposure limits for handsets

Exposure standard	Limit SAR for near-field exposure to RF fields (2000 MHz) as would apply to the head of a mobile telephone user (W/kg)	Averaging volume
ICNIRP	2	"any 10 g of contiguous tissue"
IEEE C95.1-1999 (partial-body limits incorporated into U.S. Federal Communications Commission Bulletin 65, "Evaluating Compliance with FCC Guidelines for Human Exposure to Radiofrequency Electromagnetic Fields", Washington DC 1997.)	1.6	"any 1 g of tissue, defined as a tissue volume in the shape of a cube"

The numerical limits in these two exposure standards are rather similar (1.6 vs. 2.0 W/kg). However, the different volumes of tissue over which the SAR is to be averaged has significant implications. Because the RF energy from a wireless handset is deposited near the surface of the head, mostly in the outer centimeter or so of tissue, the ICNIRP limits are much less restrictive than those of IEEE which have been incorporated into US federal regulations.

Exposures from present mobile telephone handsets are below these limits, but not by much. For example, one Web site lists the SAR values for 47 mobile handsets (averaging volume not stated) [45]. In this group of handsets, the median SAR was approximately 1 W/kg, which is approximately half of the regulatory limit.

On rare occasions handset manufacturers have exceeded these limits. For example, in December 1998, the FCC announced that Sony Electronics recalled 60,000 handsets that exceeded FCC exposure limits. Moreover, some handsets on the market (not necessarily legally) have not been tested. For example, in March 2002 the U.S. Federal Communications Commission fined two Miami firms for marketing long-range (and high powered) cordless phones that violated agency guidelines [46].

Responding to widespread public concern about the safety of mobile telephones, other countries are establishing their own regulations concerning use of mobile telephones. As of late 2002, China was considering legislation that would halve the SAR limits for mobile handsets below ICNIRP levels, and require circuits to reduce power levels after two hours of continuous use. This proposed law has been vigorously opposed by mobile telephone manufacturers (both Western and Chinese) for obvious commercial reasons [47]. In 1995, Italy passed a law requiring a minimum distance of 20 cm between a mobile phone antenna and a user's head [48]. When the law was passed, most mobile telephones were mounted in automobiles. Still on the books, the law can now be interpreted as prohibiting a user from placing a handset against his ear – a law that is obviously broken many times a day by the millions of mobile phone users in Italy. Numerous countries have, in addition, passed laws against driving while using a mobile phone – a real hazard which is not discussed in this chapter.

3.4.1 Power (SAR) to the People

Recently, governments and companies have begun to provide SAR data from mobile phones to the public, motivated in part by the ethical principle that consumers should be given information on which to make rational decisions. The SAR is usually quoted as a precise number with two or three significant digits. Beneath the seeming precision, however, remain some very large uncertainties:

1. The SAR from a handset as measured by standardized tests has very uncertain relation to the SAR actually produced in the body of a user. These tests include computer simulations and measurements on liquid-filled models of the head (called phantoms). Until recently, there has been no standard method of determining SAR, which severely limits the ability of a consumer to compare SAR from different handsets. This uncertainly has been addressed recently, as three in-

ternational standards-setting bodies (IEEE, IEC[3], and CENELEC[4]) have worked simultaneously to develop valid measurement procedures that are consistent across several countries.

These uncertainties have clear implications for industry, because the SAR produced by many handsets is close to regulatory limits, and the resulting uncertainties may be legally important. One group has reported recently that placing a handset close to the chest, for example by putting it in a shirt pocket while using a "hands-free kit", will result in higher SARs in the chest than the user would experience in the head by placing it against the ear [49]. Handsets that are in compliance with government SAR limits, when measured using standardized tests, may actually produce excessive exposure when placed against parts of the body other than the head. Health consequences aside, showing that a handset produces SAR levels above regulatory limits may have serious legal consequences for a firm

2. The actual exposure to a user of a handset in real-world conditions is determined chiefly by the quality of the network and not by design characteristics of the handset, a consequence of adaptive power control. A recent study showed that, for a person walking around the streets of Paris, most of the time his mobile phone was operating at 1% of full power [50].

3. The relation between the SAR limits and threshold for actual hazard is very uncertain. Until very recently, there have been almost no biological or toxicological studies with RF energy involving partial-body exposure. Consequently, the limits for partial-body exposure in both IEEE C95.1-1999 and ICNIRP were based on a series of ad hoc assumptions following a logical process that was, to say the least, hardly rigorous. The logical gaps, indeed, would be apparent to an engineering undergraduate.

The logical process was roughly as follows. A number of studies have shown that the peak SAR produced by far-field exposure in a number of animal and human models is as high as 20 times the SAR averaged over the whole body. This led the committee that developed an earlier version of the IEEE (ANSI C95.1-1982) to propose a peak exposure level of 8 W/kg in the body of a human, based on a whole-body limit of 0.4 W/kg. Arthur W. Guy (now emeritus from the University of Washington, Seattle), the chair of the committee that produced the 1982 standard, told me that the committee further specified that the SAR be averaged over a 1 g averaging mass. This was done for technical considerations - it corresponded to the smallest volume of tissue in which the exposure could be determined feasibly with methods available at the time. By contrast, the ICNIRP standard used cataract formation as an endpoint, and averaged the SAR over the mass of an eye.

In its 1991 update, the IEEE standard introduced a two-tiered limit, that incorporate an additional safety factor of 5 for uncontrolled exposures (roughly equivalent to exposures to the general population). This brought the limit for par-

[3] The International Electrotechnical Commission, an international standards-setting organization.

[4] The European Committee for Electrotechnical Standardization, an international standards-setting organization.

tial body exposure to 1.6 W/kg, averaged over any gram of tissue. When the FCC developed its own exposure guidelines, it adopted the 1.6-W/kg limit for gram-average exposures and applied it to mobile phones and other small transmitters.

Thus, the 1.6-W/kg limit was based on a thermal effect associated with whole-body exposure (behavioral disruption) measured in animals, and the 1-g averaging mass was based on engineering rather than toxicological considerations. This limit is certainly very conservative in guarding against thermal hazards; calculations show that extended exposures of the head at this level will cause maximum increases in tissue temperature of the order of 0.1°C, which is negligible. By contrast, studies have shown that the surface of a user's head can increase in temperature by as much as 3°C during a 20 minute phone call due to the thermal insulation effect of the handset. Placing a block of wood against the head would result in similar temperature increases.

As a result of these uncertainties, SAR data give consumers no useful information on which to base decisions. The SAR is a measure of absorbed power, i.e. heating potential. If the hazards from excessive exposure to RF energy are thermal, then reducing the SAR below levels that are thermally innocuous will not increase the "safety" of a handset any more than reducing the temperature of a cool shower will further increase its safety. If any hazard exists from mobile phones, it is certainly unrelated to heating, in which case SAR would be an insufficient measure of exposure. After decades of controversy, there is no scientific agreement that any such hazards exist, or what would be the exposure conditions that would produce them. I cannot explain to myself why anybody should be concerned about SAR (providing it is below regulatory limits), or explain to a layperson the usefulness of comparing the SAR ratings of different handsets.

Even more troubling is the expanding after-market in "shields" and other devices intended to reduce exposure to users to RF radiation from handsets. Independent tests have shown that many of these devices are ineffective in reducing exposure [51]. Some, having no plausible basis at all, are on face value blatant frauds. The US government has recently taken legal action against two manufacturers of shields for misleading advertising [52], and may take more actions in the future. Meanwhile, a consumer who wants to reduce exposure (for whatever reason) can do so by limiting the duration of calls, and by using a "hands-free" kit that moves the headset away from the body.

People can surely choose what they worry about. But I suggest that consumers should consider several major independent reviews conducted by expert groups including:

- International Commission on Non-Ionizing Radiation Protection
- European Commission Expert Group
- U.K. National Radiological Protection Board
- Royal Society of Canada Expert Group
- U.K. Independent Expert Group on Mobile Phones
- French Expert Report
- Dutch Health Council

All of these reviews have reached a similar conclusion: that the weight of evidence is that RF exposures within accepted limits (such as those of ICNIRP or the very similar US limits) do not cause any adverse health effects.

3.5 Implications for Technologists

As the above comments make clear, the possible health effects of RF energy have been debated for generations – even as health agencies have repeatedly failed to find clear scientific evidence of a problem. Technologists and industry managers should not expect science to provide a neat solution to what is in large part a human problem, the desire for certainty about risk.

In short, the wireless communications industry cannot look to health agencies for unqualified statements that their products are safe, and many citizens are convinced that they are not. There is no legal requirement for industry to prove that the emissions from its products are "safe," but rather that they meet relevant government limits.

However, the industry has a clear ethical and legal obligation to react appropriately to any new evidence for hazard. Because of an extensive research program on the brain cancer issue (supported by industry as well as governments throughout the world), the wireless communications industry may well escape the legal disaster that befell the silicone breast implant industry, which, when allegations of health problems were raised about their products, failed to mount the necessary studies to address the public's concerns. But public health concerns have led to other consequences, including restrictive regulations aimed to address public fears. These may limit the technology in a way that is more subtle but just as important as technical limitations such as channel capacity. Consider:

1. The SAR levels from mobile handsets are already close to regulatory limits. This may limit future communications technologies that require faster rates of data transmission and thus operation at higher power levels. It is hardly likely that exposure limits will increase substantially; there is strong political pressure to reduce the limits further.

2. While RF exposure levels from wireless base stations are generally far below western exposure limits (IEEE C95.1-1999 and ICNIRP), they are *not* far below limits in effect in Russia, China, several Eastern European countries, Switzerland, Italy, and other nations. The limits in Switzerland and Italy (and a few other countries) were only recently instituted in response to public concerns about possible health effects of RF energy from wireless base stations. Some critics are calling for precautionary limits in the United States as well. Such limits, undertaken in response to public fears about the safety of wireless communications technology, may have very serious and unintended consequences for other applications such as airport radar and broadcasting.

Clearly, the health issue is both sensitive and of longstanding duration. If only out of enlightened self-interest, technologists and industry managers need to take the issue seriously. There is an obvious need for further research, both to improve exposure guidelines, and to address whatever specific health concerns might arise.

But there is also a need for improved risk communication. Many engineers, in particular, have a difficult time in communicating effectively about risk, even as they seem compelled to speak in public about the risks or nonrisks of electromagnetic fields. Industry needs to educate its personnel how to communicate honestly and effectively with the public about these very sensitive issues. Government spokespeople, for their own part, need to stop making public statements that mobile phones "have not been proven safe", which is trivially true to a scientist (since no technology has been proven absolutely safe) but clearly disturbing to laypeople.

Finally, there is a need for sensitivity on the part of industry in locating facilities. Placing cellular antennas on schools or on apartment buildings in a way that appears to direct their energy into residences is sure to cause public opposition, whatever the actual risks or nonrisks might be. In part, I think that some of the public opposition to wireless base stations in Europe and elsewhere has been a result of insensitive practices of wireless companies in locating their facilities. The engineers who plan sites need to look at more than computer printouts in deciding where to locate antennas.

In the end, wireless technology will be accepted because of its great usefulness, and health fears will fade as the technology becomes commonplace. Consider the Italian experience: "(A study in 2000 showed that) eight out of every 10 Italians is concerned that cell phone antennas cause electrosmog and harm human health. But only one in 10 said that they would consider relinquishing their own *telefonino* for health concerns" [48].

References

1 Foster KR, Pickard WF (1987) Microwaves: the risks of risk research. Nature 330:531-532
2 Foster KR (2000) Thermal and nonthermal mechanisms of interaction of radiofrequency energy with biological systems. IEEE Trans Plasma Science 28:17-23
3 Michaelson SM, Lin JC (1987) Biological effects and health implications of radiofrequency radiation. Plenum, New York
4 BENER Abstracts, Information Ventures, 42 S. 15th St. Suite 700, Philadelphia PA 19102 USA
5 Alexander RC, Surrell JA, Cohle SD (1987) Microwave oven burns to children: an unusual manifestation of child abuse. Pediatrics 79:255-260
6 Ziskin MC, Adair ER, Bassen HI, et al. (2002) Medical aspects of radiofrequency radiation overexposure. Health Phys 82:387-391
7 Liu Y-H, Li X-M, Zou R-P, Li F-B (1991) Biopsies of human testes receiving multiple microwave irradiation: an histological and ultramicroscopical study. J Bioelectr 10:213-230
8 IEEE Standards Coordinating Committee 28 on Non-Ionizing Radiation Hazards: Standard for Safe Levels With Respect to Human Exposure to Radio

Frequency Electromagnetic Fields, 3 KHz to 300 GHz (ANSI/IEEE, 1999), The Institute of Electrical and Electronics Engineers, New York, 1992

9 International Commission on Non-Ionizing Radiation Protection (1996) Health issues related to the use of hand-held radiotelephones and base transmitters. Health Phys 70:587-593

10 Lilienfeld AM, Tonascia J, Tonascia S, Libauer CH, Cauthen GM, Markowitz JA, Weida S (1978) Foreign service status study: evaluation of health status of foreign service and other employees from selected Eastern European posts. NTIS Document No. PB-28B 163/9GA. Dept. of State, Washington DC, Final Report, Contract No. 6025- 619073

11 Liakouris AGJ (1998) Radiofrequency (RF) sickness in the Lilienfeld study: an effect of modulated microwaves? Arch Environ Health 53:236-238

12 Goldsmith JR (1997) Epidemiologic evidence relevant to radar (microwave) effects. Environ Health Persp 105:1579-1587, Suppl. 6

13 Zaret MM, Snyder WZ (1977) Cataracts in aviation environments (letter to editor). Lancet 1:484-485

14 Hollows FC, Douglas JB (1984) Microwave cataract in radiolinemen and controls (letter). Lancet 2:406-407

15 Lipman RM, Tripathi BJ, Tripathi RC (1988) Cataracts induced by microwave and ionizing radiation. Surv Ophthalmol 33:200-210

16 Joyner KH (1989) Microwave cataract and litigation: a case study. Health Phys 57:545-549

17 Szmigielski S (1996) Cancer morbidity in subjects occupationally exposed to high frequency (radiofrequency and microwave) electromagnetic radiation. Sci Total Environ 180:9-17

18 Independent Expert Group on Mobile Phones (2000) Mobile Phones and Health, National Radiological Protection Board (UK) 2000, see http://www.iegmp.org.uk/IEGMPtxt.htm

19 Lai H, Singh NP (1996) Single- and double-strand DNA breaks in rat brain cells after acute exposure to radiofrequency electromagnetic radiation. Int J Rad Biol 69:513-521

20 Malyapa RS, Ahern EW, Bi C, Straube WL, LaRegina M, Pickard WF, Roti Roti JL (1998) DNA damage in rat brain cells after in vivo exposure to 2450 MHz electromagnetic radiation and various methods of euthanasia. Rad Res 149:637-645

21 P. Brodeur (1977) The zapping of America: microwaves, their deadly risk, and the cover-up. Norton, New York

22 A comprehensive list of ongoing research can be found at the World Health Organization EMF Project website at http://www.who.int/peh-emf/

23 Foster KR and Moulder JE (2000) Are mobile phones safe? IEEE Spectrum 37:23-28 (August)

24 Moulder JE, Erdreich LS, Malyapa RS, et al. (1999) Cell phones and cancer: What is the evidence for a connection? Rad Res 151:513-531

25 Weinberger Z, Richter ED (2002) Cellular telephones and effects on the brain: The head as an antenna and brain tissue as a radio receiver. Med Hypotheses 59:703-5
26 Santini R, Santini P, Danze JM, LeRuz P, Seigne M (2002) Investigation on the health of people living near mobile telephone relay stations: Incidence according to distance and sex. Pathol Biol (Paris). 50:369-73
27 COST 281 (2002) Scientific comment on epidemiologic studies on the health impact of mobile communcation base stations. Available on the World Wide Web at www.cost281.org
28 Foster KR, Chou C-K, Riu P (2002) Use of "protective devices" for cellular telephones -A COMAR technical information statement. IEEE Eng. Med Biol 21:105-106.
29 McRee DI (1979) Review of Soviet/Eastern European research on health aspects of microwave radiation. Bull NY Acad Med 55:1133-1151
30 Szmigielski S (1989) Eastern European RF protection guides and rationales. In: Lin JC (ed) Electromagnetic interaction with biological systems. Plenum, New York, pp 221-244
31 Sliney DH, Wolbarsht ML, Muc AM (1985) Differing radiofrequency standards in the microwave region-implications for future research. Health Phys 49:677-683 (Editorial)
32 Gajšek P, Pakhomov AG, Klauenberg BJ (2002) Electromagnetic field standards in Central and Eastern European countries: current state and stipulations for international harmonization. Health Phys. 82:473-483
33 Grigoriev Yu (1998) The Russian standards. In: Inaugural Round Table on World EMF Standards Harmonization, 18 Nov. 1998, Zagreb, Croatia, published by the WHO International EMF Project, http://www.who.int/peh-emf/publications/standards_harmonization/zagreb_minutes_1998.pdf
34 Rubtsova N (2001) Overview of health effects of extremely low frequency electromagnetic fields, criteria for EMF standards harmonization. In: Proceedings of the Eastern European Regional EMF Meeting and Workshop, Varna, Bulgaria, 28 April 2001
35 Bielski J (1994) Bioelectrical brain activity in workers exposed to electromagnetic fields. Ann NY Acad Sci 724:435-437
36 Vasilevskii NN, Suvorov NB, Sidorov YuA, Zueva NG (1994) Neurophysiological mechanisms of ecological stability:prognosis and biorhythm correction. Vestn Ross Akad Med Nauk 2:40-44
37 Huai C (1983) Assessment of health hazard and standard promulgation in China. In: Biological effects and dosimetry of nonionizing radiation, radiofrequency and microwave energies. Grandolfo M, Michaelson SM, Rindi A (eds) Plenum, New York, pp 627-644
38 Suvorov IM (1989) Clinical variants of the disease caused by exposure to radio-frequency electromagnetic fields. Gig Tr Prof Zabol 10:19-22
39 Gluszcz M (1979) Difficulties in the certification of microwave disease. Med Przemyslowa 30:147-150

40 Djordjevic Z, Kolak A, Djokovic V, Ristic P, Kelecevic Z (1983) Results of our 15-year study into the biological effects of microwave exposure. Aviat Space Environ Med 54:539-42
41 Sanitary Rules and Norms (SanPin 2.2.4/2.1.8.055-96) (1996) Radiofrequency electromagnetic radiations (2.2.4. physical factors of industrial surroundings. 2.1.8. Physical factors of the environment) p 28, Ministry of Public Health, Russian Federation
42 Vecchia P, Foster KR (2002) Precaution and controversies. Regulating radiofrequency fields in Italy. IEEE Technology and Society Magazine 21:23–27
43 Foster KR, Vecchia P, Repacholi MH (2000) Science and the precautionary principle. Science 288:979-980
44 Pekárek L (2002) Electromagnetic Standards in the Czech Republic. In: Proceedings of the Eastern European Regional EMF Meeting and Workshop, Varna, Bulgaria, 28 April
45 GSM phone SAR levels: A-Z, at http://www.cellular.co.za/radiation_gsm_phone_a-to-z.htm
46 Silva J (2002) FAA battles illegal supercordless phones. RCR News (March 25)
47 Avioli D (2002) China may draw a sharply lower line on mobile phone radiation. IEEE Spectrum (Sept)
48 Emsden C (2000) Beware of the curse of the telefonino:smog becomes electric. Italy Daily (published with the International Herald Tribune) June 9
49 Kang G, Gandhi OP (2002). SARs for pocket-mounted mobile telephones at 835 and 1900 MHz. Phys Med Biol. 47:4301-13
50 Wiart J, Dale C, Bosisio AV, Le Cornec A (2000) Analysis of the influence of the power control and discontinuous transmission on RF exposure with GSM mobile phones. IEEE Trans. Electromag. Comp. 42:376-85
51 Oliver JP, Chou CK, Balzano Q (2003) Testing the effectiveness of small radiation shields for mobile phones. Bioelectromagnetics 24:66-69.
52 FTC: “Radiation Shields: Do They ‘Cell’ Consumers Short?” http://www.ftc.gov/bcp/conline/pubs/alerts/cellshlds.htm, Feb 20, 2002.

4 All-Planar RF Integration Approach to Millimeter-Wave Wireless Front-Ends

4.1 Introduction

Recently, indoor wireless local area network (WLAN) and broadband communication systems operating at millimeter-wave frequencies have attracted much interests because they offer the advantage of high data rates. Two distinct technologies employing non radiative dielectric (NRD) guides and rectangular waveguides have been rigorously developed to efficiently achieve stringent radio-frequency (RF) system specifications [1,2]. Yoneyama et al. reported high-performance, low-loss millimeter-wave building blocks such as filters, voltage-controlled oscillators, mixers, modulators and antennas; these components can be integrated by the NRD technology. On the other hand, a single-substrate transceiver module for 76–77 GHz autonomous cruise control (ACC) radar applications was reported, where (monolithic microwave integrated circuits) MMICs and discrete devices were mounted in a multi chip module (MCM) fashion by wire-bonding and flip-chip techniques. Both millimeter-wave module integration technologies are aimed for low-cost, high-volume and high-yield designs. The guiding technologies mentioned above perform well at millimeter-wave frequencies, however, some modules would require complicated analyses and mechanical fabrication processes, hence increasing the cost and complexity of the system integration efforts. For example, the well-known horn antennas are three-dimensional; thereby they need transitions to handle mode conversion and impedance matching from waveguides to planar circuits and vice versa [3].

This chapter reports an all-planar millimeter-wave integration approach for WLAN video transceiver modules. Innovative building blocks such as leaky-wave antenna arrays, rectangular-waveguide-stabilized oscillators, multiple-order H-plane rectangular waveguide filters and surface-mountable dual-mode packages must comply with the printed-circuit board photolithographic technology or multi layered substrate technology.

The largest component of the complete millimeter-wave RF module is often the antenna. Other non planar components, which are often integrated in the millimeter-wave RF module, are filters and dielectric-resonator oscillators (DRO). Filters implemented by finline or rectangular waveguide technologies [4, 5] show much better performance than those applying planar, photolithographic technologies. DRO, however, is the most popular technology and may replace the cavity-stabilized Gunn oscillator. Great care must be exercised for the integration of various components that are two-or three-dimensional.

We present a completely planar solution to the integration of millimeter-wave RF front-end components. Two feasibility studies are investigated. First is the 30 GHz frequency-shift keying (FSK) transceiver module; the second is the 30 GHz self-heterodyne broadcast transceiver module. Both employ the advanced coplanar strips system (ACSS) guiding structure concept. As shown in Fig. 4.1, the ACSS guiding technology is fundamentally a multi layered guiding structure that incorporates microstrips, coplanar waveguides, coplanar strips and rectangular waveguides. All of the RF front-end functional blocks, including low-profile antennas, high-power local oscillators with waveguide cavity resonators, frequency doublers, rectangular waveguide filters and sub harmonic quadrature mixers, are implemented simultaneously in a planar form by the conventional photolithographic techniques. Notice that the state-of-art planar integrated circuits using microstrips and coplanar waveguides may confront a critical restriction of large transmission losses at millimeter-wave frequencies. Therefore novel H-plane mode converters are proposed in the ACSS guiding technology to incorporate low-loss rectangular waveguides in the printed circuit board (PCB)-based circuits simultaneously; thereby the RF module performance can be greatly enhanced. Owing to its planar nature, the ACSS module is compatible with co-planar waveguide (CPW) probe testing, allowing fast evaluation cycles.

Figure 4.1 illustrates a schematic of the fundamental concepts of integrating all RF front-end building blocks into the multi layered structure. On the top layer lie the chips, packaged transistors and monolithic microwave integrated circuits

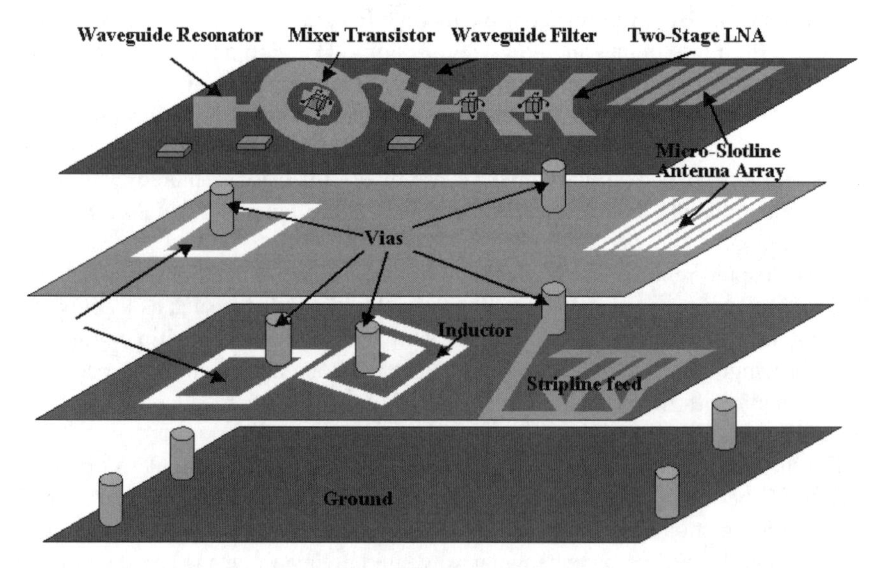

Fig. 4.1 Advanced coplanar strips system (ACSS) millimeter-wave RF module integration concept

(MMICs), which are properly wire-bonded or surface-mounted into the substrate. Antennas, filters, and resonators are all built-in by the conventional multi layered PCB (printed-circuit board) technology, which can also be in the form of LTCC (low-temperature cofired ceramics). Therefore we can design very compact millimeter-wave RF modules using familiar technology. For example, inductors can be embedded in the layered structure as seen in Fig. 4.1. In this chapter, however, we restrict our design to two-layer versions, which are less demanding PCB fabrications, but are good enough to demonstrate the concept of planar integration of millimeter-wave RF front-ends.

4.2 Ka-Band Microstrip Leaky-Mode Array Antenna for Receiver

Two all-planar millimeter-wave transceiver prototypes are built and tested. The transmitter module is installed against the ceiling and the receiver module is placed on the flat table as seen in Fig. 4.2 The broadcasted signals radiate in a cone-shaped fashion with the main beam at 48° measured from the broadside. The main beam of the receiving antenna coincides with that of the transmitting antenna. Since we are restricted to the planar integration of these antennas, challenges arise in the syntheses of antenna radiation patterns using only multi layered PCB technology.

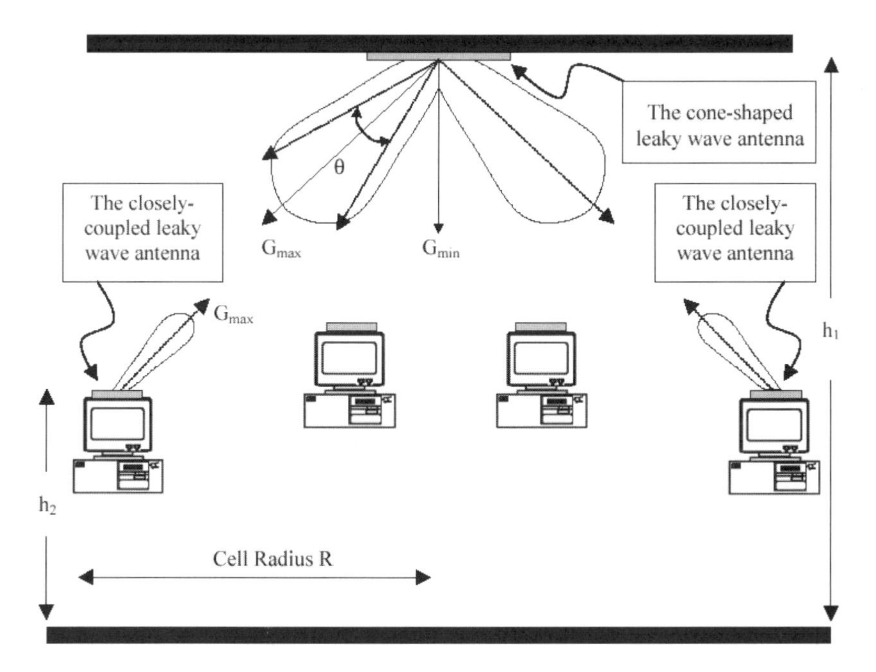

Fig. 4.2 Experimental Ka-band wireless multimedia indoor broadcast system

4.2.1 Receiving Antenna Array Design

An antenna array consisting of a highly directed beam pointing to the transmitter while occupying the least area is the most desirable antenna in the present experimental wireless indoor broadcast system. The antenna array made of closely coupled microstrips is chosen to meet such a requirement. What follows parallels the work published in [6] for the analyses and syntheses of leaky-mode microstrip arrays.

It has been shown that N mutually coupled microstrips result in N coupled modes of orders zero, one, two, …, etc. The bound quasi-TEM (transverse electromagnetic wave) modes are of order zero, designated as EH_0 modes. The higher-order microstrips are designated by EH_1, EH_2 …, etc., representing higher-order modes, which will leak below certain onset frequencies [7]. The receiving antenna array consists of eight tightly coupled microstrips with a feeding structure exciting predominantly one of the eight leaky EH_1 modes that may reside in the array. The coupled mode approach is invoked to obtain the complex propagation constants γ and the corresponding modal current distributions of the coupled microstrips [6]. This approach not only provides accurate physical insights of how coupled leaky modes work, but greatly reduces the computational time for obtaining antenna radiation patterns and makes feeding networks much easier to develop.

4.2.2 The Eigenvalue Approach for Designing Large Microstrip Leaky-Mode Arrays

To obtain the complex propagation constant of the coupled microstrips, full-wave field theory approaches such as the integrated equation method [8–10] and the mode-matching method [11] have been widely applied. These full-wave methods search for the complex roots from square matrixes, thus solving the non standard eigenvalue problem [12]. By invoking coupled-mode theory, however, we may transform such non standard eigenvalue problems into the standard eigenvalue solutions of the coupled leaky modes, which otherwise would be very difficult to obtain using the conventional non standard eigenvalue approach. The result is simply solving the following determinant expressed by

$$\det[diag(\lambda) - diag(\gamma_i) + [C_{i,j}]] = 0 \tag{4.1}$$

where λ denotes the eigenvalue of the coupled EH_1 modes; γ_i is the complex propagation constant of the ith microstrip element before coupling occurs; $C_{i,j}$ is the coupling coefficient of element i and element j. When all microstrips are of equal widths, γ_i must be equal to γ, i.e. the complex propagation constant (EH_1 mode) of a single microstrip. What interests us here is to excite the proper EH_1 mode that is leaky and carries desirable radiation characteristics of an antenna.

4.2.3 Design Procedure

Applying (4.1) for N=2, the coupling coefficient of the two symmetrical microstrips at the first higher order EH_1 leaky mode can expressed by

$$C_{1,2} = (\lambda_e - \lambda_o)/2 \tag{4.2}$$

$$\gamma = (\lambda_o + \lambda_e)/2 \tag{4.3}$$

where $\lambda_{e(o)}$ denotes the even (odd) EH_1 mode solution. The even (odd) mode shows in-phase (out-of-phase) modal current distributions along the two microstrips. Thus, a perfect electric conductor (PEC) and a perfect magnetic conductor (PMC) can be inserted along the symmetry plane for the even and odd EH_1 modes, respectively. Figures 4.3 and 4.4 display the electric and magnetic field vectors of the even and odd modes, respectively, to illustrate the symmetric properties. Once the even-mode and odd-mode full-wave analyses of the coupled microstrips are carried out, we immediately obtain the coupling coefficient C_{12} and γ. The former quantity will be used for obtaining the eigenvalues of the N-coupled microstrips. The latter can be used for checking against that obtained by a single, isolated microstrip.

We can extend the two-element study to coupled microstrip arrays of more than two elements. Given a system of N coupled microstrips of equal width W and spacing (S), we may rewrite (4.1) as

$$\begin{aligned} &\det[\, diag\,(\lambda) - diag\,(\gamma_i) + [C_{i,j}]] \\ &= \lambda^N + a_{N-1}\lambda^{N-1} + \Lambda \quad + a_1\lambda^1 + a_{01} = \prod_{i=1}^{N}(\lambda - \lambda_i) \end{aligned} \tag{4.4}$$

where λ_i is the complex propagation constant of the ith leaky mode, for i=1 to N. Expanding the determinant (det $(diag(\lambda) - diag(\gamma_i) + [C_{i,j}])$), and comparing order-by-order at both sides of (4.4), say for N=3, we obtain the following equations for solving the unknown coupling coefficients of $C_{1,2}$ and $C_{1,3}$, representing the coupling coefficients of the adjacent elements and other-than-adjacent elements, respectively. Notably, $\gamma = \gamma_1 = \gamma_2 = \gamma_3$ and $C_{1,2}$=$C_{2,3}$, since each coupled microstrip has identical width W and is positioned at equal spacing S. Thus we obtain the following system of equations for solving $C_{1,2}$ and $C_{1,3}$.

$$\begin{cases} \gamma^3 - 2C_{1,2}{}^2C_{1,3} - (2C_{1,2}{}^2 + C_{1,3}{}^2)\gamma = \lambda_1\lambda_2\lambda_3 \\ 3\gamma^2 - 2C_{1,2}{}^2 - C_{1,3}{}^2 = \lambda_1\lambda_2 + \lambda_1\lambda_3 + \lambda_2\lambda_3 \end{cases} \tag{4.5}$$

where $\gamma = (\lambda_1+\lambda_2+\lambda_3)/3$. Invoking the full-wave approach for solving the complex propagation constants λ_1, λ_2 and λ_3, respectively, we obtain the value of γ, which is the arithmetic mean of the three complex propagation constants of the coupled EH_1 modes. Such an observation can be generalized by the following expression

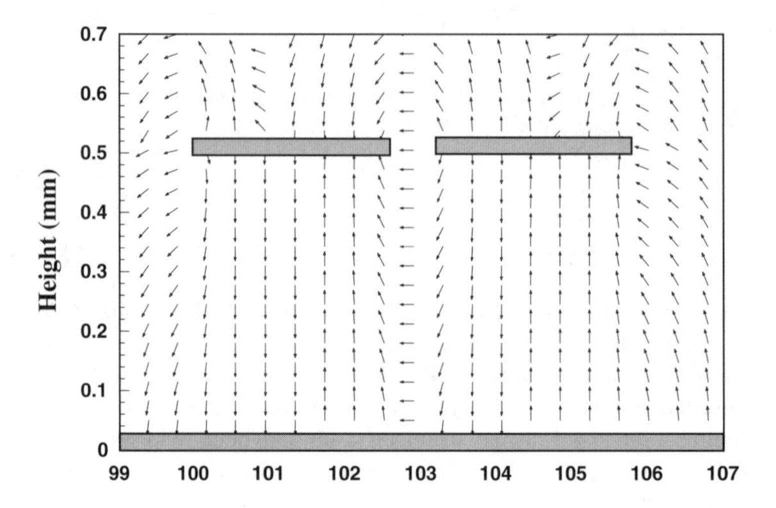

a

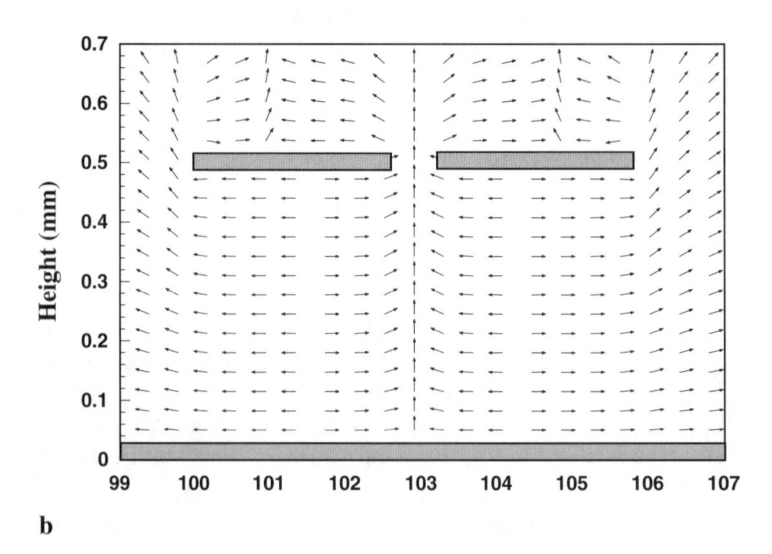

b

Fig. 4.3 Even EH_1 mode transverse field distributions. **a** E-field; **b** H-field

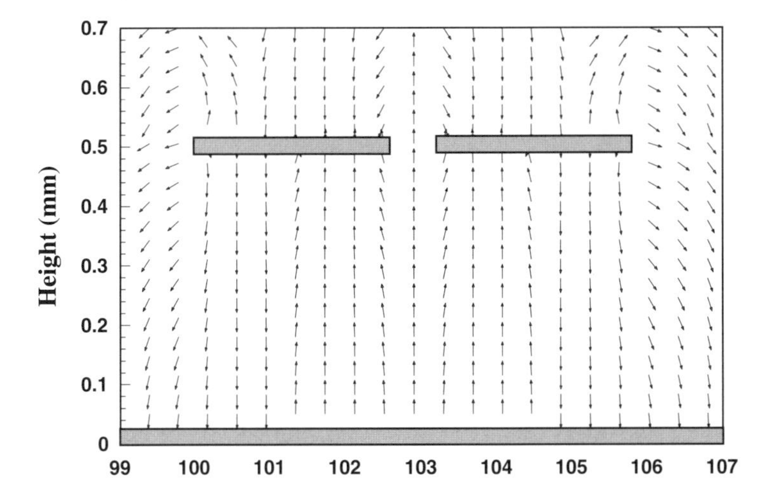

a

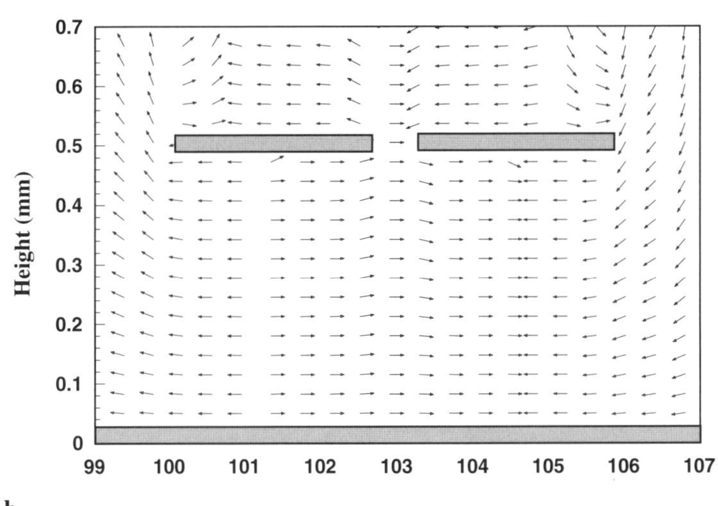

b

Fig. 4.4 Odd EH_1 mode transverse field distributions: a E-field; b H-field

$$\gamma = \sum_{i=1}^{N} \lambda_i / N \tag{4.6}$$

Similarly, for N=4, we may derive a system of equations for solving the coupling coefficients $C_{1,2}$, $C_{1,3}$, and $C_{1,4}$, noticing that $C_{1,2}=C_{2,3}=C_{3,4}$, and $C_{1,3}=C_{2,4}$.

$$\begin{aligned}
&6\gamma^2 - 3C_{1,2}^2 - 2C_{1,3}^2 - C_{1,4}^2 \\
&= \lambda_1\lambda_2 + \lambda_1\lambda_3 + \lambda_1\lambda_4 + \lambda_2\lambda_3 + \lambda_2\lambda_4 \\
&4\gamma^3 - 4C_{1,2}C_{1,3}C_{1,4} - 4C_{1,2}^2C_{1,3} - 2(3C_{1,2}^2 + 2C_{1,3}^2 + C_{1,4}^2) \\
&= (\lambda_1\lambda_2\lambda_3 + \lambda_1\lambda_2\lambda_4 + \lambda_1\lambda_3\lambda_4 + \lambda_2\lambda_3\lambda_4) \\
&\gamma^4 - (3C_{1,2}^2 + 2C_{1,3}^2 + C_{1,4}^2)\gamma^2 - 4(C_{1,2}C_{1,3}C_{1,4} + C_{1,2}^2C_{1,3})\gamma \\
&+ C_{1,2}^4 + C_{1,3}^4 + C_{1,2}^2C_{1,4}^2 - 2C_{1,2}^2C_{1,3}^2 - 2C_{1,2}^3C_{1,4} - 2C_{1,2}C_{1,3}^2C_{1,4} \\
&= \lambda_1\lambda_2\lambda_3\lambda_4
\end{aligned} \tag{4.7}$$

Following the above-mentioned design procedure, Fig. 4.5 plots the complex propagation constant of the EH_1 mode of a single microstrip (N=1), which is 2.6 mm wide printed on the substrate of thickness 0.508 mm with relative permittivity of 3.0. Figure 4.6 plots the normalized propagation constant (γ / k_0, $\gamma = \alpha + j\beta$) of the two coupled EH_1 modes (even and odd) of the symmetric, parallel microstrips separated by 0.6 mm. As expected, the arithemetic mean of λ_e and λ_o agrees closely with that of the EH_1 mode of a single microstrip attained by the full-wave analysis.

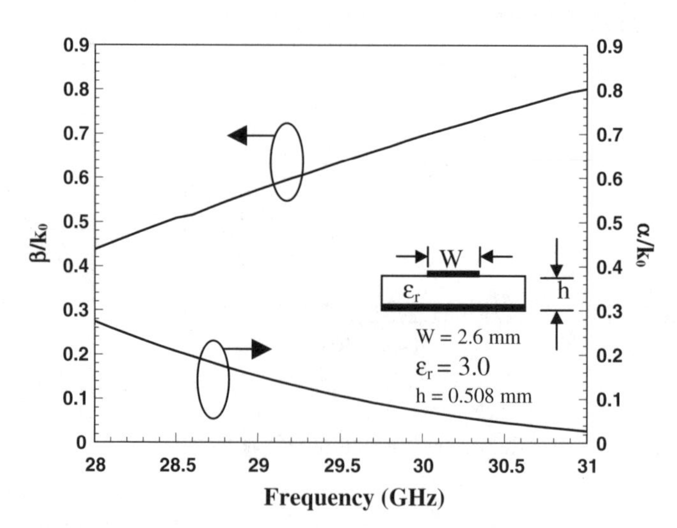

Fig. 4.5 Normalized complex propagation constant of a single microstrip at first higher order EH_1 mode

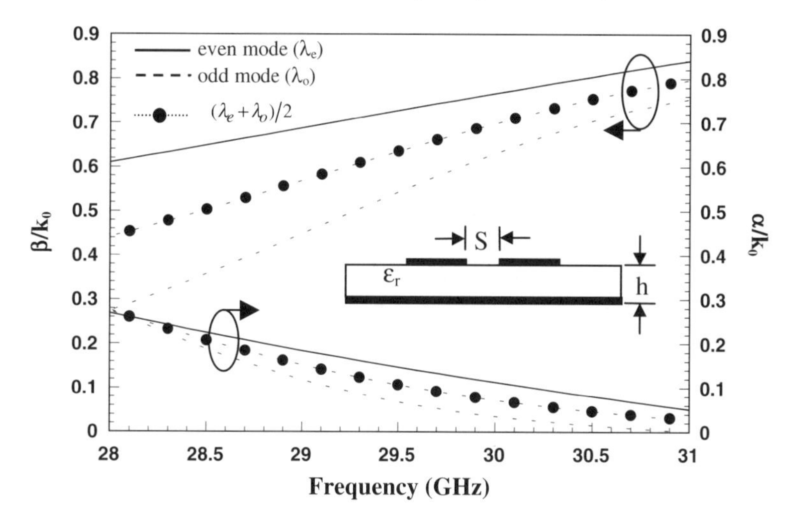

Fig. 4.6 Normalized EH_1 propagation constants λ_e and λ_o of symmetric, coupled microstrips; the arithemetic means of λ_e and λ_o are identical to those plotted in Fig. 4.5

The assessment of the magnitudes of the coupling coefficients of the closely coupled microstrip array is important for carrying out the eigenvalue analyses of the coupled EH_1 leaky modes. Figure 4.7 plots the magnitudes of the coupling coefficients for the case N=4, where each microstrip is 2.6-mm wide and is separated by only 0.6 mm. Figure 4.7 shows that the magnitude of $C_{i,i+j}$ decreases approximately in the order of $10^{-j/2}$ for j = 1, 2, …, and (N–1). In the particular case design of the Ka-band microstrip array, the coupling coefficients obtained by assuming N=3, i.e. $C_{i,k}$=0 for $|i-k|>2$, are adequate for practical accuracy.

Table 4.1 lists the eigenvalues and eigenvectors obtained by assuming N=3 for operating frequency at 30.2 GHz. Element 1 corresponds to the leftmost microstrip of the array. Element 2 is next to element 1, and so on. The rightmost column lists the eigenvalues and complex propagation constants of the eight coupled EH_1 modes.

Table 4.1 Eigenvectors $I^{(j)}$ and normalized eigenvalues (complex propagation constants) of the eight-element microstrip array at 30.2 GHz

		Element 2	Element 3	Element 4	Element 5	Element 6	Element 7	Element 8	Eigenvalues ($\gamma_i = \alpha_i + j\beta_i$)
$I^{(1)}$		0.28∠0°	0.39∠0°	0.46∠0°	0.46∠0°	0.39∠0°	0.28∠0°	0.14∠0°	0.166 +j0.77
$I^{(2)}$	0.28∠67°	0.46∠67°	0.39∠67°	0.14∠67°	0.14∠−113°	0.39∠−113°	0.46∠−113°	0.28∠−113°	0.05+j0.65
$I^{(3)}$	0.39∠−113°	0.39∠−113°	0.0∠0°	0.39∠67°	0.39∠67°	0.0∠0°	0.39∠−113°	0.39∠−113°	0.168+j0.79
$I^{(4)}$	0.46∠− 27°	0.14∠−27°	0.39∠153°	0.28∠153°	0.28∠−27°	0.39∠−27°	0.14∠153°	0.46∠153°	0.04+j0.64
$I^{(5)}$	0.46∠67°	0.14∠−120°	0.39∠−120°	0.28∠67°	0.28∠67°	0.39∠−120°	0.14∠−120°	0.46∠67°	0.172+j0.8
$I^{(6)}$	0.14∠−113°	0.28∠67°	0.39∠−113°	0.28∠67°	0.46∠−113°	0.39∠67°	0.28∠−113°	0.14∠67°	0.02+j0.81
$I^{(7)}$	0.28∠− 43°	0.46∠137°	0.39∠−43°	0.14∠0°	0.14∠137°	0.39∠−43°	0.46∠113°	0.28∠−43°	0.177+j0.61
$I^{(8)}$	0.39∠−160°	0.39∠20°	0.0∠0°	0.39∠−160°	0.39∠20°	0.0∠0°	0.39∠−160°	0.39∠20°	0.01+j0.82

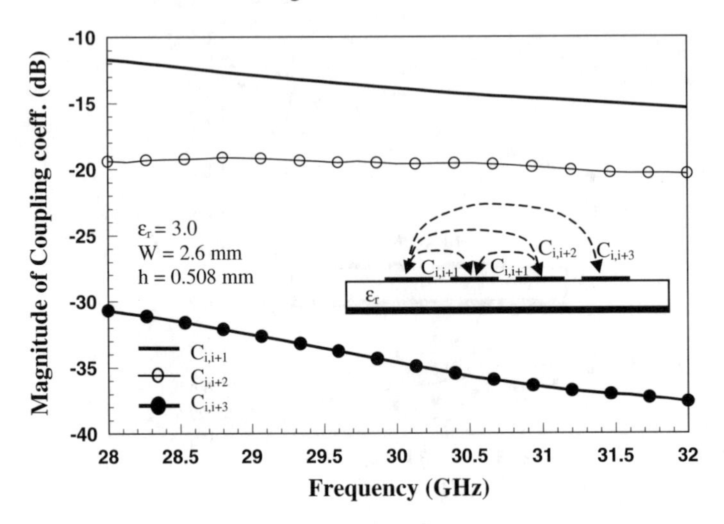

Fig. 4.7 The magnitude of the coupling coefficient in dB

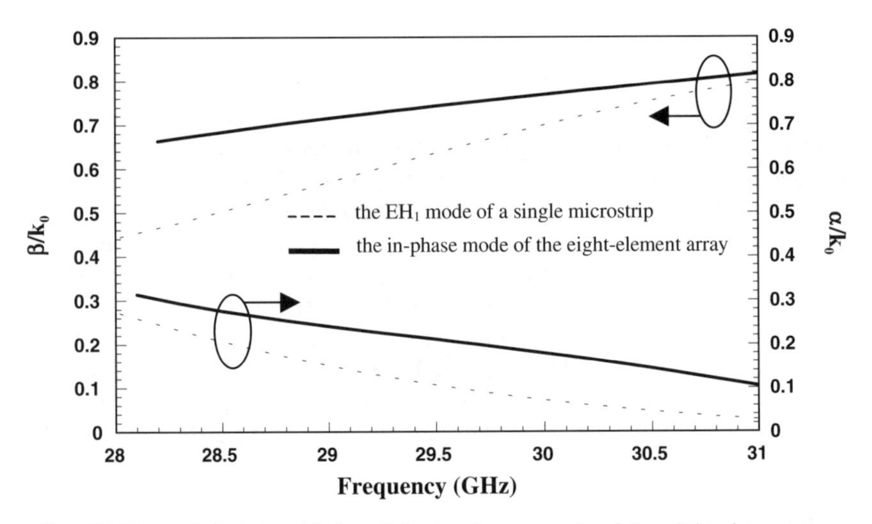

Fig. 4.8 Dispersion characteristics of the in-phaseγ_1 mode of the eight-element array and those of a single microstrip

The corresponding eigenvector for each eigenvalue, as denoted by current eigenvector $I^{(i)}$, is in the first row of the table. Notice that the first eigenvalue corresponds to the current eigenvector having the in-phase property for each element. This is the mode we wish to excite. Furthermore, as depicted in Fig. 4.8, this eigenvalue shows the normalized attenuation constant (α/k_0) of 0.166, which is greater than that of 0.06 for the EH_1 mode of a single microstrip. Figure 4.8 compares the complex propagation constants of the in-phase γ_1 mode against that of a single microstrip. This is advantageous in the sense that a smaller array length will be adequate in the antenna design.

Figure 4.9 plots the radiated powers (percent) against the above-mentioned guided modes propagating along the same distance of 10 mm (one free-space wavelength at 30 GHz). The radiated power of the in-phaseγ_1 mode is greater than 90% between 28 GHz and 30 GHz, whereas the EH_1 mode of a single microstrip shows only 56% power radiation at 30 GHz, implying that the aperture efficiency of the eight-element antenna array will be increased by using the tightly coupled leaky EH_1 mode array that excites theγ_1 mode.

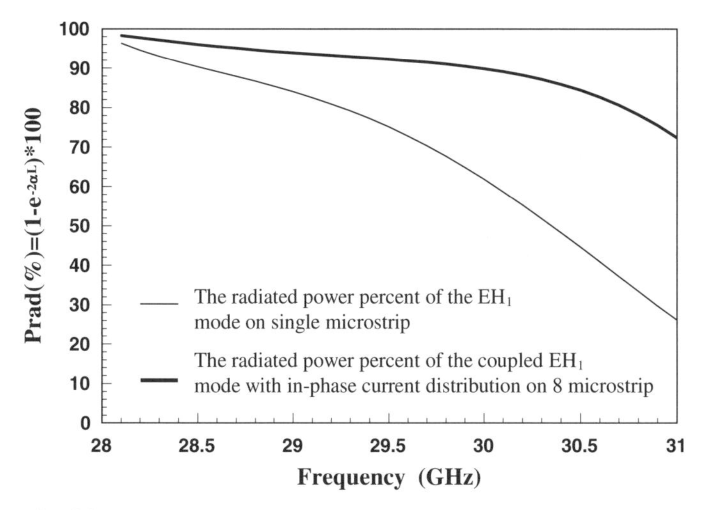

Fig. 4.9 The comparison of the radiated power (percent) of the EH_1 mode on single microstrip and the coupled EH_1 mode with in-phase modal current distribution on eight microstrip lines along a distance of 10 mm

4.2.4 Ka-Band EH_1 Leaky-Mode Antenna Array Prototype

The well-known corporate feed is adopted for its simplicity in structure at the cost of exciting multiple EH_1 modes. Each microstrip radiating element is excited differentially at both ends as seen in Fig.10, which shows the eight-element leaky-mode array prototype. The differential signals are made by a power divider with one arm half-wavelength longer than the other. The corporate feed is simply a combination of a power divider at the input, followed by two series branches. Each series branch distributes one-quarter of the power into the microstrip. In the corporate-feed condition, the current associated with all array elements is thus the superposition of the eight eigenvectors of Table 4.1. Table 4.2 lists the relative magnitude and phase (Ω_i) of each eigenvector contributing to the corporate feed total currents, showing theγ_1 mode is the major leaky EH_1 mode excited. Modes 3, 5 and 7 are also present, but play a much smaller role in the antenna array. Figure

4.11 plots the percentage of radiated power by the γ_1 mode in the 25-mm long array, showing 99.6% at 30 GHz, which is much better than that of a single microstrip leaky line. Therefore the corporate feed is a good choice for our application.

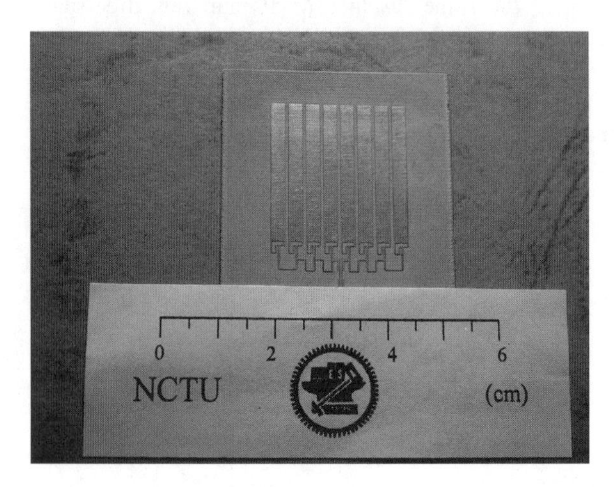

Fig. 4.10 Microstrip eight-element leaky-mode array prototype

Table 4.2 Current coefficient (Ω_i) of the ith eigenmode of the corporate-feed eight-element array

i	1	2	3	4	5	6	7	8
Ω_i	$0.951\angle 0°$	$0.00\angle 0°$	$0.24\angle 0°$	$0.00\angle 0°$	$0.12\angle 0°$	$0.00\angle 0°$	$0.05\angle 0°$	$0.00\angle 0°$

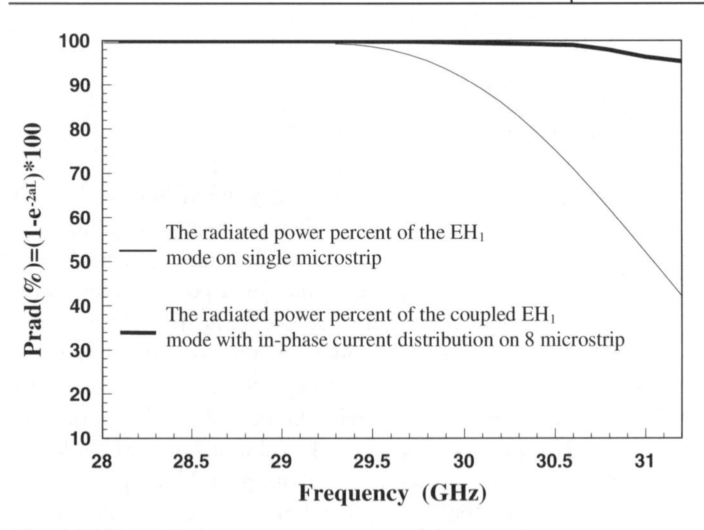

Fig. 4.11 The radiating power percentage of the γ_1 mode

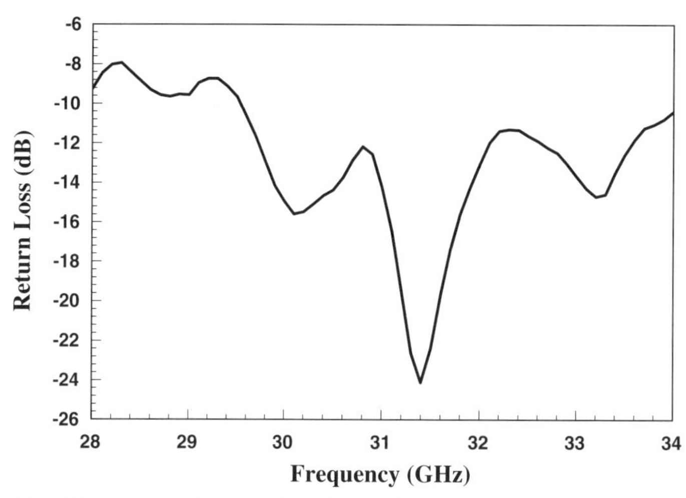

Fig. 4.12 Measured return loss (S_{11}) of the closely coupled leaky wave antenna array, showing return loss (S_{11})<□ 10dB between 29.6 GHz and 34 GHz

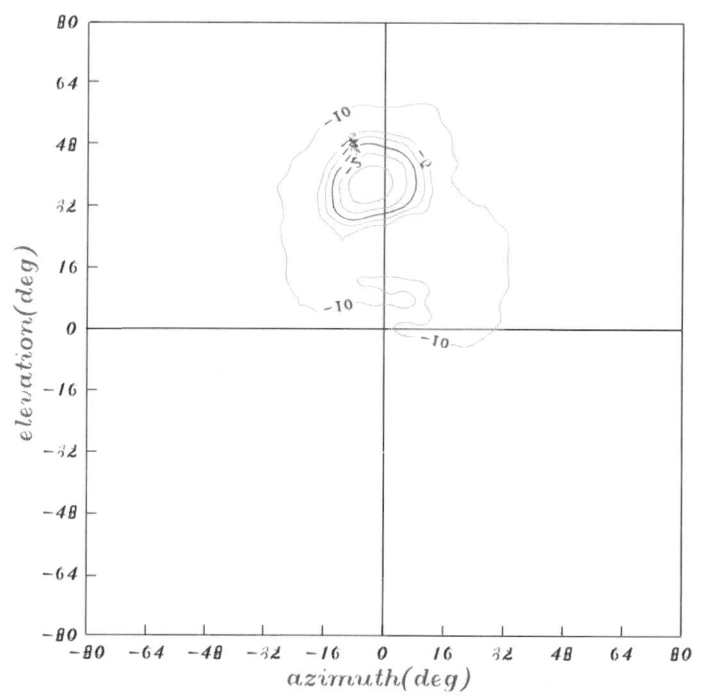

Fig. 4.13 Measured radiated contour pattern of the closely coupled leaky-wave antenna

The closely coupled leaky wave antenna prototype is built on a RO3203 substrate with a thickness of 0.508 mm, relative permittivity of 3.0 and loss tangent of 0.002. The measured return loss (S_{11}) of the microstrip array is plotted in Fig.

4.12, showing good broadband impedance matching. The loss $|S_{11}|<-10$ dB between 29.6 GHz and 34 GHz. Figure 4.13 plots the radiation contours of the antenna array, showing the main beam at 43°/94° (elevation/azimuth), the directivity of 17.3 dB, the gain of 15.8 dB and antenna efficiency of 70.8%.

4.3 Transmitting Active Integrated Antenna with Cone-Shaped Radiatoin Patterns

4.3.1 Omnidirectional Antenna and Power Amplifier Interface

Omnidirectional antennas have been widely applied for indoor WLAN (wireless local area networks). Prior works were based on various approaches such as slot arrays with cylindrical feed [13], slotted waveguide arrays [14], and so on. Although these antennas show high efficiency, the fabrications are relatively complicated and expensive [13, 14]. Furthermore, integrating millimeter-wave antenna arrays and RF modules is a nontrivial process involving careful handling of transitions and matching [1, 15].

Based on planar integration of the omnidirectional antennas, we present a Ka-band radial antenna prototype that makes the interface to millimeter-wave RF modules a straightforward practice. The antenna array is built on a low-cost RO3003 substrate of thickness 20 mils. As shown in Fig. 4.1, the antenna array consists of seven microstrips with etched holes positioned anti symmetrically. The antenna array input impedance is matched to 50 Ω, thereby allowing a direct interface to millimeter-wave RF modules by a 50 Ω microstrip. Also shown in Fig. 4.14, is part of the millimeter-wave RF module, where a power amplifier (PA) MMIC is die-attached and wire-bonded to the substrate with a neighboring bias circuit. The bias lines and the adjacent DC decoupling circuits for the PA are placed near the antenna array. The PA MMIC and its neighboring circuits destroy the symmetry of the antenna array. This, however, has negligible impact on the cone-shaped radiation pattern in the present approach, as is clear in the test results for the antenna.

4.3.2 Space Harmonics and Leaky Microstrip Lines

The radial antenna array design begins with optimizing a single leaky line. The leaky transmission line employs the concept of space-harmonic modulations on both EH_0 and EH_1 modes as the results of the perforated ground plane with anti symmetrical holes etched underneath the microstrip line. The dimension of the microstrip line is properly chosen such that the leaky EH_1 mode enters the desired operating frequencies. The perturbations caused by the etched pattern on the ground plane therefore excite both EH_0 and EH_1 modes [16]. The input EH_0 mode passes through the tapered microstrip, which acts as an impedance transformer,

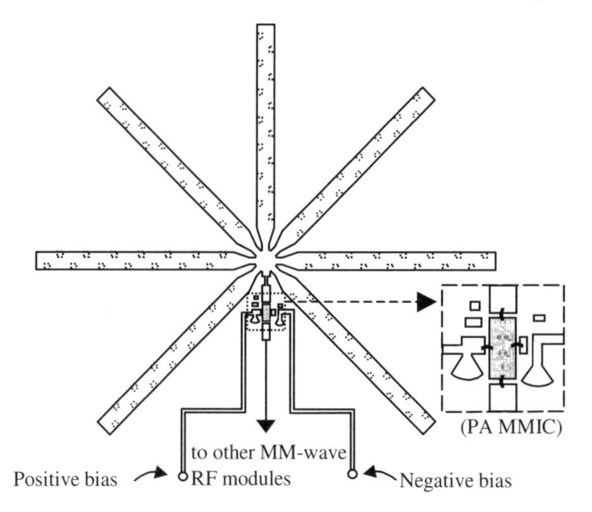

Fig. 4.14 Ka-band integrated microstrip antenna array interfaced to a PA MMIC for wireless LAN application

and then enters the leaky line. Since the perturbation of the anti symmetric is periodic, the space harmonics are excited; thus emitting forward and backward leaky waves concurrently. Figure 4.15 shows the simulated forward and backward beams pointing at – 45° and 37.5°, respectively, as measured from the broadside.

As shown in Fig. 4.15, the forward beam is approximately 4 dB less than the backward beam. This, however, will not create a problem for making the cone-shaped radiation pattern if we arrange the leaky lines in a radial fashion as shown in Fig. 4.1. These radial leaky lines are aligned collinearly except at the feed point for interfacing the array to the millimeter-wave RF module. The back-to-back

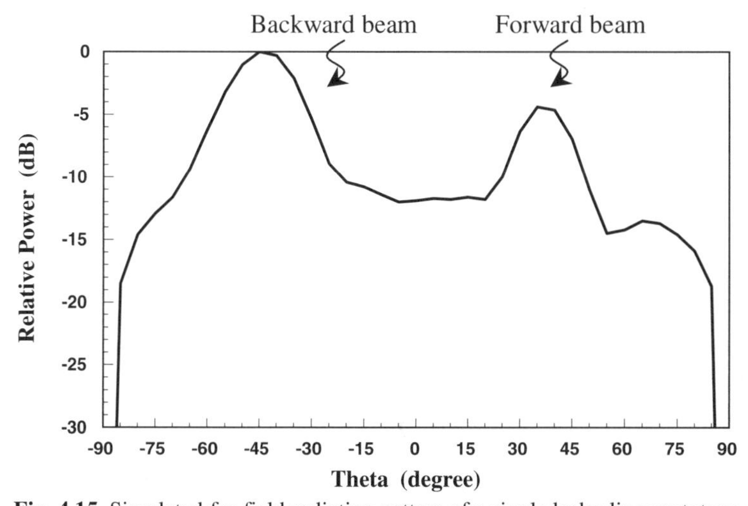

Fig. 4.15 Simulated far-field radiation pattern of a single leaky line prototype at 30.2 GHz

connection of the leaky lines along the azimuthal plane results in averaging out the radiated energies, thus forming a broadened and nearly uniform radiation pattern.

Removal of one leaky line from the otherwise perfectly symmetrical eight-line array results in one quadrant of space for placing the RF module nearby the antenna. Figure 4.14 illustrates how a PA MMIC is integrated and connected to the antenna array. The resulting seven-line antenna array is properly matched to 50 Ω using the eight-port planar circuit at the center and a series-matching circuit at the input. Figure 4.16 plots the measured return loss $|S_{11}|$ of the antenna, showing the broadband characteristics between 29.75 and 30.37 GHz for $|S_{11}| < -10$dB and very good matching at 30.10 GHz with input return loss of –22dB. Figure 4.17 shows

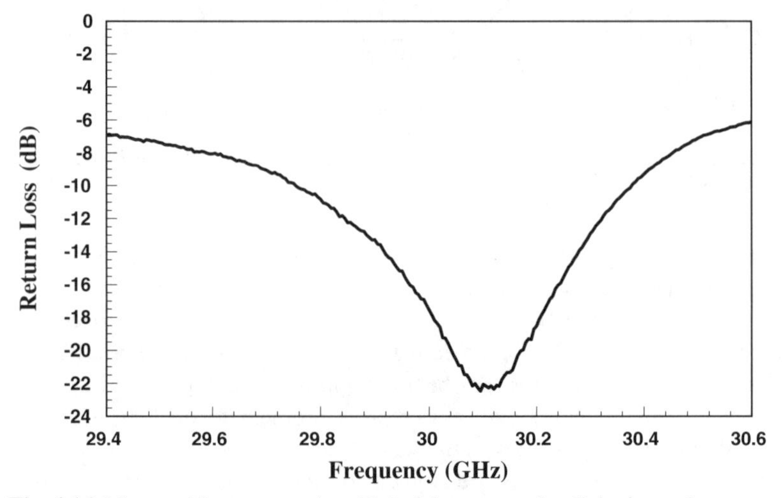

Fig. 4.16 Measured input return loss $|S_{11}|$ of the proposed radial microstrip array

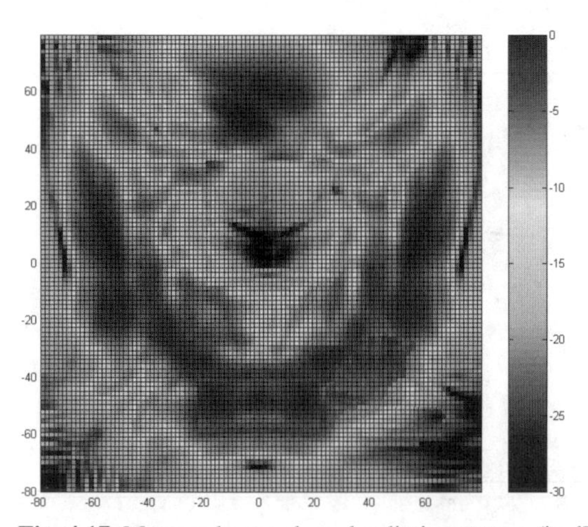

Fig. 4.17 Measured cone-shaped radiation pattern (in dB) of the proposed radial microstrip array at 30.2 GHz

the measured radiation contour at 30.2 GHz with azimuth angle and elevation angle in the horizontal and vertical planes, respectively. The radiation contour shows a circular ring of nearly uniform amplitude, signifying that the cone-shaped radiation pattern is achieved, despite the fact that one quadrant of the antenna array is used for RF interconnect and placing modules. A LMA443 PA MMIC has been successfully installed on the same antenna substrate as shown in Fig. 4.14, demonstrating the practicality of the proposed approach.

4.4 Planar Microstrip-To-Waveguide Mode Converters

The planar integrated circuits employing microstrips and coplanar waveguides are well suited for building RF front-end modules primarily because of their in easy fabrication and low cost. At millimeter-wave frequencies, however, the existence of high-order complex modes, leaky modes, surface-wave modes and radiation of these PCB based planar circuits may cause large transmission losses. Therefore, an important technology, that of (nonradiative dielectric) NRD guides, has been developed to reduce the transmission losses [17]. On the other hand, introducing the conventional metallic rectangular waveguide to integrate RFIC (radio frequency integrated circuit) provides a good alternative to achieving high-performance millimeter-wave modules [18]. Since these two well-known guiding structures are planar, transition circuits are required to convert modal energies to the planar forms. Researchers have investigated various radiators such as the quasi-Yagi antenna [19], the slot -fed patch antenna [20] and the probe-fed -type antenna [21] to convert the electromagnetic energies of the planar circuits to the rectangular waveguides, and vice versa. These transition circuits are generally three-dimensional; they make the complete millimeter-wave module a complex structure, thus making the assembly a fine art which is difficult for mass production.

To circumvent such complex non planar assembly processes for making millimeter-wave modules, we adopt the advanced coplanar strips system (ACSS) multi layered structure and design new mode converters for making good microstrip-rectangular waveguide transition circuits. Therefore low-loss rectangular waveguides can be used in the ACSS integration of millimeter-wave modules whenever necessary. Two types of planar microstrip-to-waveguide transitions have been developed for Ka-band applications.

4.4.1 Why Use Planar Rectangular Waveguides?

High-quality resonators are important for quality factor planar realization of millimeter-wave/microwave oscillators and filters. This poses a major stumbling block for our planar ACSS integration approach. Figure 4.18 plots the total quality factor (Q-factor) of a 50-Ωquarter-wavelength microstrip resonator and the

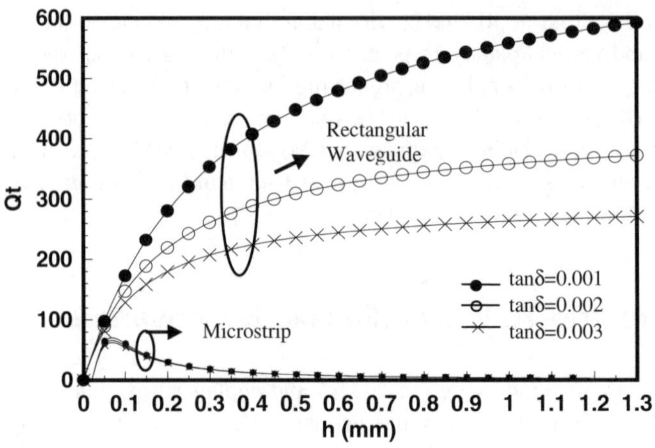

Fig. 4.18 Q-factors of microstrip and rectangular waveguide resonators versus substrate (RO3003) thickness

quarter-wavelength planar, rectangular waveguide resonator at 30 GHz (Ka-band) based on the formulas reported by Konishi [22], Bėlohoubek and Denlinger [23]. When plotting Fig. 4.18, the loss tangent of the substrate and the finite conductivity of the printed-circuit board are included in the computations. The microstrip case shows serious degradation of the Q-factor for substrate thicknesses larger than 0.2 mm and is relatively insensitive loss tangents of the substrate in the range 0.001–0.003. The rectangular waveguide case, however, is highly sensitive to the variation of substrate loss tangent, and its Q-factor improves as the substrate thickness increases. We choose a substrate thickness of 0.508 mm, which yields a total Q-factor of approx. 300 for the planar rectangular waveguide resonator. Thus we compromise the microstrip interconnection losses for better rectangular waveguide resonator performance in the proposed all-planar ACSS approach when designing the millimeter-wave transceiver module.

4.4.2 DC-Blocked Broadband Microstrip-to-Waveguide Transition

Figure 4.19 depicts the proposed planar microstrip-to-rectangular waveguide transition, where a microstrip, a tapered microstrip and a partially filled rectangular waveguide acting as a transition circuit are all aligned collinearly along the logitudinal axis (z-axis) of the rectangular waveguide [24]. The input microstrip signal line is fabricated photolithographically on layer 1, and its bottom surface shares the same ground plane as the rectangular waveguide. A Portion of the tapered microstrip is properly inserted into the partially filled rectangular waveguide to carry out field-matching and to provide the DC-blocking function. The tapered microstrip line connects the protruded microstrip on one end and the 50-Ω input port at the other end, thus acting as an impedance transformer.

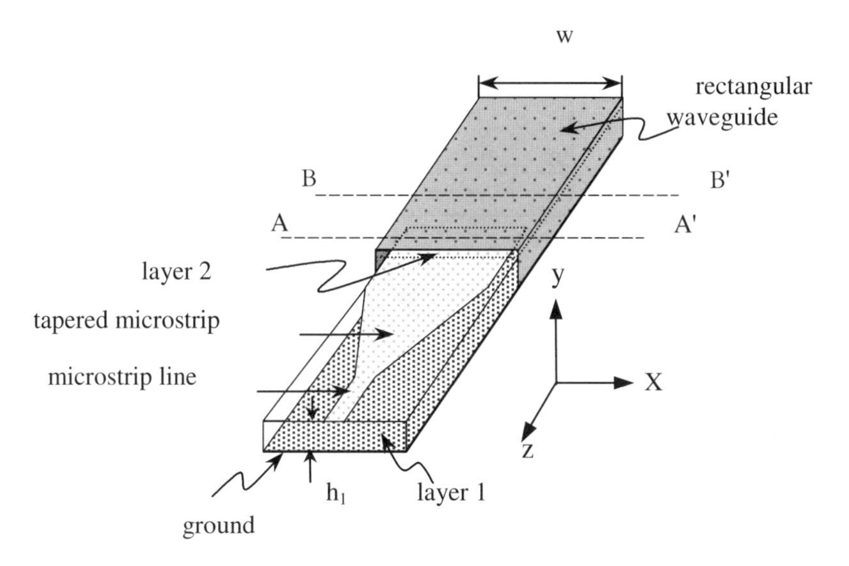

Fig. 4.19 Planar microstrip-to-waveguide transition of DC-blocked type

4.4.3 Design Principle of the Mode Converter: Field Matching

As shown in Fig.4.20, the tapered microstrip protruding into the waveguide at the A–A′ reference plane and the metallic waveguide at the B–B′ reference plane show very close agreement in the dispersion curves throughout the Ka-band (26–to–40GHz). Notice that the inserted microstrip line does not touch the waveguide side walls, thus providing a DC-blocking function. Since the definitions

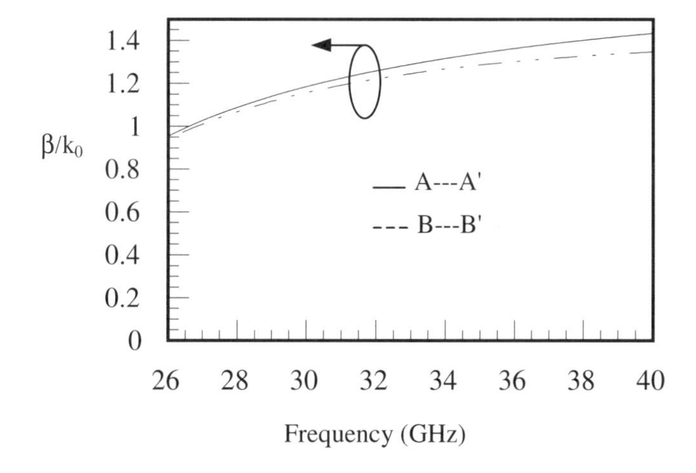

Fig. 4.20 Dispersion curve at cross section A–A' and B–B'

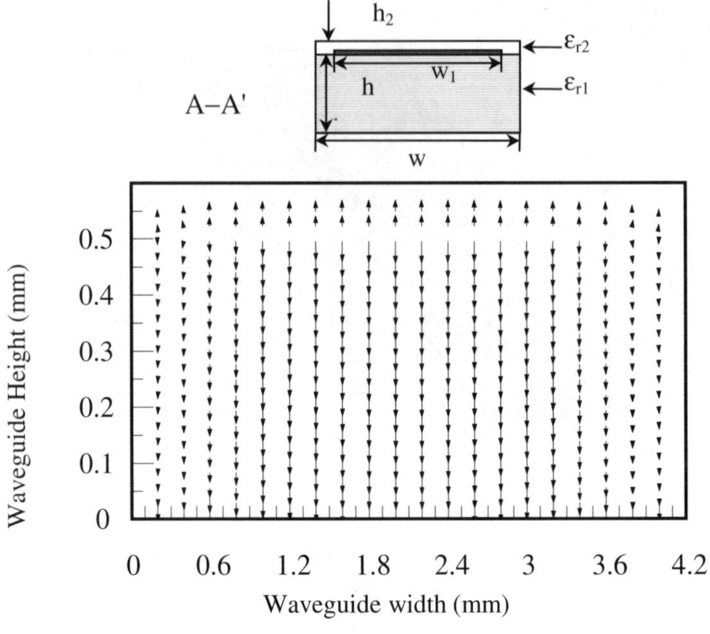

a

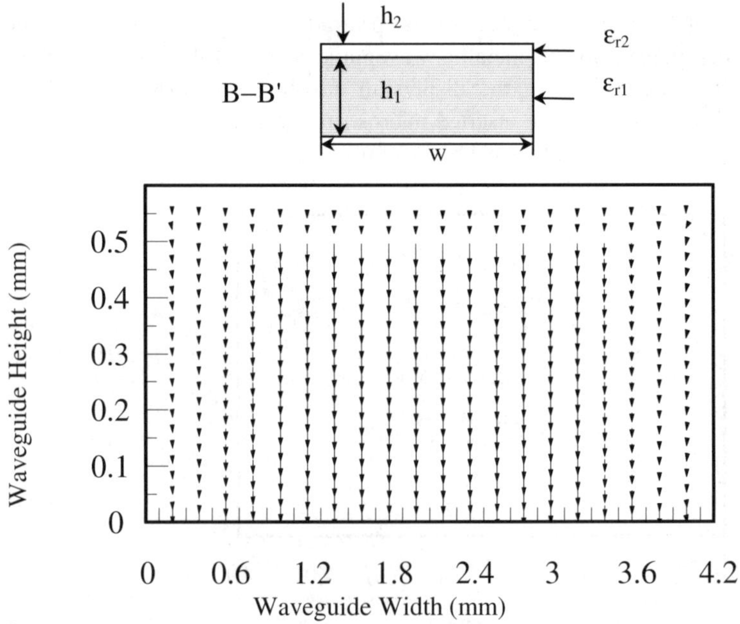

b

Fig. 4.21 Transverse electric fields. **a** at reference plane A–A′, the partially filled waveguide including the inserted microstrip; **b** at reference plane B–B′, the TE_{10}- mode partially filled waveguide

of characteristic impedance of various guiding structures such as the rectangular waveguide and microstrip are not precisely defined, the impedance matching is not initially considered. Instead, field matching between the microstrip and waveguide is investigated first. The electrical fields at the reference planes A-A′ and B-B′ are ploted in Fig. 4.21a and 4.21b, respectively. The field lines of the rectangular waveguide, including the microstrip line (cross section A–A′) resemble those of the TE_{10} mode of the partially filled waveguide (cross section B–B′). Consequently, the field matching between the two distinct guiding structures is good. Furthermore, layer 1 has a higher relative permittivity constant and thicker substrate height than layer 2, thereby bounding most of the modal energy within layer 1 and reducing the radiation loss leaking out of the aperture at the interface of the tapered microstrip and rectangular waveguide. Figure 4.22 plots the theoretical scattering parameters of the proposed microstrip-to-waveguide transition of Fig. 4.19, indicating that the return loss is better than –15 dB in the entire Ka-band and the minimum insertion loss is –0.27 dB at 33.5 GHz.

The assumed parameters in the high frequency structure simulator (HFSS) finite element method (FEM) simulator are gold metal thickness of 34 μm, relative dielectric constant of 3.0 and substrate thickness of 0.508 mm, in accordance with the RO3003 electrical specifications.

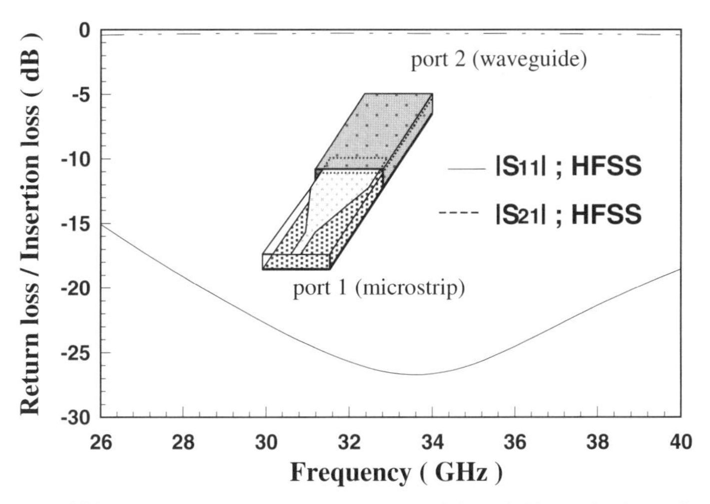

Fig. 4.22 Theoretical scattering parameters of the DC-blocked microstrip-to- waveguide transition

4.4.4 Experimental Verification

The simulated performances show in Fig. 4.22 can be hardly verified by a practical test fixture that supports the microstrip and waveguide ports simultaneously. Therefore two of the proposed microstrip-to-waveguide mode converters are

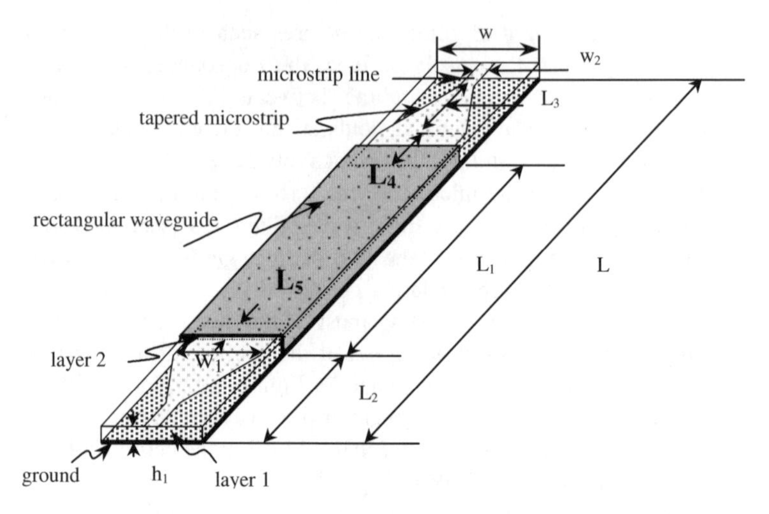

Fig. 4.23 Configuration of the back-to-back connected microstrip-to-waveguide transitions at Ka-band

connected back-to-back to make the measurement feasible. Figure 4.23 shows the configuration of the test structure, where the input and output ports are both 50-Ω microstrips, and the dominant mode of the partially filled waveguide is the TE_{10} mode with the cutoff frequency at 21.52 GHz. This planar transition employing Duroid-based substrates was also designed and verified by the HFSS simulator and network analyzer, respectively. As shown in Fig. 4.24, theoretical results show

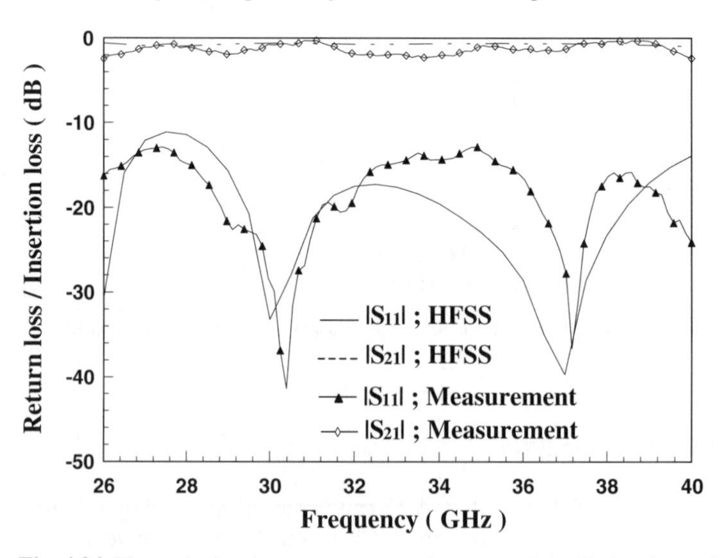

Fig. 4.24 Theoretical and measurement performance of the Ka-band transition

that the return loss is better than –10 dB from 26 GHz to 40 GHz, and the insertion losses, including radiation losses at junctions, conductor losses (σ=5.8×10^7) and dielectric losses (tanδ=0.002 for both layer 1 and layer 2) are approximately –0.6 dB at 30 GHz and 37 GHz. Figure 4.24 also plots the magnitudes of the measured scattering parameters, showing very good agreement with the theoretical ones. Based on the measurement curves, we conclude the broadband characteristic of over 40% bandwidth and average insertion loss of –1.3 dB in the entire Ka-band for the back-to-back connected mode converters plus the rectangular waveguide. The loss budgets of the mode converter computed by the HFSS simulator are 0.14 dB for conductor losses and 0.24 dB for the dielectric losses, respectively. The rest is attributed to losses associated with the tapered microstrip and the microstrip–waveguide interface, approx. 0.18 dB. According to the measurement performances described above, this mode converter is for realizing millimeter-wave planar circuits.

4.4.5 DC-Shorted Broadband Microstrip-to-Waveguide Transition

Figure 4.25 shows the second planar microstrip-to-rectangular waveguide collinear transition, where the microstrip, the rectangular waveguide and the tapered microstrip acting as the transition circuit all coincide along the axial axis of the guiding structures. The X-band microstrip-to-waveguide transition is also fabricated photolithographically on a substrate of thickness 0.508 mm and relative permittivity 3.38. Detailed dimensions are L=49.5 mm, L=19.75 mm, L_g=10 mm, W_1=1.16 mm and W_2=12.434 mm. The microstrip input and output are 50-Ω lines, and the rectangular waveguide (in the middle of Fig. 4.28) supports the TE_{10} mode with cutoff frequency at 6.56 GHz. The tapered microstrip lines, which are separately connected to the input and output ports, are the mode converters, transforming the microstrip modal energy into the waveguide, and vice versa.

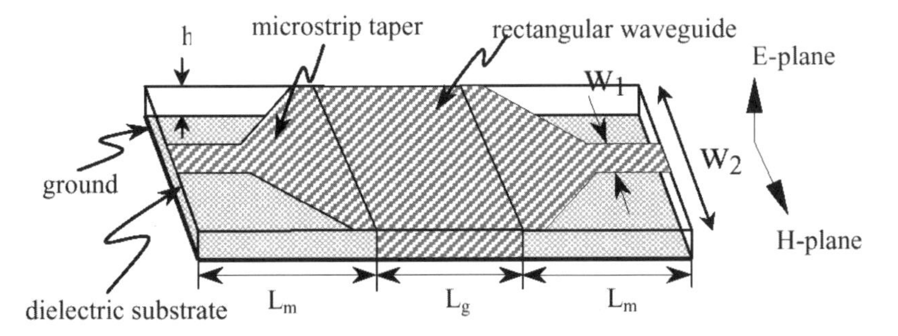

Fig. 4.25 The proposed collinear, planar microstrip-to-waveguide

As shown in Fig. 4.25, the E-plane is polarized along the E-field of the TE_{10} mode, whereas the H-plane is perpendicular to the E-plane. The tapered microstrip is obviously posing the H-plane discontinuity, thereby constituting an H-plane mode converter. The measurement results of the back-to-back connected microstrip-to-waveguide transitions show that the return loss is better than –10 dB from 9.22 to 10.19 GHz, and the insertion loss is near –0.62 dB at the center frequency of 9.78 GHz. The through measurement indicates that the bandwidth of approximately 10% is achieved. The measurement is carried out with the SMA connectors installed at both input and output ports, thus the insertion loss includes connector effects.

The performance of the proposed microstrip-to-waveguide transition is comparable to what has been achieved recently [25]. The DC-shorted type microstrip-to-waveguide mode converter has been frequently applied in our Ka-band integrating modules, which are described in the following sections.

4.5 H-Plane Planar Rectangular Waveguide Filter

Figure 4.26 illustrates a Ka-band integrated planar rectangular waveguide filter prototype, consisting of two DC-shorted H-plane mode converters and six H-plane waveguide slits. All of them are metallized on the same single-layer substrate of thickness *h*. Altogether they establish a sixth-order bandpass waveguide resonator filter. The H-plane slits control the coupling of the waveguide sections.

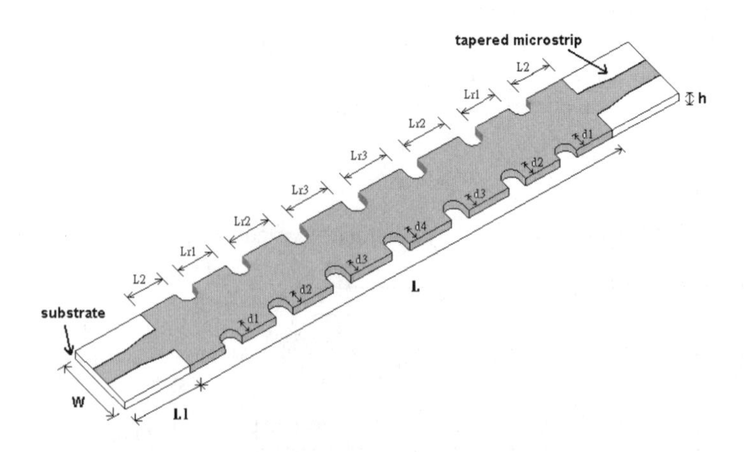

Fig. 4.26 Application of the proposed transition circuit in a H-plane waveguide filter at Ka-band

As shown in Fig. 4.26, the filter is even symmetric with sections of length 2.02 mm (Lr_1), 2.36 mm (Lr_2), 2.45 mm (Lr_3), respectively. The rectangular waveguide filter has width 4.1 mm (W), and length 3.9 mm (L_1). The total length of the planar filter is 24.86 mm (L). The step junctions, made by the symmetric waveguide slits with the lengths of 0.77 mm (d_1), 0.98 mm (d_2), 1.04 mm (d_3), 1.05 mm (d_4), respectively, and width of 1 mm (d_s) establish the necessary inductive couplings between adjacent resonators. Therefore this filter can be designed by the K-inverter equivalent circuit for procedure [26]. The mode- matching method employing the generalized scattering matrix representation of the filter circuit is well suited for the H-plane filter design [27]. The K-inverter equivalent circuit can be found by the reflection coefficient S_{11} of the GSM parameters. Figure 4.27 shows the equivalent inductance of a step junction between the two waveguides of various widths W and W'(See the inset, Fig. 4.27). The abscissa of Fig. 4.27 shows the ratio W'/W of the two waveguides (guide 1 and guide 2) under the condition $W = 0.714\lambda_0$ (λ_0 is the free-space wavelength). The ordinate of Fig. 4.27 is the normalized equivalent inductance of the step junction, defined by $X\lambda_g/2WZ_0$, where X denotes $2\pi fL$ (f is the operating frequency, and L is the equivalent inductance), Z_0,λ_g, and W are respectively the characteristic impedance, guided wavelength, and width of guide 1. Three curves in Fig. 4.27 show the convergent behavior of the mode-matching method, confirming that the mode number of 10 in the smaller waveguide is good enough to obtain the converged equivalent inductance of the step junction. Notice that the relative convergence criterion is satisfied in the above-mentioned calculations. Figure 4.28 shows the sixth-order planar realization of the H-plane rectangular waveguide filter. The printed-circuit board (RO3003) is routed by the drilling bits, thus showing rounded corners at the slits required for the inductive coupling of the adjacent resonators. These rounded

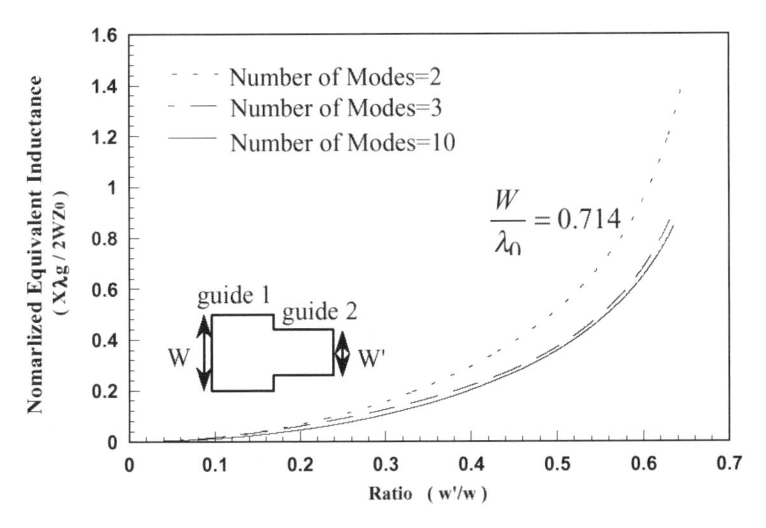

Fig. 4.27 Normalized shunt inductance of the H-plane waveguide step junction using the number of modes at the smaller waveguide as the controlling parameter

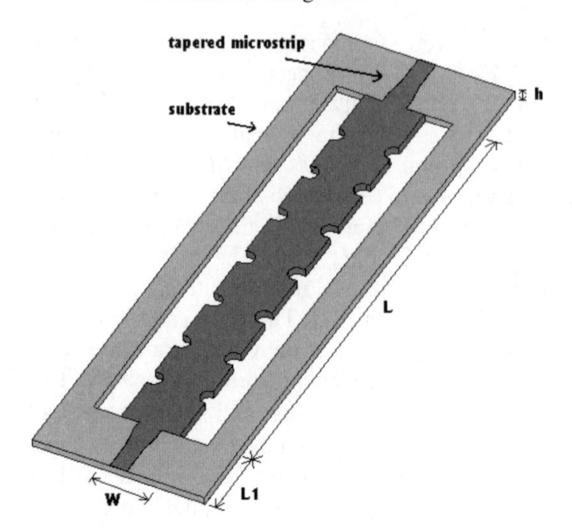

Fig. 4.28 Ka-band planar rectangular filter prototype showing rounded effects

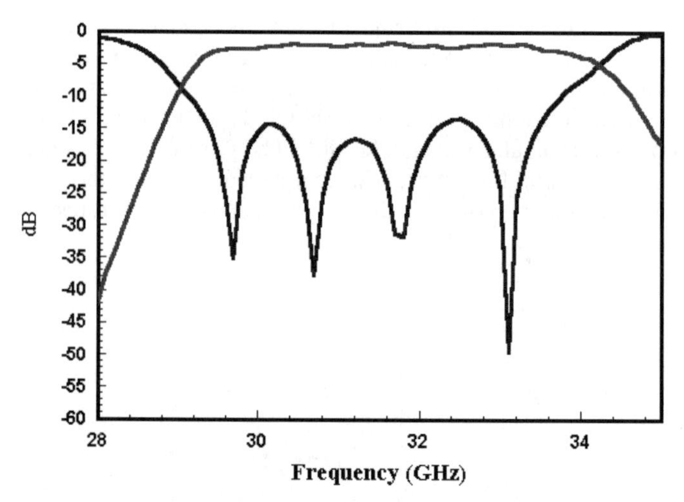

Fig. 4.29 Measured scattering parameters of a Ka-band sixth-order planar H-plane filter

effects are modeled in the FEM analyses. The filter is then fine-tuned and adjusted to the desired filter response. As seen clearly, the side walls of the rectangular filter are plated with gold. The connectorless test fixture Wiltron 3680K is applied to measure the filter. The results are shown in Fig. 4.29, indicating a 29–34-GHz pass band of maximum insertion loss of 1.7 dB at 32.4 GHz. This filter will be used for a self-hetrodyne transmitter prototype.

4.6 Low Phase-Noise Oscillator

4.6.1 Considerations of Planar Resonator

High quality factor (Q-factor) cavity-stabilized or dielectric-resonator-stabilized oscillators are often desired in microwave/millimeter-wave RF modules. Greater resonator quality means better phase-noise performance of oscillators [28, 29]. Dielectric resonators can be easily blended into the NRD-type millimeter-wave modules, thus enabling very high quality oscillator designs. This section, however, will report an alternative approach for designing planar oscillators using PCB rectangular waveguides as high-quality resonators. As shown in Sect.4.1, the total Q-factor of a microstrip a resonator decreases rapidly as the substrate thickness increases. This is largely due to the radiation losses. On the other hand, the planar rectangular waveguide resonator shows much higher total Q-factor, exceeding 300 for the case with loss tangent of 0.002 and substrate height of 0.5 mm. Such observation encourages fully planar integration of oscillators without external high-Q resonators.

4.6.2 Resonator Design for a 15 GHz Oscillator

Using the PCB microstrip resonator has the advantage of simplicity in fabrication, but its poor Q-factor limits its use for high-performance oscillator designs. To circumvent these problems, a rectangular waveguide is incorporated in the resonator design. Where before the step junction and tapered microstrip served as the mode converter, posing the H-plane field discontinuity, in this case the top surface of the rectangular waveguide lines up with the printed microstrip line and shares the same ground plane. The other three open side walls are made by routed cuts, followed by a plated through-hole process, thus forming the planar rectangular waveguide cavity. Figure 4.30 shows two different trial resonator designs for comparative studies of a 15-GHz oscillator. They are both fabricated on the same Duroid RO3003 substrate with thickness of 0.508 mm, relative permittivity of 3.38 and loss tangent of 0.022. The conductivity of the plated metal and sidewall glued silver pastes (if applied) are 5.8×10^7 and 5×10^6, respectively. Dimensions are also given to produce a parallel resonance at 15 GHz as verified by the HFSS. Simulation shows that the total Q-factor of the microstrip resonator is approx. 60.

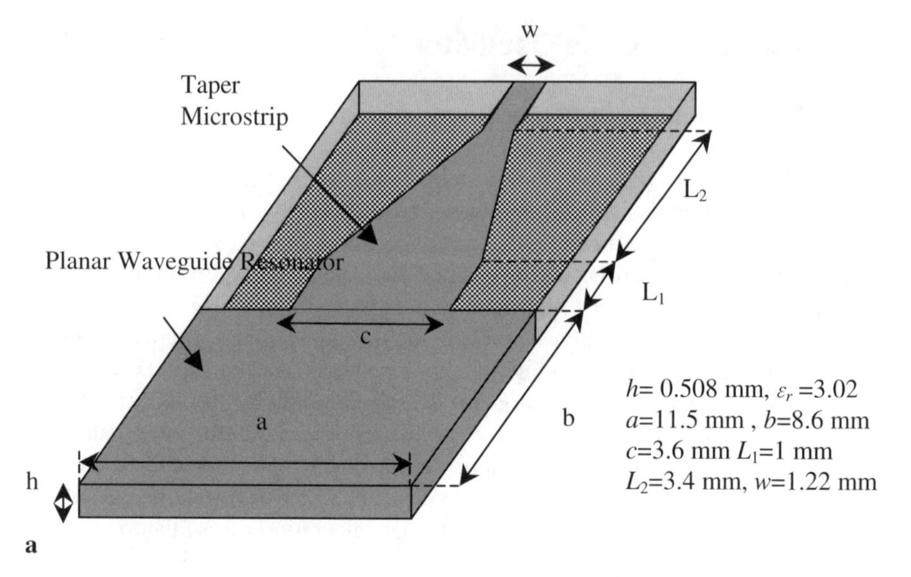

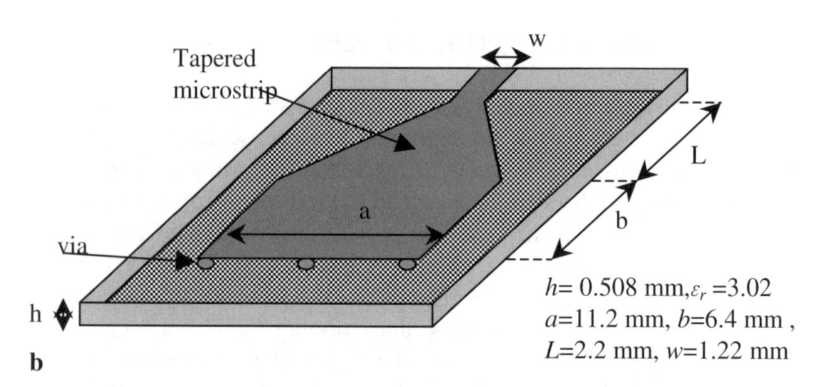

Fig. 4.30 Configuration of planar resonators. **a** Planar rectangular waveguide resonator; **b** conventional short-circuited $\lambda/4$ short-end printed microstrip resonator

4.6.3 All-Planar Microstrip Oscillator

The oscillator was designed using a packaged 0.2-μm NEC 32584C, pseudomorphic high electron mobility transistor (PHEMT), with a typical noise figure of 0.45 dB and associated gain of 12.5 dB at 12 GHz. Since PHEMT technology shows much lower frequency noise than MESFET and HEMT devices [30], thus reducing the low- frequency noise upconversion near the carrier [29, 31, 32]. Curves fitting in the linear and nonlinear models are carried out using the measurement data. The resultant oscillator schematic diagram is shown in Fig. 4.31, illustrating an open-gate topology with a waveguide resonator connected to the source port. It is interesting to note that the short-circuited waveguide resonator provides a series feedback and DC current path simultaneously. Thus only a single bias voltage is applied to drain through the RF choke. The second oscillator applies a series microstrip line connected between the source and microstrip resonator that is adjusted to make the same transistor oscillating at the same frequency. The outputs of the two oscillators are simply a series of a short section of microstrip and $\lambda/4$ -coupled lines tuned at 15 GHz.

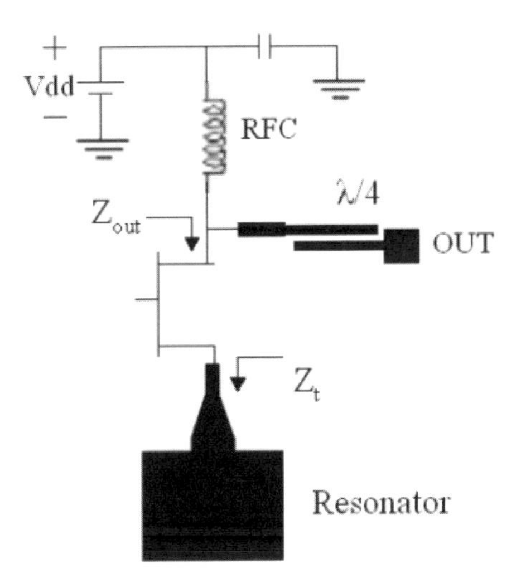

Fig. 4.31 A 15-GHz open-gate, source-feedback, rectangular-waveguide cavity-stabilized oscillator

Measurements of the oscillation frequency, output power and phase noise are carried out by using the test fixture (Wiltron 3680K) and spectrum analyzer (HP 8565E). Figure 4.32 plots the measured power spectrum of the 15 GHz oscillator using the planar rectangular waveguide resonator, showing a stable output power of 14 dBm, phase noise of 98 dBc/Hz at 100 kHz offset from carrier, under 5 V single bias. The second oscillator applying the conventional microstrip resonator at the source exhibits a phase noise of –85 dBc/Hz at 100 kHz offset. The experiments clearly demonstrate that the oscillator with the planar rectangular waveguide resonator outperforms the oscillator with conventional microstrip resonator by 13 dB phase-noise improvement. Furthermore, the open-gate, series feedback oscillator topology provides noise immunity against gate loading. This oscillator needs only a single supply since the pHEMT attempts to oscillate while building up gate bias automatically. Experiments show that more than 10-dB power control and better phase noise can be achieved by varying the drain voltage, which results in superior phase noise [33].

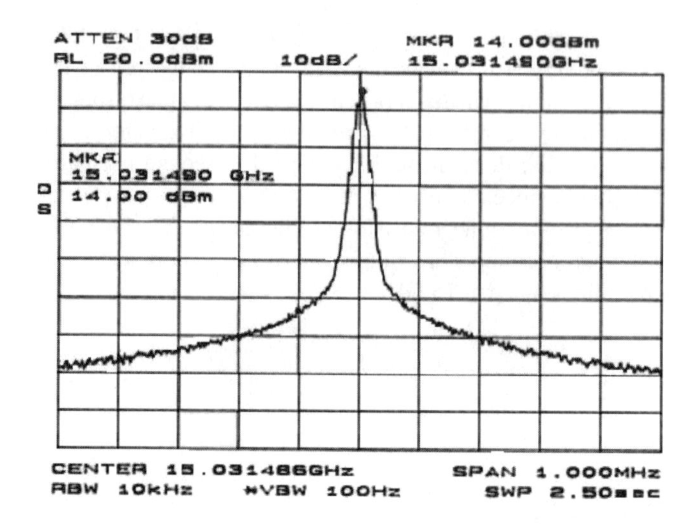

Fig. 4.32 Measured spectrum of the 15 GHz planar waveguide oscillator over a 1 MHz span (phase noise: –98 dBc/Hz at 100 KHz offset)

4.6.4 Frequency Doubler

Followed by the oscillator, a common-source frequency doubler [34] is designed using 0.25-μm PHEMT packaged by a dual-mode configuration that supports both microstrip and coplanar waveguide modes [35], as illustrate in Fig. 4.33. The dual-mode package is modeled by a field solver. The biasing point of the doubler is chosen to be 0 V at the gate and 3 V at the drain, respectively. A $\lambda/4$-shunt open stub followed by a drain serves as a trap for suppressing the 15 GHz fundamental signal and passes the 30-GHz second harmonic to the load. A linear simulation was first performed to match the input at 15 GHz, then utilizing the nonlinear simulation to match the output for maximum output power at 30 GHz [36]. The $\lambda/4$-coupled lines section at the output also acts as a DC-block element. Combining the oscillator with the designed 15 GHz/30 GHz doubler, a millimeter-wave frequency source at 30 GHz is realized. Figure 4.34 plots the measured power spectrum at 30 GHz, showing a power level of 9 dBm and phase noise of 100 dBc/Hz at 1 MHz offset from carrier. The fundamental frequency suppression of the doubler is equal to 29 dBc. The measured results show the great potential of applying standard PCB processes to develop millimeter wave modules without using external components such as dielectric resonators.

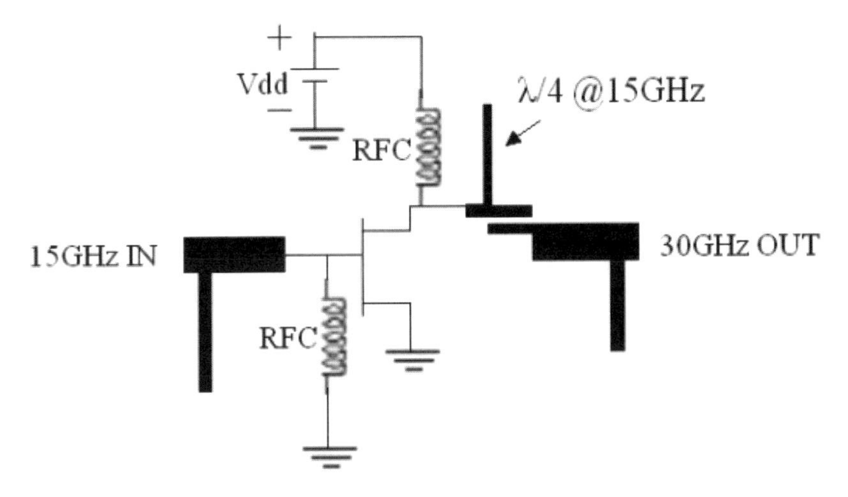

Fig. 4.33 A 15 GHz–30 GHz frequency doubler using a PHEMT package

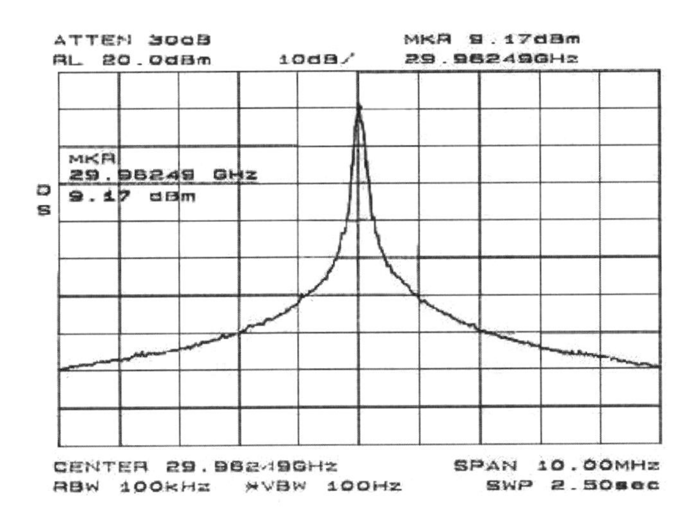

Fig. 4.34 Measured output spectrum of the 30 GHz frequency doubler over a 10 MHz span (phase noise: –103 dBc/Hz @ 1MHz offset)

4.7 All-Planar Sub harmonic Mixer

In modern RF communication systems, the most stringent characteristic for an up converter circuits is related to the local oscillator (LO)-to-RF isolation since the local oscillator frequency can be close to the desired RF band. When the circuit is operating at millimeter-wave frequencies, the local oscillator is difficult to realize using the conventional PCB technology. Sub harmonic mixing provides a remedy for these problems. A conventional sub harmonic mixer consists of an antiparallel diode pair and a waveguide type filter [37, 38]. Rectangular waveguides made by plastic injection-molding have been rigorously pursued for designing high-performance millimeter-wave modules. However, these guiding structures are not planar and need transition circuits to interconnect devices in either discrete or monolithic forms.

The planar rectangular waveguide filter described in Sect. 4.5 provides an alternative approach to new sub harmonic circuit design. Here we report a sub harmonic mixer utilizing a 29–34 GHz planar rectangular waveguide filter to achieve good RF-to-LO isolation, and integrate all the necessary components in one PCB module.

Figure 4.35 shows the schematic of the sub harmonic mixer. The mixer consists of an antiparallel diode pair (APDP), the parallel coupled lines, an LO-matching network, a planar rectangular waveguide image-rejection bandpass filter and several quarter-wavelength open stubs. The $\lambda_{g,LO}/4$ ($\approx \lambda_{g,RF}/2$) open-circuit stub on the RF side terminates the diode pair with a short circuit at the LO frequency and with an open circuit at the RF frequency. Therefore, the RF signal can pass undisturbed. The RF rectangular waveguide bandpass filter is DC short-circuited and provides a DC/IF return path to ground. It also provides isolation between RF port and LO/IF ports. The LO signal is coupled to the diode pair through the parallel coupled lines, which are nearly open for the IF signal and provide the LO port isolation to the IF port. The series $\lambda_{g,RF}/4$ transmission line and the shunt

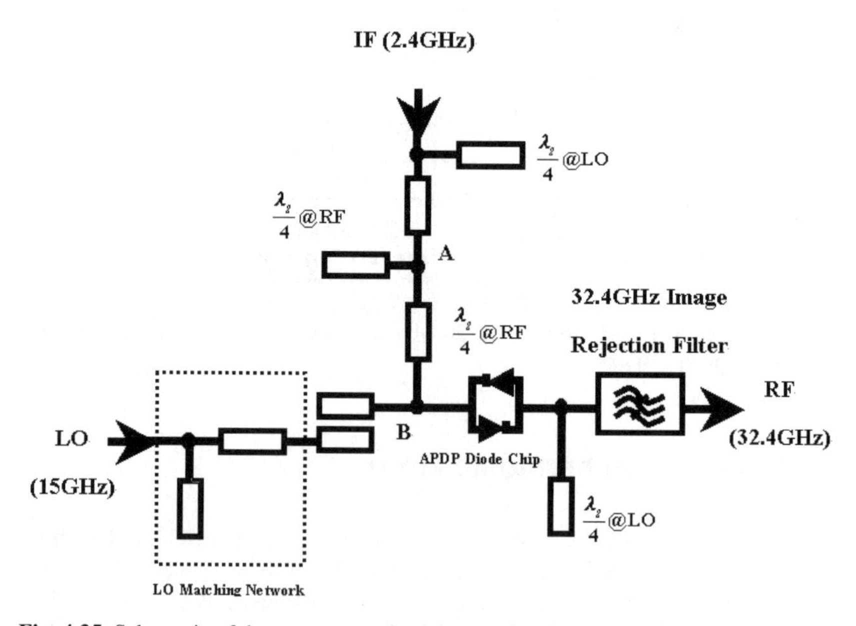

Fig. 4.35 Schematic of the up converted sub harmonic mixer

$\lambda_{g,RF}/4$ open-circuit stub at the IF side provide an open circuit for the RF signal. The additional series transmission line together with the shunt $\lambda_{g,LO}/4$ open-circuit stub make a desired impedance Z^* at node A, which subsequently makes node B see an open circuit for the LO signal. Thus the LO signal can only pass undisturbed into the APDP and the IF port isolation to the LO port is achieved simultaneously. The up converted RF signal is filtered by the planar rectangular with passband between 29 and 34 GHz, which rejects the image signal at 27.6 GHz. A sixth-order Chebyshev filter is designed for good image rejection, resulting in nearly 46 dB rejection (Fig. 4.36).

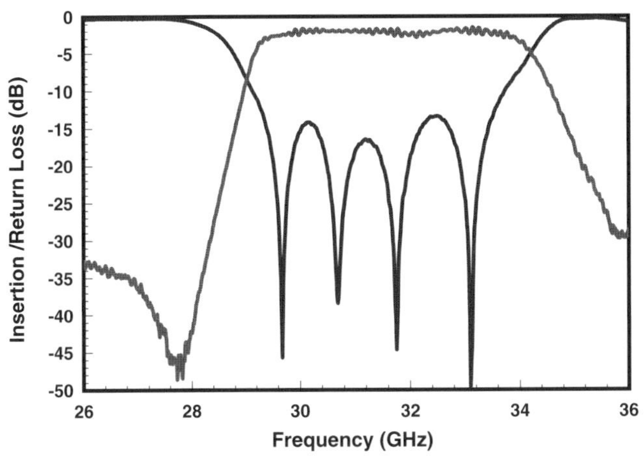

Fig. 4.36 Measured scattering parameter of the sixth-order image-rejection filter made by planar rectangular waveguide

Figure 4.37 shows the sub harmonic mixer prototype implemented on the RO3003 substrate of thickness 0.508 mm, relative dielectric constant 3.0, metal thickness 17μm and loss tangent 0.002. The M/A-COM MA4E2039 anti parallel beam lead diode pair is soldered to the substrate.

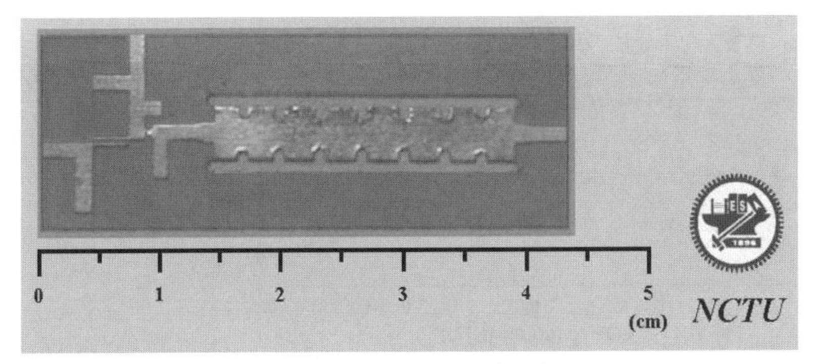

Fig. 4.37 Photograph of the fabricated sub harmonic mixer

Figure 4.38 shows the measured single side band (SSB) up conversion loss versus IF frequency at the LO power corresponding to the minimum conversion loss. In the upper side band (USB) between 31.5 and 33 GHz the conversion loss of the mixer is between 11 and 12.5 dB, whereas the lower side band (LSB) shows conversion loss between 19 and 43 dB at the LO power of 4 dBm. The lower side band signal is suppressed by the image-rejection rectangular waveguide filter. Figure 4.39 shows the measured port-to-port isolation. The RF/LO and RF/IF isolation are better than 35 and 75 dB, respectively when varying LO frequencies from 13–16 GHz. On the other hand, the RF/2LO isolation are better than 42 dB. The measured data of the sub harmonic mixer using all-planar integration technology also show great potential for making compact millimeter-wave modules.

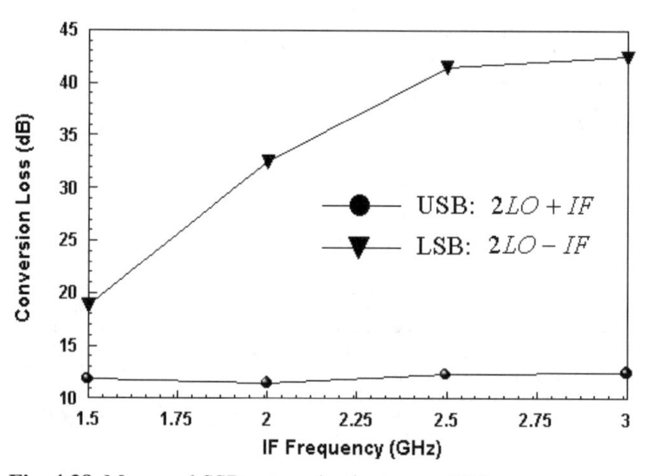

Fig. 4.38 Measured SSB conversion loss versus IF frequency

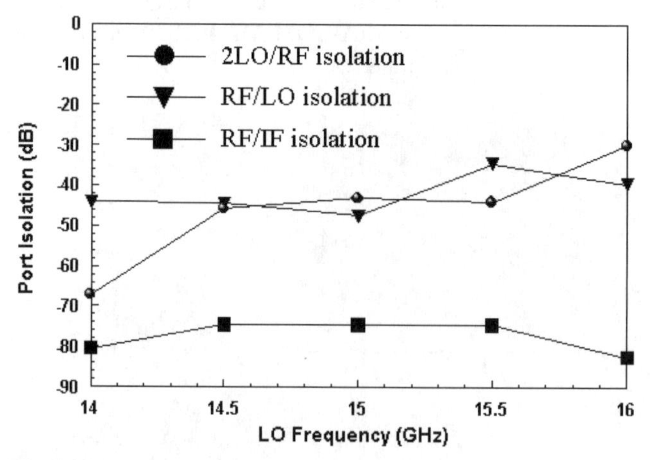

Fig. 4.39 Measured port isolation versus LO frequency

4.8 Applications

The all-planar advanced coplanar strips system (ACSS) integration approach to millimeter-wave RF front-end modules as described in the previous sections can be extended to various types of transceivers. This section reports two transceiver prototypes: one is the frequency-shift keying (FSK) transceiver module, followed by a self-hetrodyne transceiver for broadcast applications. Both examples are developed mainly to demonstrate the practicality of the all-planar ACSS integration approach.

4.8.1 FSK (Frequency-Shift Keying)

Frequency-shift keying (FSK) is a popular digital modulation technique in modern RF communication systems. The FSK modulation has been applied primarily in lower RF frequencies; however, in microwave or millimeter-wave frequencies, FSK modulation is also common in various wireless systems. The design of a FSK modulator at millimeter-wave frequencies is a non trivial matter. Figure 4.40 shows the block diagram of the FSK modulator; the FSK signals are created at half the RF frequency, followed by a doubler. The oscillator and doubler circuits have been reported in Sects. 6.3, and 6.4, respectively.

Figure 4.41 shows the schematic of the transistor-controlled FSK modulator [39]. A surface-mounted PHEMT oscillator with a floating gate is designed to generate high power at the drain output using a single supply. Applying a high-Q rectangular waveguide resonator through an H-plane microstrip-to-waveguide transition at the source terminal can result in a low phase-noise oscillator. Additionally, a voltage-controlled PHEMT switch is shunted across the planar waveguide resonator at a proper position. When digital input data, alternating between 0 and –5 V, are applied to the gate terminal of the switching PHEMT transistor, the source impedance of the oscillating transistor changes accordingly, thereby switching the carrier frequencies.

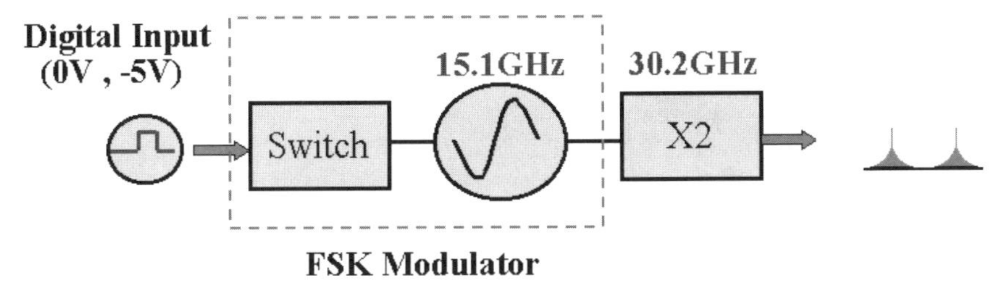

Fig. 4.40 The schematic of the millimeter-wave FSK modulator

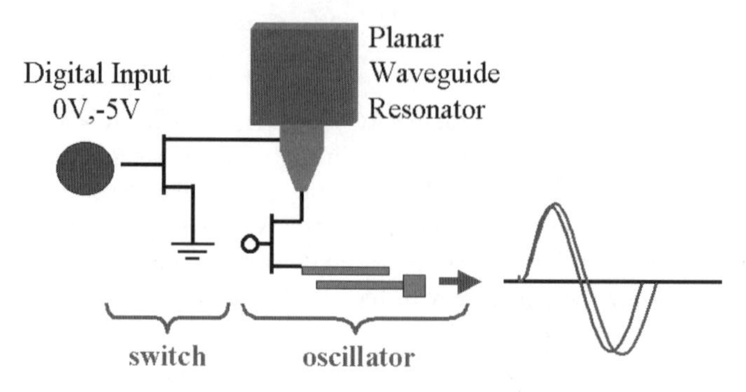

Fig. 4.41 The schematic of the transistor-controlled FSK modulator

Figure 4.42 shows a more complete layout of the FSK transmitter, where the FSK modulator is followed by a MMIC power amplifier and the planar radial array antenna emitting a cone-shaped radiation pattern. The FSK transmitter is obviously all planar and measures 140 by 100 mm^2 in size. The measured FSK signals at output alternate at 30.18 GHz and 30.28 GHz, respectively. When data rates

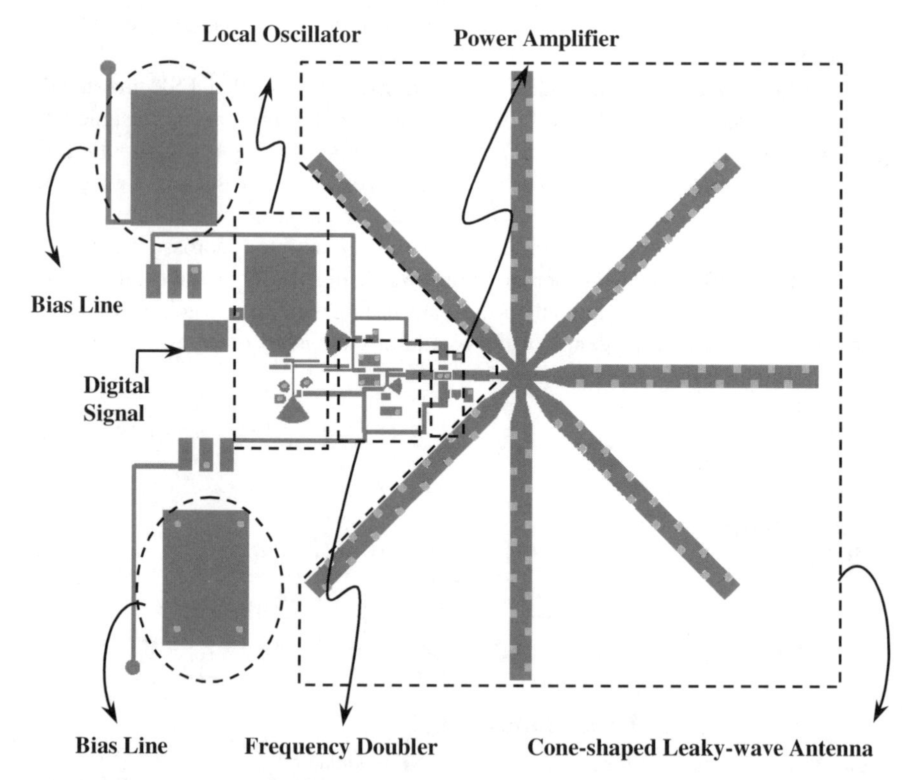

Fig. 4.42 Layout of the 30 GHz FSK transmitter

reach 10 Mbps (duty cycles 50%), the FSK signals are still distinguishable on the spectrum analyzer, indicating that the fast FSK switching speed at the Ka-band can be achieved. The measured radiation pattern of the radial antenna array is nearly omni directional, with the main beam pointing to 48° measured from broadside and a 3-dB beam width of 22°. Figure 4.43 plots the spectrum of a modulated signal with 2 Mb/s pulse train with duty cycle 50% at the output port of the frequency doubler.

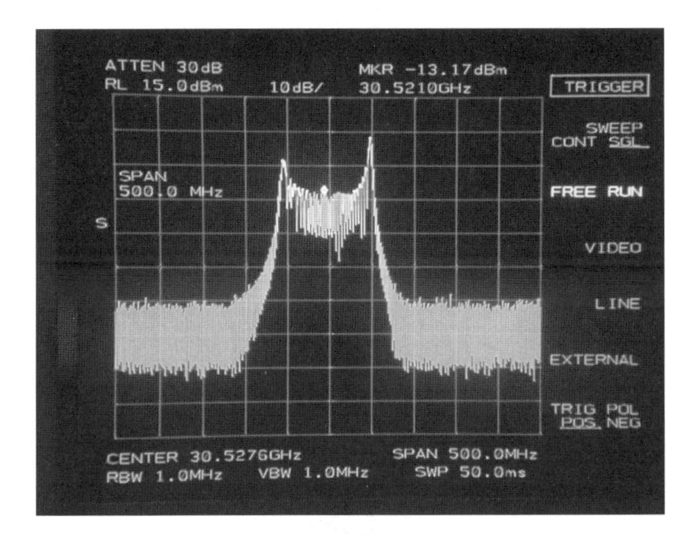

Fig. 4.43 The output signal of FSK modulator after a frequency doubler with input digital data rate, 2 Mb/s after a frequency doubler

4.8.2 FSK Direct-Conversion Receiver

Figure 4.44 shows the architecture of the direct-conversion FSK receiver with its layout illustrated in Fig. 4.45. The leaky-mode antenna array of 2.5×3.5cm^2 in size (Sect. 4.2) exhibits an aperture efficiency over 70%. Since the main beam of this

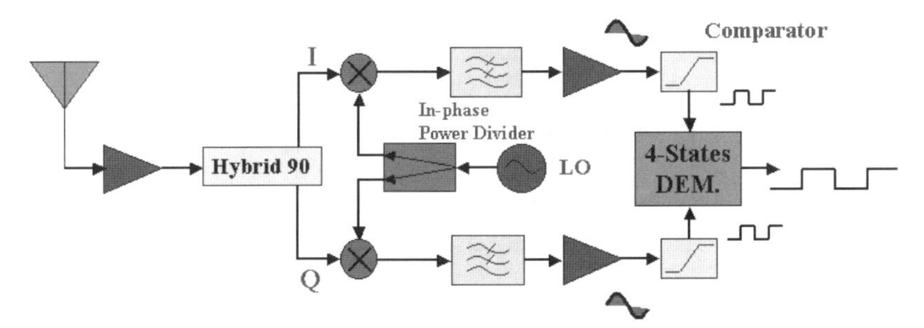

Fig. 4.44 The architecture of the Ka-band FSK direction-conversion receiver

receiving antenna points at the same elevated angle of 43°, one can install the receiver module horizontally to gather the maximum power of –37 dBm at the distance of 1 m away from the transmitter module.

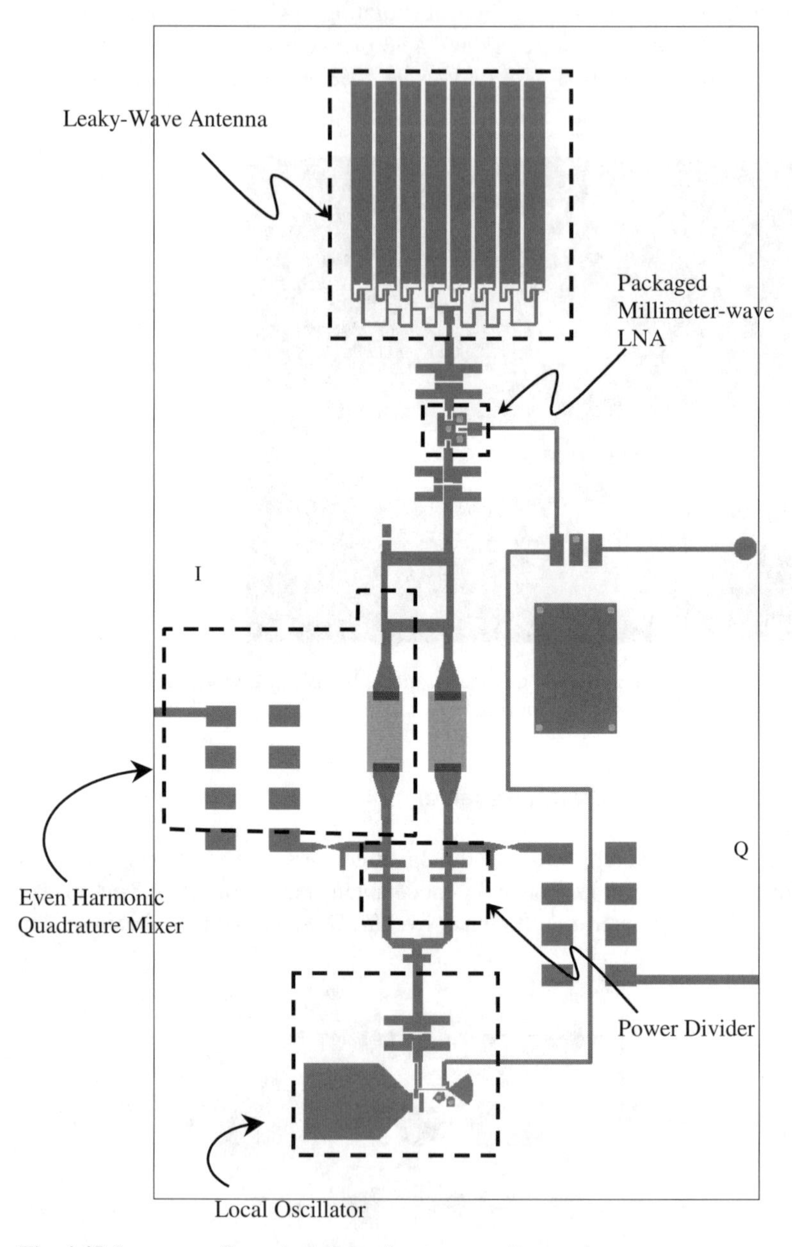

Fig. 4.45 Layout configuration of the direct-conversion receiver

To eliminate the disadvantages of direct conversion such as second-order intermodulation (IM2) and LO noise, an even harmonic quadrature mixer, including a 90° branch coupler, an in-phase power divider and an anti parallel diode pair (APDP) is incorporated into the downconverter. The LO frequency is half of the RF frequency; therefore a simple diplexer approach is used to reduce the area size of the receiver module [40]. A low-loss planar rectangular waveguide and step-impedance microstrips acting as the high-pass and low-pass filters, respectively, can easily separate the RF and LO signals with isolation greater than 30 dB. The mixer conversion loss of 12.3 dB can be observed, where the corresponding power levels of RF and LO are –22 and 11.5 dBm. Received RF signals at 30.28 and 30.18 GHz are directly converted to the baseband at 50 MHz by the even harmonic quadrature mixer. Figure 4.46 reports the measured demodulated waveforms at I/Q outputs, showing a 90.2°phase difference and illustrating the feasibility of the proposed design approach.

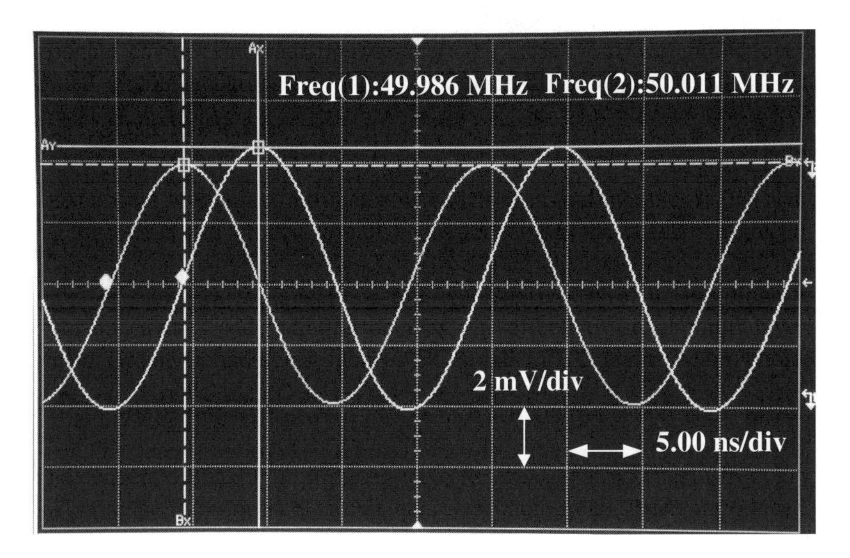

Fig. 4.46 Output waveforms at I channel and Q channel

4.9 Conclusions

The concept of making millimeter-wave RF front-end modules using planar guiding structures and printed circuit board technology has been introduced and its potential explored. Key elements for successful planarization of the millimeter-wave RF modules are mastering the available modes of the guiding structures, resonant cavities and radiating devices; the effective use of multi layered integrated circuit technologies such as PCB and LTCC and a better choice of RF front-end architectures to meet specific needs. The antenna and filter, for instance, are no longer considered as separate external entities. They are, in fact, parts of the integrated modules in the so-called advanced coplanar strip system (ACSS), which takes full advantage of the rapid advances in microelectronics that both III–V-based and silicon-based monolithic technologies offer to continuously improve RF integrated circuits. The great leap in microelectronics, however, demands less stringent requirements on guiding losses of the planar modules. Various design examples have been discussed in the millimeter-wave regime. The migration of the presented design examples near 30 GHz to lower and higher frequencies can be equally challenging. Nevertheless, the ACSS concept is one approach for making low-cost RF front-ends at large volume.

References

[1] F. Kuroki, M. Sugioka, S. Masukawa, K. Ikeda, T. Yoneyama (1998) High-speed ASK transceiver based on the NRD-guide technology at 60-GHz band IEEE Trans on Microwave Theory and Tech. 46: 806-810

[2] I. Gresham, N. Jain, N. T. Budka, A. Alexanian, N. Kinayman, B. Ziegner, S. Brown, P. Staecker (2001) A compact manufacturable 76-77 Radar module for commercial ACC applications. IEEE Trans on Microwave Theory and Tech 49: 44-58

[3] K. Hamaguchi, Y. Shoji, H. Ogawa, H. Sato, E. Kawakami, Y. Hirachi, T. Iwasaki, A. Akeyama, Y. Shimomichi, T. Kizawa I. Kuwana (2001) Development of millimetre-Wave video transmission system II. In: Proc TSMMW2001, March, pp 191-214

[4] Y. C. Shin, T. Itoh, L. Q. Bui (1983) Computer-aided design of millimeter-wave E-plane filters. IEEE Trans on Microwave Theory and Tech 31: 135-142

[5] J. Dittloff, F. Arndt (1989) Rigorous field theory design of millimeter-wave E-plane integrated circuit multiplexer. IEEE Trans on Microwave Theory and Tech 37: 340-350

[6] C. N. Hu, C. K. C. Tzuang (2001) Analysis and design of large leaky-mode array employing the coupled–mode approach. IEEE Trans on Microwave Theory and Tech 49: 629-636

[7] C. C. Lin, C. K. C. Tzuang (2000) A dual-beam micro-CPW leaky-mode antenna. IEEE Trans. on Antennas and Propagation 48: 310-316

[8] A. Oliner (1987) Leakage from higher modes on microstrip line with application to antennas. Radio Science: 907-912

[9] K. A. Michalski, D. Zheng (1989) On the leaky modes of open microstrip lines. Microwave Opt Tech Lett 2: 6-8

[10] J. S. Bagby, C. H. Lee, D. P. Nyquist, Y. Yuan (1993) Identification of propagation regimers on integrated microstrip transmission line. IEEE Trans on Microwave Theory and Tech 41: 1887-1894

[11] K. F. Fuh, C. K. C. Tzuang (1995) The effects of covering on complex wave propagation in gyromagnetic slotlines. IEEE Trans on Microwave Theory and Tech 43: 1100-1105

[12] C. K. C. Tzuang, T. Itoh (1986) Finite element analysis of a slow-wave Schottky contact printed line. IEEE Trans on Microwave Theory and Tech 34: 1483-1489

[13] S. Qi, K. Wu (1998) Leakage and resonance characteristics of radiating cylindrical dielectric structure suitable for use as a feeder for high-efficient omnidirectional/sectorial antenna. IEEE Trans. on Microwave Theory and Tech 46: 1767-1773

[14] M. Ando, J. Hirokawa, T. Yamamoto, A. Akiyama, Y. Kimura, N. Goto (1998) Novel single-layer waveguides for high-efficiency millimeter-wave arrays. IEEE Trans. on Microwave Theory and Tech 46: 792-799

[15] W. Menzel, D. Pilz, R. Leberer (1999) A 77-GHz FM/CW radar front-end with a low-profile low-loss printed antenna. IEEE Trans. on Microwave Theory and Tech 47: 2237-2241

[16] K. C. Chen, C. K. C. Tzuang, Y. Qian, T. Itoh (1999) Leaky properties of microstrip above a perforated ground plane. In: MTT-S International Microwave Symposium Digest, pp. 69-72

[17] T. Yoneyama, S. Nishida (1981) Nonradiative dielectric waveguide for millimeter-wave integrated circuits," IEEE Trans on Microwave Theory and Tech 29: 1188-1192

[18] W. Menzel, J. Kassner (1999) Filter and circuit design for a mm-wave communication module. In: Proc 29th European Microwave Conference.

[19] N. Kaneda, Y. Qian, T. Itoh (1999) Broadband microstrip-to-waveguide transition using quasi-Yagi antenna. IEEE Trans on Microwave Theory and Tech 47: 2562-2567

[20] W. Grapher, B. Hudler, W. Menzel (1994) Microstrip to waveguide transition compatible with MM-wave integrated circuits. IEEE Trans. on Microwave Theory and Tech 42: 1842-1843

[21] D. M. Pozar (1996) Microstrip transmission line transition structure having an integral slot and antenna coupling arrangement. In: US Patent no 5,793, 263

[22] Y. Konishi (1998) Microwave electronic circuit technology. Dekker, New York , chapter 4

[23] E. Bėlohoubek, E. Denlinger (1975) Loss considerations for microstrip resonators. IEEE Trans on Microwave Theory and Tech 23: 522-526

[24] C. J. Lee, H. S. Wu, and C. K. C. Tzuang (2001) A broadband microstrip-to-waveguide mode converter. In: (ISSSE01) International Symposium on Signal, Systems, and Electronics Digest, pp. 331-334.

[25] C. K. C. Tzuang, K. C. Chen, C. J. Lee, C. C. Ho, H. S. Wu (2000) H-plane mode conversion and application in printed microwave integrated circuit. In: Digest of the 25th European Microwave Conference, October, p. 32

[26] G. L. Matthaei, L. Young, E.M.T. Jones (1980) Microwave filters, impedance-matching networks, and coupling structures. Artech House.
[27] Q. Zhang, T. Itoh(1988) Computer-aided design of evanescent-mode waveguide filter with nontouching E-Plane fins. IEEE Trans on Microwave Theory and Tech 36: 404-412
[28] D. B. Leeson (1966) A simple model of feedback oscillator noise spectrum. In: Proc IEEE, vol. 54, Feb. pp. 329–330.
[29] A. Hajimiri, T. Lee (1998) A general theory of phase noise in electrical oscillators. IEEE Journal of Solid-State Circuits 33: 179–194
[30] R. Plana, L. Escotte, O. Llopis, H. Amine, T. Parra, M. Gayral, J. Graffeuil (1993) Noise in AlGaAs/InGaAs/GaAs pseudomorphic HEMT's from 10 Hz to 18 GHz. IEEE Trans. Electron Devices 40:852–858
[31] H. J. Siweris, B. Schiek (1985) Analysis of noise upconversion in microwave FET oscillators. IEEE Trans. on Microwave Theory and Tech 33: 233–242
[32] H. Rohdin, C. Y. Su, C. Stolte (1984) A study of the relation between device low frequency noise and oscillator phase noise. IEEE Trans. on Microwave Theory and Tech 32: 267–269.
[33] T. Kashiwa, T. Ishida, T. Katoh, H. Kurusu, H. Hoshi, Y. Mitsui (1998) V band high-power low phase-noise monolithic oscillators and investigation of low phase-noise performance at high drain bias. IEEE Trans on Microwave Theory and Tech 46: 1559–1565.
[34] S. A. Maas (1988) Nonlinear Microwave Circuits. Artech House, Norwood, Mass
[35] C. K. Tzuang (1998) Dual mode Micrometer/Millimeter wave integrated circuit package. US Patent no 5,783,847
[36] D. G. Thomas, Jr., G. R. Branner (1996) Optimization of active microwave frequency multiplier performance utilizing harmonic terminating impedances. IEEE Trans. on Microwave Theory and Tech 44: 659–662.
[37] M. Cohn, J. E. Degenford, B. A. Newman (1975) Harmonic mixing with an antiparallel diode pair. IEEE Trans. on Microwave Theory and Tech 23: 667-673
[38] J. Matreci, F. K. David (1983) Unbiased subharmonic mixer for millimeter wave spectrum analyzers. In: IEEE MTT-S International Microwave Symposium Digest, pp.130-132
[39] C. J. Lee, T. Hung, S. C. Lin, H. S. Wu, C. Y. Tsai, K. F. Hung, Y. C. Chen, W. C. Lee, C. K. C. Tzuang (2001) A fully all-planar integrated Ka-band FSK transceiver module. In: Proc Asia-Pacific Microwave Conference: 1084–1087.
[40] M. Shimozawa, T. Katsura, N. Suematsu, K. Itoh, Y. Isota, O. Ishida (2000) A passive-type even harmonic quadrature mixer using simple filter configuration for direct conversion receiver. In: MTT-S Int. Microwave Symposium Digest, pp. 517-520

5. Recent Progress in Time-Slotted CDMA

5.1 Introduction

As a vital issue when designing mobile radio systems, the multiple-access (MA) scheme underlying the air interface has to be defined. There are three basic MA schemes available, namely frequency-division MA (FDMA), time-division MA (TDMA), and code-division MA (CDMA). Each of these schemes has its specific pros and cons, as discussed in detail in [1, 2]. In the first generation (1G) mobile radio systems the transmission was analog, and as a consequence, these systems were restricted to the use of FDMA. The second generation (2G) mobile radio systems introduced in the early 1990s and the forthcoming third generation (3G) systems apply digital transmission technology, which allows the use of TDMA and CDMA as well as combinations of the three basic MA schemes. The idea behind such hybrids consists of retaining the advantages of the contained basic MA schemes and simultaneously circumventing their disadvantages in the combinations. Virtually all modern mobile radio systems apply hybrid MA schemes, with all those schemes comprising an FDMA component, which facilitates frequency planning and enables cluster sizes larger than one with a view to reduce intercell and adjacent channel multiple-access interference (MAI). For instance, the MA scheme of the de facto 2G world standard, GSM (Global System for Mobile Communications), is the hybrid FDMA+TDMA.

Especially attractive with respect to flexibility and spectral efficiency are hybrids incorporating all three basic MA schemes. For such hybrids the term time-slotted CDMA has been coined, which indicates that, in addition to the already used scheme FDMA, the schemes TDMA and CDMA are being utilized. About 10 years ago investigations of time-slotted CDMA were initialized with a view to apply this MA hybrid in 3G mobile radio systems [3]. As a special feature of this MA approach, intracell MAI and intersymbol interference (ISI), both typical of conventional CDMA [4], can be simultaneously and efficiently combatted in the receivers by detection schemes termed joint detection (JD) [5, 6]. Simple JD schemes apply linear algorithms, and they are similar to linear multiuser detectors (MUD), which focus on the elimination only of MAI and not of ISI [5]. More advanced JD schemes exploit a priori knowledge about the transmitted signals as, for instance, the knowledge of the value discreteness of the data symbols and of the structure of the applied forward error correction (FEC) codes [7–9]. Such schemes are nonlinear, and their implementation is more complex than that of linear JD. In the meantime, as part of the efforts towards 3G systems, time-slotted CDMA has been verified by extensive computer simulations [10] as well as by laboratory experiments and field tests performed with trial systems [11]. In time-slotted CDMA, the pros of CDMA, namely frequency diversity, interferer diversity and flexibility [12], come together with the following major benefits of TDMA:

- Signals of different TDMA time slots are perfectly separated.
- Thanks to the TDMA component, relatively few user signals are simultaneously active, which facilitates the application of interference-combatting techniques like JD [6] and of adaptive antennas [13].
- In the case of dynamic channel allocation (DCA) the assignment of TDMA time slots to users can be chosen in such a way that situations characterized by a severe impact of MAI are a priori avoided [13].
- The TDMA component easily lends itself to the implementation of packet transmission schemes, which will be a major issue in future mobile radio systems.
- TDMA fits nicely together with time division duplexing (TDD), which, in turn is highly advantageous if different data rates have to be supported in uplink (UL) and downlink (DL). Such asymmetries are typical of future wireless Internet and multimedia applications. There is no way to offer spectrally efficient, highly asymmetric services based on frequency division duplexing (FDD). As another major advantage of TDD enabled by a TDMA component, channel impulse response estimates obtained in the UL reception can be used as a priori information when generating the DL transmission signals at the base station (BS).
- In contrast to the FDD scheme the application of TDD facilitated by TDMA does not require paired frequency bands for UL/DL transmission. Therefore, the reassignment of frequency bands made free by discarded systems, like analog mobile radio systems, as well as the allocation of new frequency bands is easier than in the case of FDD.
- The mobile stations (MS) have to listen to the BS to which they are assigned only in certain TDMA time slots. During other time slots they can probe the propagation conditions to other BS and initiate handover if advisable. Such mobile assisted handovers help to significantly reduce the transmit powers and, consequently, the MAI caused in other radio connections of the system.

The Third-Generation Partnership Project (3GPP) has selected time-slotted CDMA as a key technology for the air interfaces of 3G mobile radio systems. Time Division–CDMA (TD–CDMA) technology has become a constituent part of the 3G standards IMT–2000 and UMTS, where TD–CDMA is designated for the TDD radio bands [14]. Also the world standard TD–SCDMA, launched by China, where S stands for synchronous, is based on this concept [14]. The state of the art already achieved by the time-slotted CDMA systems presently being developed by industry meets the demands on 3G mobile radio systems. Nevertheless, further performance enhancements are desirable in order to increase the system capacity and performance. Promising approaches in this direction, which are presently pursued in academic research and industrial development, concern novel concepts in the fields of antenna array processing for the UL and the DL, advanced JD algorithms, intercell MAI reduction and transmit signal optimization for the DL. In the present chapter, after an introduction into the basics of time-slotted CDMA, these novel concepts are reviewed and illustrated by means of examples including some simulation results.

Chapter 5 is structured as follows: In Sect. 5.2 the basics of time-slotted CDMA, including the mathematical description of this MA technique, are recapitulated. In Sects. 5.3 and 5.4 the application of adaptive antennas at the BS is considered for the UL and the DL, respectively. Section 5.5 concerns advanced JD algorithms, which avail themselves of the iterative (turbo) principle. In Sect. 5.6 methods to mitigate intercell MAI are presented. Section 5.7 deals with an approach to utilize transmit signals in the DL, which have minimum power and, thus, cause minimum MAI to other links of the system, and simultaneously allow very simple receiver implementations at the MS. Finally, Sect. 5.8 summarizes the information in this chapter.

5.2 Time-Slotted CDMA

5.2.1 Frame and Burst Structure

In time-slotted CDMA, K_{s} CDMA signals, which will be termed burst signals in the following, are simultaneously active in the same frequency slot and time slot, with each of these signals using a signal-specific CDMA code to allow signal separation at the receiver [15]. The frame structure of this concept with its FDMA, TDMA and CDMA components is illustrated in Fig. 5.1 [15], where B, T_{fr}, N_{fr}, T_{bu} and K_{s} denote the bandwidth of an FDMA frequency slot, the duration of a TDMA frame, the number of bursts, that is, time slots per TDMA frame, the burst duration and the number of user signals per FDMA frequency slot and time slot, respectively. As shown in Fig. 5.2 the signals of each burst consist of two data blocks α and β separated by a user-specific midamble, which is transmitted to allow channel estimation at the receiver, and a guard interval. Each of the two data blocks of a burst signal contains N data symbols, and each such data symbol is spectrally spread by a CDMA code of Q chips. The midamble consists of L_{m} chips, and the guard interval comprises L_{g} chips. The transmission capacity of a single burst signal, that is, of a single CDMA signal per frame, is termed one resource. Consequently, a total of $N_{\mathrm{fr}}K_{\mathrm{s}}$ resources is available per FDMA frequency slot. Except for the CDMA component, the frame and burst signal

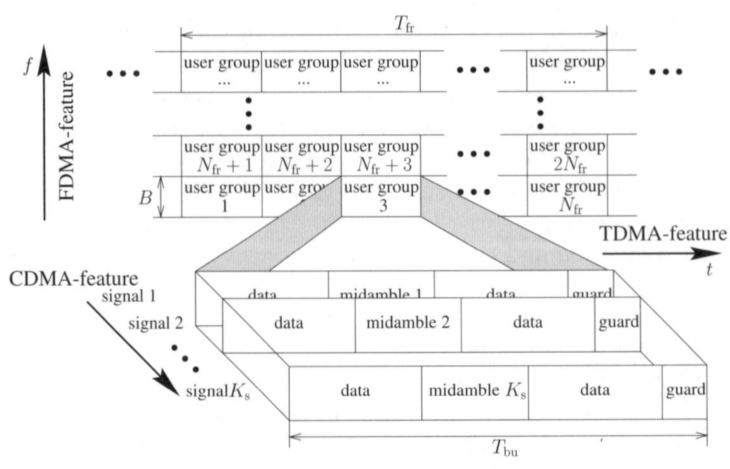

Fig. 5.1. Frame structure of time-slotted CDMA [15]

structures shown in Figs. 5.1 and 5.2, respectively, are similar to those known from GSM and would facilitate backward compatibility with this de facto 2G world standard.

5.2.2 Generic Transmission Model

The goal of a generic transmission model consists in quantitatively visualizing how the receiver input signal results from the transmitted data symbols under consideration of the relevant system parameters. In the case of time-slotted CDMA the transmission model has to include the spectral spreading of the data symbols at the transmitters, the transmission over the radio channels, the superposition of the output signals of the radio channels at the receive antennas and the inclusion of noise, which results from intercell MAI and receiver noise.

Figure 5.3 shows a generic transmission model developed for time-slotted CDMA [6], which is adopted in the following. In this model the time-discrete equivalent low-pass representation of signals and impulse responses is chosen. Consequently, signals and impulse responses are written as complex column vectors or matrices, respectively, which are both printed in bold face. Let us now consider one of the two data blocks of the simultaneously transmitted bursts, for instance block α. Then K_{s} data blocks consisting of N data symbols $\underline{d}_n^{(k_{\mathrm{s}})}$, $n=1,\ldots,N$ each are simultaneously transmitted. Each of these data blocks can be written as a data vector

$$\underline{\mathbf{d}}^{(k_{\mathrm{s}})} = \left(\underline{d}_1^{(k_{\mathrm{s}})} \ldots \underline{d}_N^{(k_{\mathrm{s}})}\right)^{\mathrm{T}}, \; k_{\mathrm{s}}=1,\ldots,K_{\mathrm{s}}, \tag{5.1}$$

termed partial data vector. The N data symbols $\underline{d}_n^{(k_{\mathrm{s}})}$ in $\underline{\mathbf{d}}^{(k_{\mathrm{s}})}$ of (5.1) only take on values from a finite data symbol set

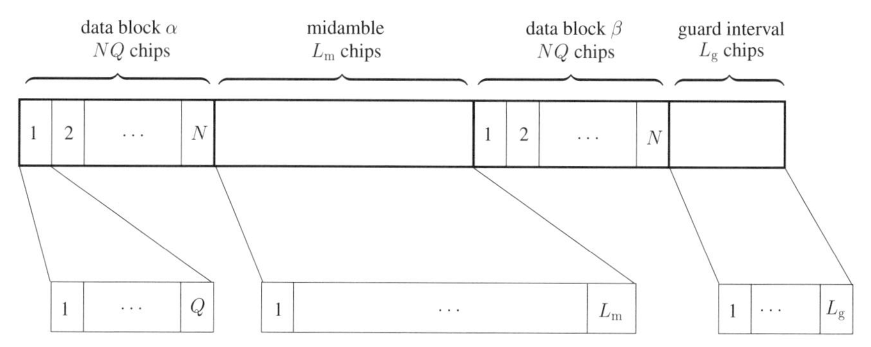

Fig. 5.2. Structure of a burst signal of time-slotted CDMA [15]

$$\mathbb{V}_{\mathrm{d}} = \left\{\underline{V}_{\mathrm{d},1} \cdots \underline{V}_{\mathrm{d},M_{\mathrm{d}}}\right\} \tag{5.2}$$

of cardinality M_{d}. They are coded data symbols originating in the FEC encoder by processing the uncoded data bits to be transmitted.

Prior to transmission each data symbol $\underline{d}_n^{(k_{\mathrm{s}})}$ is spectrally spread by the data vector-specific CDMA code

$$\underline{\mathbf{c}}^{(k_{\mathrm{s}})} = \left(\underline{c}_1^{(k_{\mathrm{s}})} \ldots \underline{c}_Q^{(k_{\mathrm{s}})}\right)^{\mathrm{T}}, \; k_{\mathrm{s}}=1,\ldots,K_{\mathrm{s}}, \tag{5.3}$$

of dimension Q, where Q is the spreading factor. The chips $\underline{c}_q^{(k_{\mathrm{s}})}$, $q=1,\ldots,Q$ of $\underline{\mathbf{c}}^{(k_{\mathrm{s}})}$, $k_{\mathrm{s}}=1,\ldots,K_{\mathrm{s}}$ of (5.3) are taken from the finite chip set

$$\mathbb{V}_{\mathrm{c}} = \left\{\underline{V}_{\mathrm{c},1} \cdots \underline{V}_{\mathrm{c},M_{\mathrm{c}}}\right\} \tag{5.4}$$

of cardinality M_{c}. Usually, the K_{s} CDMA codes $\underline{\mathbf{c}}^{(k_{\mathrm{s}})}$ form an orthogonal set. The N spectrally spread data symbols $\underline{d}_n^{(k_{\mathrm{s}})}\underline{\mathbf{c}}^{(k_{\mathrm{s}})}$ are fed into a radio channel, each with the impulse response

$$\underline{\mathbf{h}}^{(k_{\mathrm{s}})} = \left(\underline{h}_1^{(k_{\mathrm{s}})} \ldots \underline{h}_W^{(k_{\mathrm{s}})}\right)^{\mathrm{T}}, \; k_{\mathrm{s}}=1,\ldots,K_{\mathrm{s}}, \tag{5.5}$$

of dimension W. The elements $\underline{h}_w^{(k_{\mathrm{s}})}$, $w = 1,\ldots,W$ of $\underline{\mathbf{h}}^{(k_{\mathrm{s}})}$ from (5.5) are termed channel coefficients. It is assumed that by the channel impulse responses $\underline{\mathbf{h}}^{(k_{\mathrm{s}})}$ the effects of the usually limited system bandwidth are also taken into account [16]. A band limitation leads to correlations between the elements $\underline{h}_w^{(k_{\mathrm{s}})}$ of the channel impulse response $\underline{\mathbf{h}}^{(k_{\mathrm{s}})}$. The antenna receives a superposition of the output signals of the K_{s} radio channels plus noise, symbolized by the noise vector $\underline{\mathbf{n}}$. In this way the receiver input signal symbolized by the vector $\underline{\mathbf{e}}$ is finally obtained. Later in this section we will see that both $\underline{\mathbf{n}}$ and $\underline{\mathbf{e}}$ have the dimension $NQ+W-1$.

CDMA codes $\underline{\mathbf{c}}^{(k_{\mathrm{s}})}$ of (5.3) with different superscripts k_{s} have to be chosen in such a way that they differ from each other, because they constitute the basis for performing signal separation at the receiver. In the case of the UL, the K_s partial data vectors $\underline{\mathbf{d}}^{(k_\mathrm{s})}$ of (5.1) originate in different MS unless CDMA code pooling is applied (Sect. 5.4), and, consequently, the K_s channel impulse responses $\underline{\mathbf{h}}^{(k_\mathrm{s})}$ of (5.5) generally differ from each other. In the case of the DL, all K_s data vectors $\underline{\mathbf{d}}^{(k_\mathrm{s})}$ of (5.1) originate in one and the same BS, and, consequently, the K_s channel impulse responses $\underline{\mathbf{h}}^{(k_\mathrm{s})}$ of (5.5) are all equal, unless adaptive transmit antennas are utilized (Sect. 5.4).

In a fundamental treatise on time-slotted CDMA [6] Klein showed that the transmission model of Fig. 5.3 can be described by a concise mathematical formalism. In this formalism, with $\underline{\mathbf{c}}^{(k_\mathrm{s})}$ of (5.3) and $\underline{\mathbf{h}}^{(k_\mathrm{s})}$ of (5.5), first the K_s composite channel impulse responses

$$\begin{aligned}\underline{\mathbf{b}}^{(k_\mathrm{s})} &= \underline{\mathbf{c}}^{(k_\mathrm{s})} * \underline{\mathbf{h}}^{(k_\mathrm{s})} \\ &= \left(\underline{b}_1^{(k_\mathrm{s})} \dots \underline{b}_{Q+W-1}^{(k_\mathrm{s})}\right)^\mathrm{T}, \; k_\mathrm{s}=1,\dots,K_\mathrm{s},\end{aligned} \tag{5.6}$$

valid for the data vectors $\underline{\mathbf{d}}^{(k_\mathrm{s})}$ of (5.1) are introduced. Then, with $\underline{\mathbf{b}}^{(k_\mathrm{s})}$ of (5.6), the matrices

$$\underline{\mathbf{A}}^{(k_\mathrm{s})} = \left(\underline{A}_{i,n}^{(k_\mathrm{s})}\right), \; i=1,\dots,NQ+W-1, \; n=1,\dots,N,$$

$$\underline{A}_{i,n}^{(k_\mathrm{s})} = \begin{cases} \underline{b}_{i-(n-1)Q}^{(k_\mathrm{s})}, & \text{if } 1 \le i-(n-1)Q \le Q+W-1, \\ 0, & \text{else,} \end{cases} \tag{5.7}$$

and

$$\underline{\mathbf{A}} = \left(\underline{\mathbf{A}}^{(1)} \dots \underline{\mathbf{A}}^{(K_\mathrm{s})}\right) \tag{5.8}$$

are established. The dimensions of $\underline{\mathbf{A}}^{(k_\mathrm{s})}$ and $\underline{\mathbf{A}}$ are $(NQ+W-1)\times N$ and $(NQ+W-1)\times NK_\mathrm{s}$, respectively. The matrices $\underline{\mathbf{A}}^{(k_\mathrm{s})}$ and $\underline{\mathbf{A}}$ are termed partial and total system matrices, respectively. By stacking the K_s data vectors $\underline{\mathbf{d}}^{(k_\mathrm{s})}$ of (5.1), we obtain the total data vector

$$\underline{\mathbf{d}} = \left(\underline{\mathbf{d}}^{(1)\mathrm{T}} \dots \underline{\mathbf{d}}^{(K_\mathrm{s})\mathrm{T}}\right)^\mathrm{T} \tag{5.9}$$

of dimension NK_s. Now, the vector $\underline{\mathbf{e}}$ representing the signal at the receiver antenna output and at the receiver input can be expressed by $\underline{\mathbf{A}}$ of (5.8), $\underline{\mathbf{d}}$ of (5.9) and the noise vector $\underline{\mathbf{n}}$:

$$\underline{\mathbf{e}} = \underline{\mathbf{A}}\,\underline{\mathbf{d}} + \underline{\mathbf{n}}. \tag{5.10}$$

Equation (5.10) shows that, as already mentioned, both $\underline{\mathbf{e}}$ and $\underline{\mathbf{n}}$ have the dimension $NQ+W-1$.

The elements of the total data vector $\underline{\mathbf{d}}$ of (5.9) and of the noise vector $\underline{\mathbf{n}}$ appearing in (5.10) are assumed to be samples of wide-sense stationary processes [17] in what follows. Then, by averaging over an infinite series of

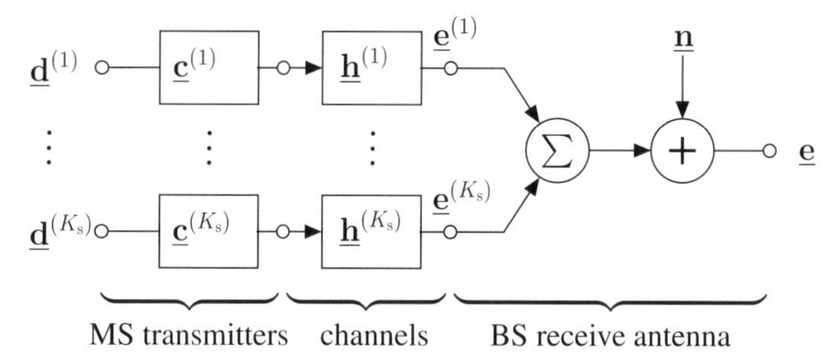

Fig. 5.3. Generic transmission model for the UL of time-slotted CDMA [15]

bursts, the temporal correlation properties of $\underline{\mathbf{d}}$ and $\underline{\mathbf{n}}$ can be statistically characterized by their covariance matrices

$$\underline{\mathbf{R}}_{\mathrm{d}} = \mathrm{E}\left\{\underline{\mathbf{d}}\,\underline{\mathbf{d}}^{*\mathrm{T}}\right\}, \tag{5.11}$$

and

$$\underline{\mathbf{R}}_{\mathrm{n}} = \mathrm{E}\left\{\underline{\mathbf{n}}\,\underline{\mathbf{n}}^{*\mathrm{T}}\right\}, \tag{5.12}$$

respectively. If FEC coding with sufficiently deep interleaving is applied when generating $\underline{\mathbf{d}}$ of (5.9) from the uncoded data in the transmitters, then $\underline{\mathbf{R}}_{\mathrm{d}}$ of (5.11) is a diagonal matrix. If $\underline{\mathbf{n}}$ in (5.12) is white noise, (5.12) takes the form

$$\underline{\mathbf{R}}_{\mathrm{n}} = 2\sigma^2\,\mathbf{I}^{(NQ+W-1)}. \tag{5.13}$$

5.2.3 Data Detection

The data detector located at the receiver must determine an estimate $\widehat{\underline{\mathbf{d}}}$ of the transmitted total data vector $\underline{\mathbf{d}}$ of (5.9) by evaluating the received signal $\underline{\mathbf{e}}$ of (5.10). This estimate should be as accurate as possible. Optimum detectors would be the maximum likelihood sequence estimator (MLSE), or the maximum a posteriori probability sequence estimator (MAPSE) [5]. In the case of the MLSE, with the probability $\mathrm{P}(\underline{\mathbf{d}}|\underline{\mathbf{e}})$ of $\underline{\mathbf{d}}$ of (5.9) conditioned on $\underline{\mathbf{e}}$ of (5.10), the estimate of $\underline{\mathbf{d}}$ is determined, with $\mathbb{V}_{\mathrm{d}}$ of (5.2), by

$$\widehat{\underline{\mathbf{d}}}_{\mathrm{MLSE}} = \arg\max_{\underline{\mathbf{d}}'\in\mathbb{V}_{\mathrm{d}}} \left[\mathrm{P}(\underline{\mathbf{e}}|\underline{\mathbf{d}}')\right]. \tag{5.14}$$

The MLSE is a nonlinear detector because it takes into account that the data symbols $\underline{d}_n^{(k_s)}$ of (5.1) are discrete valued, see (5.2). It is well known that the MLSE is prohibitively expensive in the case of CDMA systems, even if one resorts to complexity-saving implementations like the Viterbi algorithm [18]. Therefore, one has to look for less complex detectors, which, of course, are suboptimum as compared to the MLSE. A class of lower complexity suboptimum detectors are the linear detectors, which determine continuous valued estimates $\widehat{\underline{d}}_n^{(k_\mathrm{s})}$ of the data symbols $\underline{d}_n^{(k_\mathrm{s})}$ of (5.1). These estimates could then be discretized, or they could be used to obtain soft inputs for the subsequent FEC decoder [5, 19].

The theory of linear detectors for CDMA signals is presented in detail in [6]. A linear detector for the received signal $\underline{\mathbf{e}}$ of (5.10) can be generally described by a demodulator matrix $\underline{\mathbf{D}}$ of dimensions $NK_\mathrm{s} \times (NQ+W-1)$. Then with this matrix and $\underline{\mathbf{e}}$ of (5.10) we obtain the linear estimate

$$\widehat{\underline{\mathbf{d}}} = \underline{\mathbf{D}}\,\underline{\mathbf{e}} \tag{5.15}$$

of $\underline{\mathbf{d}}$ of (5.9). Substituting $\underline{\mathbf{e}}$ of (5.10) in (5.15) yields

$$\begin{aligned}\widehat{\underline{\mathbf{d}}} &= \underline{\mathbf{D}}\,\underline{\mathbf{A}}\,\underline{\mathbf{d}} + \underline{\mathbf{D}}\,\underline{\mathbf{n}} \\ &= \underbrace{\mathrm{diag}\left(\underline{\mathbf{D}}\,\underline{\mathbf{A}}\right)\underline{\mathbf{d}}}_{\text{desired data}} + \underbrace{\overline{\mathrm{diag}}\left(\underline{\mathbf{D}}\,\underline{\mathbf{A}}\right)\underline{\mathbf{d}}}_{\substack{\text{ISI and intracell}\\ \text{MAI}}} + \underbrace{\underline{\mathbf{D}}\,\underline{\mathbf{n}}}_{\text{noise}}\,.\end{aligned} \tag{5.16}$$

As indicated on the right side of (5.16), the estimate $\widehat{\underline{\mathbf{d}}}$ generated by the linear detector consists of the following three components:

- a useful component representing, up to factors given by the diagonal elements of the matrix $\underline{\mathbf{D}}\,\underline{\mathbf{A}}$, the desired data symbols $\underline{d}_n^{(k_\mathrm{s})}$;
- a component representing the disturbances of the estimate $\widehat{\underline{d}}_n^{(k_\mathrm{s})}$ of $\underline{d}_n^{(k_\mathrm{s})}$ by all other data symbols $\underline{d}_{n'}^{(k'_\mathrm{s})}$, $n' \neq n$ and/or $k'_\mathrm{s} \neq k_\mathrm{s}$. Disturbances caused by other data symbols of the data vector $\underline{\mathbf{d}}^{(k_\mathrm{s})}$ to which $\underline{d}_n^{(k_\mathrm{s})}$ belongs constitute the ISI already mentioned in Sect. 5.1. Disturbances by data symbols belonging to data vectors other than $\underline{\mathbf{d}}^{(k_\mathrm{s})}$ constitute the intracell MAI;
- a component representing the impact of the received noise.

As shown in [6], with $\underline{\mathbf{R}}_\mathrm{d}$ of (5.11) and $\underline{\mathbf{R}}_\mathrm{n}$ of (5.12), we can express the signal-to-noise-and-interference ratio (SNIR) of the estimate $\widehat{\underline{d}}_n^{(k_\mathrm{s})}$ of $\underline{d}_n^{(k_\mathrm{s})}$ obtained by (5.16), with $[\cdot]_{j,j}$ designating the jth diagonal element of the matrix in brackets, as

$$\begin{aligned}\gamma_n^{(k_\mathrm{s})} = {}& \left(\left|\left[\underline{\mathbf{D}}\,\underline{\mathbf{A}}\right]_{j,j}\right|^2 \mathrm{E}\left\{\left[\underline{\mathbf{R}}_\mathrm{d}\right]_{j,j}\right\}\right) \Big/ \left(\left[\underline{\mathbf{D}}\,\underline{\mathbf{A}}\,\underline{\mathbf{R}}_\mathrm{d}\left(\underline{\mathbf{D}}\,\underline{\mathbf{A}}\right)^{*\mathrm{T}}\right]_{j,j}\right. \\ & -2\mathrm{Re}\left\{\left[\underline{\mathbf{D}}\,\underline{\mathbf{A}}\,\underline{\mathbf{R}}_\mathrm{d}\right]_{j,j}\left[\underline{\mathbf{D}}\,\underline{\mathbf{A}}\right]_{j,j}^*\right\} \\ & \left. + \left|\left[\underline{\mathbf{D}}\,\underline{\mathbf{A}}\right]_{j,j}\right|^2 \mathrm{E}\left\{\left[\underline{\mathbf{R}}_\mathrm{d}\right]_{j,j}\right\} + \left[\underline{\mathbf{D}}\,\underline{\mathbf{R}}_\mathrm{n}\underline{\mathbf{D}}^{*\mathrm{T}}\right]_{j,j}\right), \\ & j = n + N(k_\mathrm{s}-1).\end{aligned} \tag{5.17}$$

Let us assume that $\underline{\mathbf{A}}$ of (5.8), $\underline{\mathbf{R}}_\mathrm{d}$ of (5.11) and $\underline{\mathbf{R}}_\mathrm{n}$ of (5.12) are known at the receiver. Then when designing a time-slotted CDMA system, an important question is the selection of a suitable demodulator matrix $\underline{\mathbf{D}}$ based on the knowledge of $\underline{\mathbf{A}}$, $\underline{\mathbf{R}}_{\mathrm{d}}$ and $\underline{\mathbf{R}}_{\mathrm{n}}$. This selection depends on the criterion to be fulfilled by the estimate $\widehat{\underline{\mathbf{d}}}$ of $\underline{\mathbf{d}}$. Well-known linear detectors are [6] the

- decorrelating matched filter (DMF);
- zero forcing block linear equalizer (ZF–BLE);
- minimum mean square error block linear equalizer (MMSE–BLE).

In order to formulate the criterion to be fulfilled by the DMF, with $\underline{\mathbf{e}}$ and $\underline{\mathbf{n}}$ appearing in (5.10) the vectors

$$\widetilde{\underline{\mathbf{e}}}^{(n)} = \left(\underline{e}_{(n-1)Q+1} \cdots \underline{e}_{nQ+W-1}\right)^{\mathrm{T}} \tag{5.18}$$

and

$$\widetilde{\underline{\mathbf{n}}}^{(n)} = \left(\underline{n}_{(n-1)Q+1} \cdots \underline{n}_{nQ+W-1}\right)^{\mathrm{T}} \tag{5.19}$$

of dimension $Q+W-1$ and the partial covariance matrices

$$\widetilde{\underline{\mathbf{R}}}_{\mathrm{n}}^{(n)} = \mathrm{E}\left\{\widetilde{\underline{\mathbf{n}}}^{(n)}\widetilde{\underline{\mathbf{n}}}^{(n)^{*\mathrm{T}}}\right\} \tag{5.20}$$

of dimensions $(Q+W-1)\times(Q+W-1)$ are introduced. Then, the criterion to be fulfilled by the DMF can be written as

$$\widehat{\underline{d}}_n^{(k_{\mathrm{s}})} = \arg \min_{\underline{d}' \in \mathbb{C}} \left| \left(\widetilde{\underline{\mathbf{e}}}^{(n)} - \underline{\mathbf{b}}^{(k_{\mathrm{s}})}\underline{d}'\right)^{*\mathrm{T}} \widetilde{\underline{\mathbf{R}}}_{\mathrm{n}}^{(n)^{-1}} \left(\widetilde{\underline{\mathbf{e}}}^{(n)} - \underline{\mathbf{b}}^{(k_{\mathrm{s}})}\underline{d}'\right)\right|,$$
$$\forall\, n=1,\ldots,N,\ \forall\, k_{\mathrm{s}}=1,\ldots,K_{\mathrm{s}}. \tag{5.21}$$

In the case of the ZF–BLE and the MMSE–BLE, the criteria read

$$\widehat{\underline{\mathbf{d}}} = \arg \min_{\underline{\mathbf{d}}' \in \mathbb{C}^{NK_{\mathrm{s}}}} \begin{cases} \left(\underline{\mathbf{e}} - \underline{\mathbf{A}}\,\underline{\mathbf{d}}'\right)^{*\mathrm{T}} \underline{\mathbf{R}}_{\mathrm{n}}^{-1} \left(\underline{\mathbf{e}} - \underline{\mathbf{A}}\,\underline{\mathbf{d}}'\right), & \text{(ZF–BLE)}, \\ \mathrm{E}\left\{\|\underline{\mathbf{d}}' - \underline{\mathbf{d}}\|^2\right\}, & \text{(MMSE–BLE)}. \end{cases} \tag{5.22}$$

In order to fulfill the criteria (5.21) and (5.22), respectively, the demodulator matrix has to be chosen as follows:

$$\underline{\mathbf{D}} = \begin{cases} \left[\operatorname{diag}\left(\underline{\mathbf{A}}^{*\mathrm{T}}\underline{\mathbf{R}}_{\mathrm{n}}^{-1}\underline{\mathbf{A}}\right)\right]^{-1} \underline{\mathbf{A}}^{*\mathrm{T}}\underline{\mathbf{R}}_{\mathrm{n}}^{-1}, & \text{(DMF)}, \\ \left(\underline{\mathbf{A}}^{*\mathrm{T}}\underline{\mathbf{R}}_{\mathrm{n}}^{-1}\underline{\mathbf{A}}\right)^{-1} \underline{\mathbf{A}}^{*\mathrm{T}}\underline{\mathbf{R}}_{\mathrm{n}}^{-1}, & \text{(ZF–BLE)}, \\ \left(\underline{\mathbf{A}}^{*\mathrm{T}}\underline{\mathbf{R}}_{\mathrm{n}}^{-1}\underline{\mathbf{A}} + \underline{\mathbf{R}}_{\mathrm{d}}^{-1}\right)^{-1} \underline{\mathbf{A}}^{*\mathrm{T}}\underline{\mathbf{R}}_{\mathrm{n}}^{-1}, & \text{(MMSE–BLE)}. \end{cases} \tag{5.23}$$

The DMF maximizes $\gamma_n^{(k_{\mathrm{s}})}$ of (5.17) but totally neglects the impact of ISI and intracell MAI. The ZF–BLE totally eliminates ISI and MAI. However, as compared to the DMF, the impact of the noise $\underline{\mathbf{n}}$ is enhanced. This enhancement can be quantitatively expressed by the SNR degradation [6]

$$\delta_n^{(k_s)} = \left[\left(\underline{\mathbf{A}}^{*\mathrm{T}}\underline{\mathbf{R}}_\mathrm{n}^{-1}\underline{\mathbf{A}}\right)^{-1}\right]_{j,j}\left[\underline{\mathbf{A}}^{*\mathrm{T}}\underline{\mathbf{R}}_\mathrm{n}^{-1}\underline{\mathbf{A}}\right]_{j,j} \geq 1, \quad j = n + N(k_\mathrm{s}-1). \tag{5.24}$$

A quite good approximation of $\delta_n^{(k_s)}$ of (5.24) empirically found by Weckerle [20] reads

$$\delta_n^{(k_s)} \approx \frac{Q+1}{Q-K_\mathrm{s}+1}, \tag{5.25}$$

see also [19, 5]. The ratio K_s/Q is called the system load with a ratio of $K_\mathrm{s}/Q = 1$ representing full or 100% system load.

In the case of white noise $\underline{\mathbf{n}}$, the SNR degradations $\delta_n^{(k_s)}$ of (5.24) take minimum values if $\underline{\mathbf{A}}^{*\mathrm{T}}\,\underline{\mathbf{A}}$ is a diagonal matrix. This is the case if the K_s composite channel impulse responses $\underline{\mathbf{b}}^{(k_s)}$ of (5.6) were structured in such a way that their $K_\mathrm{s}(K_\mathrm{s}-1)$ cross-correlation coefficients forming the off-diagonal elements of $\underline{\mathbf{A}}^{*\mathrm{T}}\underline{\mathbf{A}}$ vanish and their K_s auto correlation coefficients forming the diagonal elements of $\underline{\mathbf{A}}^{*\mathrm{T}}\underline{\mathbf{A}}$ are nonzero. The MMSE–BLE maximizes $\gamma_n^{(k_s)}$ of (5.17) under consideration of ISI, intracell MAI and noise. In order to implement the DMF or the ZF–BLE, $\underline{\mathbf{R}}_\mathrm{n}$ of (5.12) has to be known only up to an arbitrary factor. In the case of the MMSE–BLE, $\underline{\mathbf{R}}_\mathrm{n}$ of (5.12) must be known exactly, which requires the knowledge of the noise level at the receiver input. In all three detectors of (5.23), $\underline{\mathbf{R}}_\mathrm{n}$ could be replaced, in the case of the MMSE–BLE again up to a factor, by the unit matrix, a step that would degrade the detector performance in the case of correlated noise.

5.2.4 Channel Estimation

When detecting the data as described in Sect. 5.2.3 the knowledge of the composite channel impulse responses $\underline{\mathbf{b}}^{(k_s)}$, $k_\mathrm{s}=1,\ldots,K_\mathrm{s}$ of (5.6) is indispensable at the receiver, because these channel impulse responses determine the total system matrix $\underline{\mathbf{A}}$, see (5.7, 5.8), which in turn is needed to form the demodulator matrix $\underline{\mathbf{D}}$, see (5.23). According to (5.6), the $\underline{\mathbf{b}}^{(k_s)}$, $k_\mathrm{s}=1,\ldots,K_\mathrm{s}$ are constituted by the CDMA codes $\underline{\mathbf{c}}^{(k_s)}$, $k_\mathrm{s}=1,\ldots,K_\mathrm{s}$ of (5.3) and the channel impulse responses $\underline{\mathbf{h}}^{(k_s)}$, $k_\mathrm{s}=1,\ldots,K_\mathrm{s}$ of (5.5). As far as the utilized CDMA codes $\underline{\mathbf{c}}^{(k_s)}$ are concerned, these can be assumed to be a priori known at the receiver. However, this is not true for the channel impulse responses $\underline{\mathbf{h}}^{(k_s)}$. Therefore, in order to make the composite channel impulse responses $\underline{\mathbf{b}}^{(k_s)}$, $k_\mathrm{s}=1,\ldots,K_\mathrm{s}$ available at the receiver, first the channel impulse responses $\underline{\mathbf{h}}^{(k_s)}$ have to be estimated. As already mentioned in Sect. 5.2.1, channel estimation in the considered time-slotted CDMA systems relies on the midambles, which are embedded between the two data blocks of the transmitted bursts (Fig. 5.2). In this section, the midamble-based channel estimation following the approach developed in [21] will be briefly considered.

The midamble of the burst signal k_s can be described by a midamble code

$$\underline{\mathbf{m}}^{(k_s)} = \left(\underline{m}_1^{(k_s)} \ldots \underline{m}_{L_\mathrm{m}}^{(k_s)}\right)^\mathrm{T} \tag{5.26}$$

of dimension L_m. The elements $\underline{m}_{l_\mathrm{m}}^{(k_\mathrm{s})}$, $l_\mathrm{m}=1,\ldots,L_\mathrm{m}$ of $\underline{\mathbf{m}}^{(k_\mathrm{s})}$ from (5.26) are taken from the finite set

$$\mathbb{V}_\mathrm{m} = \left\{\underline{V}_{\mathrm{m},1} \cdots \underline{V}_{\mathrm{m},M_\mathrm{m}}\right\} \tag{5.27}$$

of cardinality M_m. With the channel impulse response $\underline{\mathbf{h}}^{(k_\mathrm{s})}$ valid for burst signal k_s, the transmission of $\underline{\mathbf{m}}^{(k_\mathrm{s})}$ of (5.26) leads to the receiver input signal

$$\widetilde{\underline{\mathbf{e}}}_\mathrm{m}^{(k_\mathrm{s})} = \underline{\mathbf{m}}^{(k_\mathrm{s})} * \underline{\mathbf{h}}^{(k_\mathrm{s})}. \tag{5.28}$$

Because $\underline{\mathbf{m}}^{(k_\mathrm{s})}$ is located between the two data blocks α and β (Fig. 5.2), the head and tail of $\widetilde{\underline{\mathbf{e}}}_\mathrm{m}^{(k_\mathrm{s})}$ of (5.28) are superposed by contributions caused by the transmitted data blocks α and β. This impact of the data hampers the usefulness of $\widetilde{\underline{\mathbf{e}}}_\mathrm{m}^{(k_\mathrm{s})}$ from (5.28) for channel estimation. However, if, with W defined in (5.5), the dimension of the midamble code $\underline{\mathbf{m}}^{(k_\mathrm{s})}$ of (5.26) fulfills the condition

$$L_\mathrm{m} > W - 1, \tag{5.29}$$

then a central section of $\widetilde{\underline{\mathbf{e}}}_\mathrm{m}^{(k_\mathrm{s})}$ of (5.28) is free from the influence of the data. This useful section of $\widetilde{\underline{\mathbf{e}}}_\mathrm{m}^{(k_\mathrm{s})}$ is termed $\underline{\mathbf{e}}_\mathrm{m}^{(k_\mathrm{s})}$ and has the dimension

$$L = L_\mathrm{m} - W + 1 \geq 1. \tag{5.30}$$

In what follows, we assume that (5.29) is fulfilled. With the midamble code $\underline{\mathbf{m}}^{(k_\mathrm{s})}$ of (5.26) we can establish the $L\times W$ matrix

$$\begin{aligned}\underline{\mathbf{G}}^{(k_\mathrm{s})} &= \left(\underline{G}_{i,j}^{(k_\mathrm{s})}\right), i=1,\ldots,L, \quad j=1,\ldots,W,\\ \underline{G}_{i,j}^{(k_\mathrm{s})} &= \underline{m}_{W+i-j}^{(k_\mathrm{s})},\end{aligned} \tag{5.31}$$

termed the partial midamble matrix. Now, we can express $\underline{\mathbf{e}}_\mathrm{m}^{(k_\mathrm{s})}$ by $\underline{\mathbf{h}}^{(k_\mathrm{s})}$ of (5.5) and $\underline{\mathbf{G}}^{(k_\mathrm{s})}$ of (5.31) as follows:

$$\underline{\mathbf{e}}_\mathrm{m}^{(k_\mathrm{s})} = \underline{\mathbf{G}}^{(k_\mathrm{s})}\,\underline{\mathbf{h}}^{(k_\mathrm{s})}. \tag{5.32}$$

By stacking the K_s channel impulse responses $\underline{\mathbf{h}}^{(k_\mathrm{s})}$ of (5.5) we obtain the total channel impulse response

$$\underline{\mathbf{h}} = \left(\underline{\mathbf{h}}^{(1)\mathrm{T}} \ldots \underline{\mathbf{h}}^{(K_\mathrm{s})\mathrm{T}}\right)^\mathrm{T} = \left(\underline{h}_1 \ldots \underline{h}_{K_\mathrm{s}W}\right)^\mathrm{T} \tag{5.33}$$

of dimension $K_\mathrm{s}W$. The K_s midamble matrices $\underline{\mathbf{G}}^{(k_\mathrm{s})}$ of (5.31) are arranged to the total midamble matrix

$$\underline{\mathbf{G}} = \left(\underline{\mathbf{G}}^{(1)} \ldots \underline{\mathbf{G}}^{(K_\mathrm{s})}\right) \tag{5.34}$$

of dimension $L\times K_\mathrm{s}W$. Now, the total midamble-induced receiver input signal not disturbed by the transmitted data but including received noise, which is described by the noise vector

$$\underline{\mathbf{n}}_{\mathrm{m}} = \left(\underline{n}_{\mathrm{m},1} \cdots \underline{n}_{\mathrm{m},L}\right)^{\mathrm{T}} \tag{5.35}$$

of dimension L, can be expressed as

$$\underline{\mathbf{e}}_{\mathrm{m}} = \underline{\mathbf{G}}\,\underline{\mathbf{h}} + \underline{\mathbf{n}}_{\mathrm{m}}. \tag{5.36}$$

At the receiver $\underline{\mathbf{e}}_{\mathrm{m}}$ and $\underline{\mathbf{G}}$ are known, whereas $\underline{\mathbf{h}}$ and $\underline{\mathbf{n}}_{\mathrm{m}}$ are unknown. Therefore, (5.36) can be considered as a system of L equations that has to be fulfilled by the $K_{\mathrm{s}}W$ unknown elements $\underline{h}_j$, $j=1,\ldots,K_{\mathrm{s}}W$ of $\underline{\mathbf{h}}$ from (5.33), and by the unknown components $\underline{n}_{\mathrm{m},l}$ of $\underline{\mathbf{n}}_{\mathrm{m}}$ of (5.35). For the estimation of the components of $\underline{\mathbf{h}}$ of (5.33) by (5.36), at least $K_{\mathrm{s}}W$ equations are required, which, together with (5.30), leads to the condition

$$L = L_{\mathrm{m}} - W + 1 \geq K_{\mathrm{s}}W. \tag{5.37}$$

From (5.37) we obtain the condition

$$L_{\mathrm{m}} \geq (K_{\mathrm{s}} + 1)W - 1 \tag{5.38}$$

for the dimension of the midamble codes $\underline{\mathbf{m}}^{(k_{\mathrm{s}})}$ of (5.26).

Let us now assume that the noise $\underline{\mathbf{n}}_{\mathrm{m}}$ of (5.35) is wide-sense stationary [17] and has the covariance matrix

$$\underline{\mathbf{R}}_{\mathrm{m}} = \mathrm{E}\left\{\underline{\mathbf{n}}_{\mathrm{m}}\,\underline{\mathbf{n}}_{\mathrm{m}}^{*\mathrm{T}}\right\}. \tag{5.39}$$

Then, we obtain from (5.36) the unbiased minimum variance estimate [21]

$$\widehat{\underline{\mathbf{h}}} = \left(\underline{\mathbf{G}}^{*\mathrm{T}}\underline{\mathbf{R}}_{\mathrm{m}}^{-1}\underline{\mathbf{G}}\right)^{-1}\underline{\mathbf{G}}^{*\mathrm{T}}\underline{\mathbf{R}}_{\mathrm{m}}^{-1}\underline{\mathbf{e}}_{\mathrm{m}} \tag{5.40}$$

for $\underline{\mathbf{h}}$ from (5.33). If $\underline{\mathbf{n}}_{\mathrm{m}}$ from (5.35) is Gaussian, $\widehat{\underline{\mathbf{h}}}$ from (5.40) is also the maximum-likelihood estimate of $\underline{\mathbf{h}}$ of (5.33). If $\underline{\mathbf{R}}_{\mathrm{m}}$ is not known, this matrix can be substituted by the unit matrix $\mathbf{I}^{(L\times L)}$ in (5.40), which leads to a performance degradation in the case of correlated noise $\underline{\mathbf{n}}_{\mathrm{m}}$. By using (5.40), estimates of all K_{s} channel impulse responses $\underline{\mathbf{h}}^{(k_{\mathrm{s}})}$, $k_{\mathrm{s}} = 1,\ldots,K_{\mathrm{s}}$ of (5.5) are simultaneously generated. Therefore, the approach described by (5.40) is termed joint channel estimation (JCE). Similar to (5.24) with $\underline{\mathbf{G}}$ from (5.34) and $\underline{\mathbf{R}}_{\mathrm{m}}$ from (5.39), a SNR degradation

$$\tilde{\delta}_w^{(k_{\mathrm{s}})} = \left[\left(\underline{\mathbf{G}}^{*\mathrm{T}}\underline{\mathbf{R}}_{\mathrm{m}}^{-1}\underline{\mathbf{G}}\right)^{-1}\right]_{j,j}\left[\underline{\mathbf{G}}^{*\mathrm{T}}\underline{\mathbf{R}}_{\mathrm{m}}^{-1}\underline{\mathbf{G}}\right]_{j,j} \geq 1, \quad j = w + W(k_{\mathrm{s}}-1), \tag{5.41}$$

can be introduced for the elements $\widehat{\underline{h}}_j$ of the estimate $\widehat{\underline{\mathbf{h}}}$ from (5.40). The SNR degradations $\tilde{\delta}_w^{(k_{\mathrm{s}})}$ of (5.41) depend on the chosen midamble codes $\underline{\mathbf{m}}^{(k_{\mathrm{s}})}$ from (5.26). By computer search midamble code families $\underline{\mathbf{m}}^{(k_{\mathrm{s}})}$, $k_{\mathrm{s}}=1,\ldots,K_{\mathrm{s}}$ leading to rather small SNR degradations $\tilde{\delta}_w^{(k_{\mathrm{s}})}$ can be found [21]. This search can be significantly facilitated if all K_{s} midamble codes $\underline{\mathbf{m}}^{(k_{\mathrm{s}})}$, $k_{\mathrm{s}}=1,\ldots,K_{\mathrm{s}}$ are derived by taking shifted sections of dimension L_{m} of a single basic midamble code of dimension $L_{\mathrm{m}}+(K_{\mathrm{s}}-1)W$ [22]. Such an approach also helps to reduce the implementation complexity of the channel estimator. Further, said basic midamble code can be chosen periodic with a period P equal to L from (5.30). Then the channel estimator can be simply implemented as a cyclic correlator, and we obtain the low-cost Steiner channel estimator described in [22].

5.2.5 Assignment of CDMA Codes to Mobile Stations

In Sects. 5.2.1–5.2.4 time-slotted CDMA is described in a rather general way, which does not take into account how the $N_{\mathrm{fr}}K_{\mathrm{s}}$ resources per frequency slot, cf. Fig. 5.1, are utilized. Let us now again consider a single TDMA time slot and assume that the K_{s} simultaneous burst signals, each based on one of the K_{s} CDMA codes $\underline{\mathbf{c}}^{(k_{\mathrm{s}})}$, $k_{\mathrm{s}}=1,\ldots,K_{\mathrm{s}}$, are assigned to K UL or DL connections between a BS and K MS μ_k, $k=1,\ldots,K$ to be served by this BS. In the most simple case, the number K of MS is equal to the number K_{s} of CDMA codes, and one of the K_{s} CDMA codes $\underline{\mathbf{c}}^{(k_{\mathrm{s}})}$, $k_{\mathrm{s}}=1,\ldots,K_{\mathrm{s}}$ is uniquely allotted to each of the K MS μ_k, $k=1,\ldots,K$. Then, each of the K MS would use one of the K_{s} resources of the considered TDMA time slot, and each MS could transmit with the same data rate. In a more general case, the K_{s} CDMA codes $\underline{\mathbf{c}}^{(k_{\mathrm{s}})}$, $k_{\mathrm{s}}=1,\ldots,K_{\mathrm{s}}$ are not equally distributed among the K MS μ_k, $k=1,\ldots,K$. Rather, MS demanding larger data rates are assigned more CDMA codes than MS demanding lower data rates. Such an unequal distribution of the CDMA codes is possible if K_{s} is larger than K. Assigning more than one CDMA code per MS is termed CDMA code pooling. The assignment of CDMA codes to MS within a TDMA time slot can be described by a relation [23]

$$k_{\mathrm{s}} \longmapsto z(k_{\mathrm{s}}),\; z(k_{\mathrm{s}}) \in \{1\ldots K\}, k_{\mathrm{s}}=1,\ldots,K_{\mathrm{s}}, \tag{5.42}$$

which uniquely assigns the CDMA code $\underline{\mathbf{c}}^{(k_{\mathrm{s}})}$ to the MS $\mu_{z(k_{\mathrm{s}})}$. In the case of CDMA code pooling, by virtue of (5.42), several values $k_{\mathrm{s}} \in \{1\ldots K_{\mathrm{s}}\}$ lead to the same value $z(k_{\mathrm{s}})$. With $z(k_{\mathrm{s}})$ of (5.42) all CDMA codes that are used to generate the burst signals intended for MS μ_k constitute the set

$$\mathbb{C}^{(k)} = \bigcup_{\forall k_{\mathrm{s}} \,|\, z(k_{\mathrm{s}})=k} \underline{\mathbf{c}}^{(k_{\mathrm{s}})},\; k=1,\ldots,K. \tag{5.43}$$

5.3 Uplink with Adaptive BS Antennas

5.3.1 Transmission Model

It is well known that the performance of mobile radio systems can be enhanced by utilizing multielement antennas instead of single antennas [2, 10].

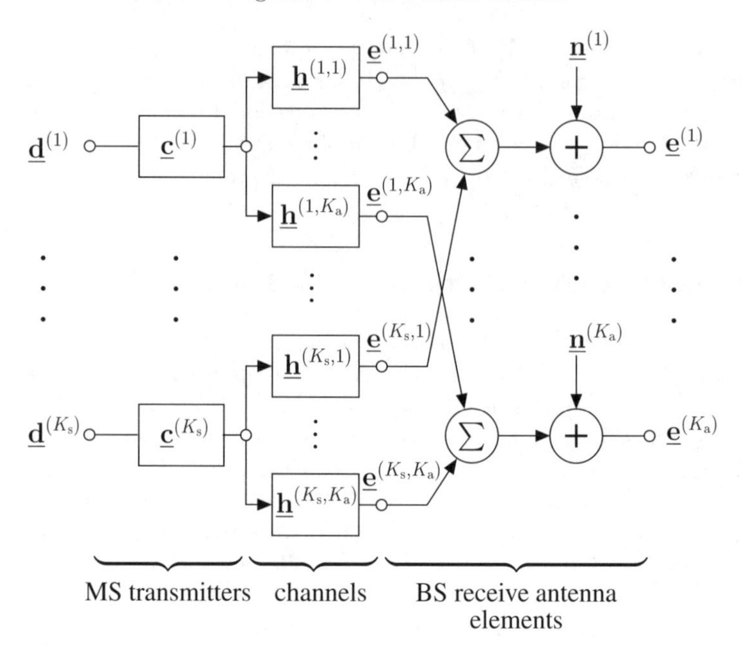

Fig. 5.4. Uplink transmission model of time-slotted CDMA systems with multielement BS antenna

In this section we consider the uplink with multielement antennas at the BS. The presented considerations apply for both single- and multielement transmit antennas at the MS. However, because of the limited space available at the MS, today single-element antennas are the usual solution at the MS, which is also adopted in the following. Nevertheless, the new services to be offered by 3G mobile radio systems may require MS with displays and keyboards larger then those of today's MS. Then space may be available to also install multielement antennas at the MS.

In this section the transmission model of Fig. 5.3 will be adapted and extended for the case of an UL with K_{a} receiving antenna elements at the BS. This leads to the transmission model shown in Fig. 5.4. Let us assume that in the considered UL a number of $K \leq K_{\mathrm{s}}$ MS μ_k, $k=1,\ldots,K$ are active in the same TDMA time slot, and that the ensemble of these MS utilize all K_{s} available CDMA codes $\underline{\mathbf{c}}^{(k_{\mathrm{s}})}$, $k_{\mathrm{s}}=1,\ldots,K_{\mathrm{s}}$. From each of the K MS exists a mobile radio channel to each of the K_{a} BS antenna elements, that is, we have a total of KK_{a} mobile radio channels. The channel impulse response of the channel leading from the antenna input of MS μ_k, $k=1,\ldots,K$ to the output of the BS antenna element k_{a}, $k_{\mathrm{a}}=1,\ldots,K_{\mathrm{a}}$ is termed

$$\underline{\tilde{\mathbf{h}}}^{(k,k_{\mathrm{a}})} = \left(\underline{\tilde{h}}_1^{(k,k_{\mathrm{a}})} \ldots \underline{\tilde{h}}_W^{(k,k_{\mathrm{a}})}\right)^{\mathrm{T}} . \tag{5.44}$$

All burst signals originating in a certain MS μ_k travel over the same K_a mobile radio channels to the BS antenna elements. Now, with $\underline{\tilde{\mathbf{h}}}^{(k,k_\mathrm{a})}$ of (5.44) and with the assignment relation of (5.42), the channel impulse responses valid for the burst signal k_s are given by

$$\underline{\mathbf{h}}^{(k_\mathrm{s},k_\mathrm{a})} = \underline{\tilde{\mathbf{h}}}^{(z(k_\mathrm{s}),k_\mathrm{a})}. \tag{5.45}$$

The superscripts k_s and k_a of $\underline{\mathbf{h}}^{(k_\mathrm{s},k_\mathrm{a})}$ indicate for which CDMA code and for which antenna element, respectively, this channel impulse response is valid; this assignment is unique. At the output of the antenna element k_a each of the k_s transmitted burst signals generates a signal $\underline{\mathbf{e}}^{(k_\mathrm{s},k_\mathrm{a})}$, $k_\mathrm{s} = 1,\ldots,K_\mathrm{s}$. In addition, noise that is described by K_a partial noise vectors $\underline{\mathbf{n}}^{(k_\mathrm{a})}$ occurs. Therefore, we obtain the output signal

$$\underline{\mathbf{e}}^{(k_\mathrm{a})} = \sum_{k_\mathrm{s}=1}^{K_\mathrm{s}} \underline{\mathbf{e}}^{(k_\mathrm{s},k_\mathrm{a})} + \underline{\mathbf{n}}^{(k_\mathrm{a})},\ k_\mathrm{a}=1,\ldots,K_\mathrm{a}, \tag{5.46}$$

of the antenna element k_a. The K_a signals $\underline{\mathbf{e}}^{(k_\mathrm{a})}$ of (5.46) are termed the partial received signals. Like $\underline{\mathbf{e}}$ of (5.10), the signals $\underline{\mathbf{e}}^{(k_\mathrm{a})}$ of (5.46) also have the dimension $NQ+W-1$. By stacking these signals we obtain the total received signal

$$\underline{\mathbf{e}} = \left(\underline{\mathbf{e}}^{(1)^\mathrm{T}} \ldots \underline{\mathbf{e}}^{(K_\mathrm{a})^\mathrm{T}}\right)^\mathrm{T} \tag{5.47}$$

observed over all K_a BS antenna elements. With the partial noise vectors $\underline{\mathbf{n}}^{(k_\mathrm{a})}$, $k_\mathrm{a}=1,\ldots,K_\mathrm{a}$ the noise contained in $\underline{\mathbf{e}}$ of (5.47) is given by the total noise vector

$$\underline{\mathbf{n}} = \left(\underline{\mathbf{n}}^{(1)^\mathrm{T}} \ldots \underline{\mathbf{n}}^{(K_\mathrm{a})^\mathrm{T}}\right)^\mathrm{T}. \tag{5.48}$$

Both $\underline{\mathbf{e}}$ of (5.47) and $\underline{\mathbf{n}}$ of (5.48) have the dimension $K_\mathrm{a}(NQ+W-1)$. The covariance matrix $\underline{\mathbf{R}}_\mathrm{n}$ of $\underline{\mathbf{n}}$ of (5.48) characterizes the noise correlations both in the spatial and in the temporal domain. As a special case close to practical situations, $\underline{\mathbf{n}}$ originates in independent impinging noise waves with each of these waves having the same spectral shape, and the K_a antenna elements are so closely spaced that each antenna element is hit by the same noise waves. Then, the $K_\mathrm{a}\times K_\mathrm{a}$ matrix $\underline{\mathbf{R}}_\mathrm{s}$ describing the spatial correlations and the $(NQ+W-1)\times(NQ+W-1)$ matrix $\underline{\mathbf{R}}_\mathrm{t}$ describing the temporal correlations of $\underline{\mathbf{n}}$ of (5.48) can be introduced [20, 24]. With these two matrices the covariance matrix of $\underline{\mathbf{n}}$ can be written as the Kronecker product [20, 24]

$$\underline{\mathbf{R}}_\mathrm{n} = \underline{\mathbf{R}}_\mathrm{s} \otimes \underline{\mathbf{R}}_\mathrm{t}. \tag{5.49}$$

With the CDMA codes $\underline{\mathbf{c}}^{(k_\mathrm{s})}$ of (5.3) and the channel impulse responses $\underline{\mathbf{h}}^{(k_\mathrm{s},k_\mathrm{a})}$ of (5.45), the composite channel impulse responses

$$\underline{\mathbf{b}}^{(k_\mathrm{s},k_\mathrm{a})} = \underline{\mathbf{c}}^{(k_\mathrm{s})} * \underline{\mathbf{h}}^{(k_\mathrm{s},k_\mathrm{a})},\ k_\mathrm{s}=1,\ldots,K_\mathrm{s},\ k_\mathrm{a}=1,\ldots,K_\mathrm{a}, \tag{5.50}$$

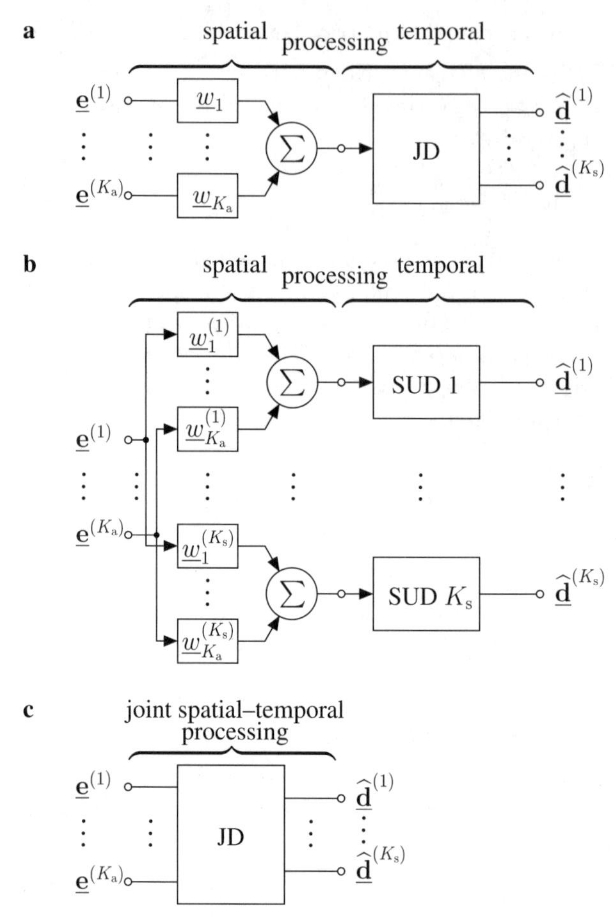

Fig. 5.5. Data detection schemes for the UL of time-slotted CDMA systems with adaptive BS antennas: **a** separate spatial and temporal processing; temporal processing by JD; **b** separate spatial and temporal processing; temporal processing by SUD; **c** joint spatial and temporal processing by JD

can be formed in analogy to (5.6). Then, with $\underline{\mathbf{b}}^{(k_\mathrm{s},k_\mathrm{a})}$ from (5.50), the $K_\mathrm{s}K_\mathrm{a}$ partial system matrices $\underline{\mathbf{A}}^{(k_\mathrm{s},k_\mathrm{a})}$, $k_\mathrm{s}=1,\ldots,K_\mathrm{s}$, $k_\mathrm{a}=1,\ldots,K_\mathrm{a}$ can be established in analogy to (5.7) just by introducing the second superscript k_a. Each of the matrices $\underline{\mathbf{A}}^{(k_\mathrm{s},k_\mathrm{a})}$ has the dimension $(NQ+W-1)\times N$. By combining the $K_\mathrm{s}K_\mathrm{a}$ partial system matrices $\underline{\mathbf{A}}^{(k_\mathrm{s},k_\mathrm{a})}$, finally the total system matrix

$$\underline{\mathbf{A}} = \begin{pmatrix} \underline{\mathbf{A}}^{(1,1)} & \ldots & \underline{\mathbf{A}}^{(K_\mathrm{s},1)} \\ \vdots & & \vdots \\ \underline{\mathbf{A}}^{(1,K_\mathrm{a})} & \ldots & \underline{\mathbf{A}}^{(K_\mathrm{s},K_\mathrm{a})} \end{pmatrix} \tag{5.51}$$

is obtained. Matrix $\underline{\mathbf{A}}$ has the dimensions $K_\mathrm{a}(NQ+W-1)\times NK_\mathrm{s}$. Now, with $\underline{\mathbf{d}}$ of (5.9), $\underline{\mathbf{A}}$ of (5.51) and $\underline{\mathbf{n}}$ of (5.48) the total antenna output signal $\underline{\mathbf{e}}$ of (5.47) can be expressed as

$$\underline{\mathbf{e}} = \underline{\mathbf{A}}\,\underline{\mathbf{d}} + \underline{\mathbf{n}}. \qquad (5.52)$$

5.3.2 Data Detection

By processing of the K_a partial received signals $\underline{\mathbf{e}}^{(k_\mathrm{a})}$, $k_\mathrm{a}=1,\ldots,K_\mathrm{a}$ of (5.46) under consideration of the properties of the channel impulse responses $\underline{\mathbf{h}}^{(k_\mathrm{s},k_\mathrm{a})}$ of (5.45) and of the noise $\underline{\mathbf{n}}$ of (5.48), the BS multielement antenna becomes an adaptive antenna. A variety of signal processing schemes is conceivable. Figure 5.5 shows three such schemes.

In the scheme of Fig. 5.5a the K_a partial received signals $\underline{\mathbf{e}}^{(k_\mathrm{a})}$ of (5.46) are first processed spatially by forming their weighted sum $\sum_{k_\mathrm{a}=1}^{K_\mathrm{a}} \underline{w}_{k_\mathrm{a}}\,\underline{\mathbf{e}}^{(k_\mathrm{a})}$. Then this sum is processed in the time domain in the sense of JD with the goal to obtain an estimate $\widehat{\underline{\mathbf{d}}}$ of the transmitted total data vector $\underline{\mathbf{d}}$ of (5.9). In the scheme of Fig. 5.5b, first K_s CDMA code-specific weighted sums $\sum_{k_\mathrm{a}=1}^{K_\mathrm{a}} \underline{w}_{k_\mathrm{a}}^{(k_\mathrm{s})}\,\underline{\mathbf{e}}^{(k_\mathrm{a})}$ of the partial received signals $\underline{\mathbf{e}}^{(k_\mathrm{a})}$, $k_\mathrm{a}=1,\ldots,K_\mathrm{a}$ of (5.46) are formed. Then, each of these K_s sums is temporally processed in the sense of single-user detection (SUD) [5]. At the output of each SUD an estimate $\widehat{\underline{\mathbf{d}}}^{(k_\mathrm{s})}$ of a partial data vector $\underline{\mathbf{d}}^{(k_\mathrm{s})}$ of (5.1) is obtained. A variety of approaches exist for determining the weights $\underline{w}_{k_\mathrm{a}}$ and $\underline{w}_{k_\mathrm{a}}^{(k_\mathrm{s})}$, respectively, with a view to suppressing the undesired signals, while simultaneously enhancing the desired signal components by adding them constructively. All schemes following Fig. 5.5a, b fall into the category of antenna beam-forming with subsequent temporal processing and are suboptimal [16].

In the scheme of Fig. 5.5c the total received signal $\underline{\mathbf{e}}$ of (5.47) is immediately fed into a unit performing spatial–temporal signal JD. In this unit $\underline{\mathbf{e}}$ of (5.47) and the covariance matrix $\underline{\mathbf{R}}_\mathrm{n}$ of $\underline{\mathbf{n}}$ of (5.48) are utilized to determine an estimate $\widehat{\underline{\mathbf{d}}}$ of the transmitted total data vector $\underline{\mathbf{d}}$ of (5.9). Thanks to the formal identity of (5.10) and (5.52), the spatial–temporal JD can be performed by immediately resorting to (5.15) and (5.23). This approach is optimum with respect to the criteria (5.21) and (5.22), respectively. In the case of $K_\mathrm{a}>1$, we obtain the maximum ratio combiner (MRC) instead of the DMF [16].

For a given input SNR per antenna element, the quality of the estimate $\widehat{\underline{\mathbf{d}}}$ of $\underline{\mathbf{d}}$ from (5.9) obtained by applying (5.15) and (5.23), that is, the SNR of $\widehat{\underline{\mathbf{d}}}$, increases with the number K_a of antenna elements [13]. The degree of this increase depends on various factors as, for instance, the propagation scenario, the velocity v of the MS, and so on. The main reasons behind this improvement of the estimation quality are:

- The effective power of the desired received signal relative to the effective power of the received noise has a tendency to grow linearly with K_a, because the desired partial signals $\underline{\mathbf{e}}^{(k_a)}$ of (5.46) are constructively added, which is not the case for the partial noise signals $\underline{\mathbf{n}}^{(k_a)}$.

- The number of utilized radio channels between a MS and the BS is equal to K_a. The larger this number, that is, the larger the diversity, the better the chances of maintaining a good connection even if some of these radio channels are of poor quality.
- The SNR degradations $\delta_n^{(k_s)}$ of (5.24) decrease with increasing K_a. This decrease can be roughly quantified by substituting Q by $K_a Q$ in (5.25). Therefore, doubling the number K_a of antenna elements is approximately equivalent to doubling the spreading factor Q [20].

In general, when going from one antenna element to $K_a > 1$ antenna elements at the BS, the SNR gain exceeds the factor of K_a [16].

In the case of a multielement receive antenna at the BS, the total system matrix $\underline{\mathbf{A}}$ of (5.51) and the covariance matrix $\underline{\mathbf{R}}_n$ of $\underline{\mathbf{n}}$ of (5.48) incorporate directional information on the desired signals and the noise signals, respectively, impinging at the BS. Thanks to this information the receiver scheme of Fig. 5.5c has the inherent potential of adapting itself to the signal-plus-noise situation in such a way that the data detection is optimum with respect to the criteria (5.21) and (5.22), respectively, and under consideration of the directions of incidence of the desired and noise signals. In order to avail oneself of this potential, the knowledge of the matrices $\underline{\mathbf{A}}$ and $\underline{\mathbf{R}}_n$ is required at the BS. The knowledge of the matrix $\underline{\mathbf{A}}$ is of fundamental importance for the JD process and results from the a priori determined CDMA codes $\underline{\mathbf{c}}^{(k_s)}$ of (5.3) and of the channel impulse responses $\underline{\mathbf{h}}^{(k_s,k_a)}$ of (5.45), which have to be estimated at the BS receiver (Sect. 5.3.3). However, the knowledge of the matrix $\underline{\mathbf{R}}_n$ is not readily available at the BS receiver, and it requires a special effort to estimate this matrix. A proposal how $\underline{\mathbf{R}}_n$ can be estimated and how the estimate $\widehat{\underline{\mathbf{R}}}_n$ can be beneficially utilized at the BS receiver [25] is illustrated in Fig. 5.6. The ability to estimate $\underline{\mathbf{R}}_n$ includes the ability to estimate also $\underline{\mathbf{R}}_m$ of (5.39) and to improve channel estimation by considering the estimate $\widehat{\underline{\mathbf{R}}}_m$ in (5.40). In the receiver scheme shown in Fig. 5.6 the processing of the received signal is performed in the following steps:

- JCE and JD under the assumptions

$$\underline{\mathbf{R}}_m = 2\sigma^2 \mathbf{I}^{(K_a L)} \tag{5.53}$$

and

$$\underline{\mathbf{R}}_n = 2\sigma^2 \mathbf{I}^{(K_a(NQ+W-1))} \tag{5.54}$$

in order to obtain first estimates $\widehat{\underline{\mathbf{d}}}^{(k_s)}$, $k_s = 1, \ldots, K_s$ of the coded data $\underline{\mathbf{d}}^{(k_s)}$, $k_s = 1, \ldots, K_s$.
- FEC decoding in order to obtain first estimates $\widehat{\mathbf{u}}^{(k_s)}$, $k_s = 1, \ldots, K_s$ of the uncoded data $\mathbf{u}^{(k_s)}$, $k_s = 1, \ldots, K_s$ corresponding to the coded data vectors $\underline{\mathbf{d}}^{(k_s)}$ of (5.1).

- Approximate reconstruction of the received desired signal $\underline{\mathbf{e}} - \underline{\mathbf{n}}$ based on the estimates $\widehat{\underline{\mathbf{u}}}^{(k_s)}$, $k_s = 1, \ldots, K_s$, on the utilized FEC code, on the a priori given CDMA codes $\underline{\mathbf{c}}^{(k_s)}$, $k_s = 1, \ldots, K_s$ of (5.3) and on the estimates $\widehat{\underline{\mathbf{h}}}^{(k_s,k_a)}$, $k_s = 1, \ldots, K_s$, $k_a = 1, \ldots, K_a$ of the channel impulse responses $\underline{\mathbf{h}}^{(k_s,k_a)}$, $k_s = 1, \ldots, K_s$, $k_a = 1, \ldots, K_a$ from (5.45).
- Subtraction of the reconstructed desired received signal from the total received signal $\underline{\mathbf{e}}$ in order to obtain an estimate $\widehat{\underline{\mathbf{n}}}$ of $\underline{\mathbf{n}}$ of (5.48).
- Determination of estimates $\widehat{\underline{\mathbf{R}}}_n$ of $\underline{\mathbf{R}}_n$ and $\widehat{\underline{\mathbf{R}}}_m$ of $\underline{\mathbf{R}}_m$ based on $\widehat{\underline{\mathbf{n}}}$.
- Once more performing JCE and JD, this time under consideration of the estimates $\widehat{\underline{\mathbf{R}}}_m$ and $\widehat{\underline{\mathbf{R}}}_n$, when applying (5.40) and (5.23), respectively. This second processing of the received total signal $\underline{\mathbf{e}}$ yields improved coded and uncoded data estimates $\widehat{\underline{\mathbf{d}}}^{(k_s)}$, $k_s = 1, \ldots, K_s$ and $\widehat{\underline{\mathbf{u}}}^{(k_s)}$, $k_s = 1, \ldots, K_s$, respectively.

In practical applications interleaving over several frames (Fig. 5.1) is performed in conjunction with FEC coding. For the sake of simplicity this aspect is omitted in the above description of the receiver concept of Fig. 5.6. The concept was evaluated by simulations for the parameters listed in Table 5.1. Concerning the noise $\underline{\mathbf{n}}$, which represents, as already stated, intercell MAI plus receiver noise, the two-dimensional interference scenario depicted in Fig. 5.7 was assumed. In this scenario 80% of the noise power impinges from a single discrete direction, whereas 20% of this power impinges homogeneously distributed over the azimuth. The noise $\underline{\mathbf{n}}$ is assumed to be directionally uncorrelated, in which case $\underline{\mathbf{R}}_n$ can be expressed by a spatial component $\underline{\mathbf{R}}_s$ and a temporal component $\underline{\mathbf{R}}_t$, as shown in (5.49). The radio channels between the MS and the BS were chosen according to the Indoor Office B channel model defined by the international telecommunications union (ITU) [26], and line-of-sight propagation was assumed. After every fourth frame the K MS were again randomly distributed over the considered cell. In Fig. 5.8 the obtained simulation results are shown. In this figure the coded bit error rate (BER) is depicted versus the carrier-to-interference ratio C/I per antenna element. The two sets of curves are valid for the cases of perfectly known and estimated channel impulse responses $\underline{\mathbf{h}}^{(k_s,k_a)}$ of (5.45), respectively. As explained by the legend of Fig. 5.8, different scenarios concerning $\underline{\mathbf{R}}_s$ are considered. The lowest BER is obtained if the channel impulse responses $\underline{\mathbf{h}}^{(k_s,k_a)}$ were perfectly known at the BS, and if $\underline{\mathbf{R}}_s$ is considered in the data detection, see curve I. The BER becomes largest if estimates $\widehat{\underline{\mathbf{h}}}^{(k_s,k_a)}$ of the channel impulse responses are used instead of the exact $\underline{\mathbf{h}}^{(k_s,k_a)}$, and if $\underline{\mathbf{R}}_s$ is substituted by the unity matrix in both channel estimation and data detection, see curve IV.

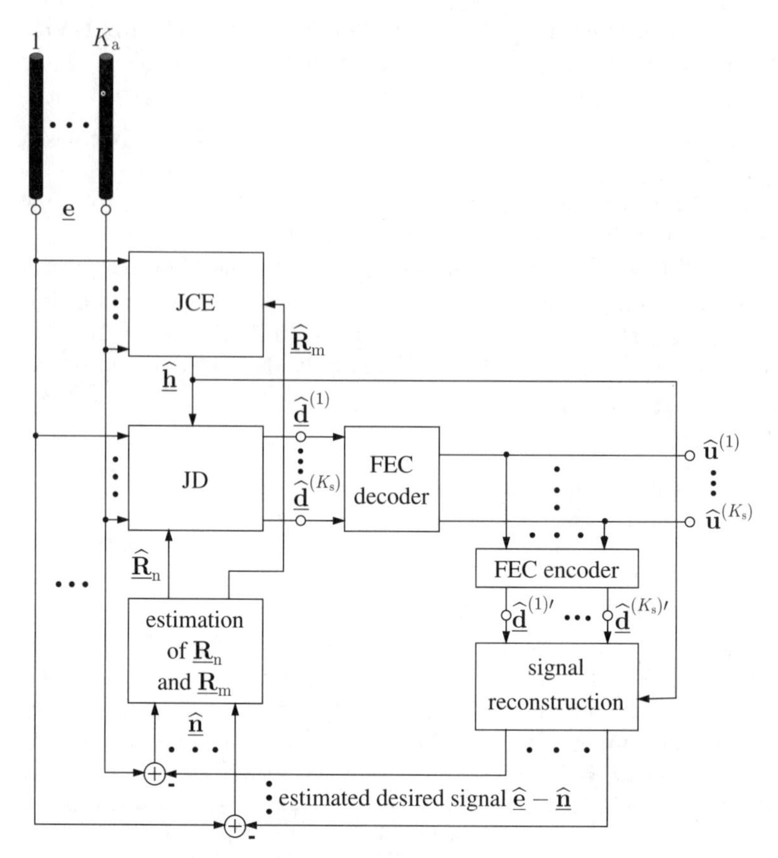

Fig. 5.6. BS receiver concept incorporating the estimation and exploitation of the noise covariance matrices $\underline{R}_m$ and $\underline{R}_n$

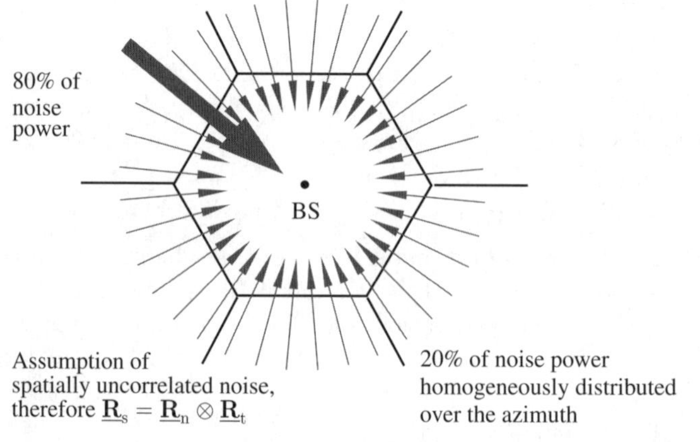

Fig. 5.7. Intercell MAI scenario considered in the simulations of the BS receiver concept of Fig. 5.6

Table 5.1. Parameters chosen in the simulations of the scheme of Fig. 5.6

Number of CDMA codes and MS	$K_\mathrm{s}=K=4$ per cell
Spreading factor	$Q=16$
Data symbols per burst signal	$2N=56$
Midamble dimension	$L_\mathrm{m}=296$
Bursts per TDMA frame	$N_\mathrm{fr}=8$
Chip impulse	GMSK
Chip rate	$1/T_\mathrm{c}=2.167$ Mchip/s
MS velocity	$v=3$ km/h
Carrier frequency	$f_\mathrm{c}=1800$ MHz
FEC code rate	$R_\mathrm{c}=1/2$
FEC code constraint length	$K_\mathrm{c}=5$
Interleaving and averaging to estimate $\underline{\mathbf{R}}_\mathrm{m}$ and $\underline{\mathbf{R}}_\mathrm{n}$	4 bursts
BS antenna	Square array, $K_\mathrm{a}=4$
Power control	$C=\mathrm{const}$

5.3.3 Channel Estimation

As explained in Sect. 5.3.2, when increasing the number of BS antenna elements from 1 to values $K_\mathrm{a}>1$, the dimension NK_s of the total data vector $\underline{\mathbf{d}}$ of (5.9), that is, the number of data symbols $\underline{d}_\mathrm{n}^{(k_\mathrm{s})}$, $n=1,\ldots,N$, $k_\mathrm{s}=1,\ldots,K_\mathrm{s}$, unknown at the receiver is not increased. However, the number of equations available to determine these unknowns grows by a factor of K_a from $NK_\mathrm{s}+W-1$ to $K_\mathrm{a}(NK_\mathrm{s}+W-1)$, cf. (5.10) and (5.52). As already stated in Sect. 5.3.2, this increase helps to improve the quality of the estimates $\hat{\underline{\mathbf{d}}}_n^{(k_\mathrm{s})}$ of $\underline{\mathbf{d}}_n^{(k_\mathrm{s})}$ when using a multielement receive antenna instead of a single antenna. Concerning channel estimation, again the number of equations available for determining estimates of the channel coefficients is increased from $L_\mathrm{m}-W+1$ to $K_\mathrm{a}(L_\mathrm{m}-W+1)$, if K_a receive antenna elements are utilized instead of only one. This is because at each of the K_a receive antenna elements a channel estimator relying on the midamble-induced sections of the received signal has to be provided; each of these K_a channel estimators utilizes $L_\mathrm{m}-W+1$ equations for performing channel estimation, see (5.29), (5.34) and (5.36). However, the number of unknowns given by $K_\mathrm{s}W$, see (5.5), in the case $K_\mathrm{a}=1$ and by $K_\mathrm{a}K_\mathrm{s}W$, see (5.45), in the case of $K_\mathrm{a}>1$ is increased by the same factor K_a as the number of equations. Therefore, the channel estimation quality is basically not improved when going from a single- to a multielement antenna. This means that in receivers with multielement antennas the problem of channel estimation, in contrast to the problem of data detection, is not mitigated when going from single antennas to multielement antennas. Fortunately, the problem of channel estimation can be, nevertheless, mitigated under certain conditions by exploiting the structural properties of the channel impulse responses $\underline{\mathbf{h}}^{(k_\mathrm{s},k_\mathrm{a})}$ of (5.45). If the K_a antenna elements are sufficiently closely spaced, it is possible to express the $K_\mathrm{s}K_\mathrm{a}$ channel impulse

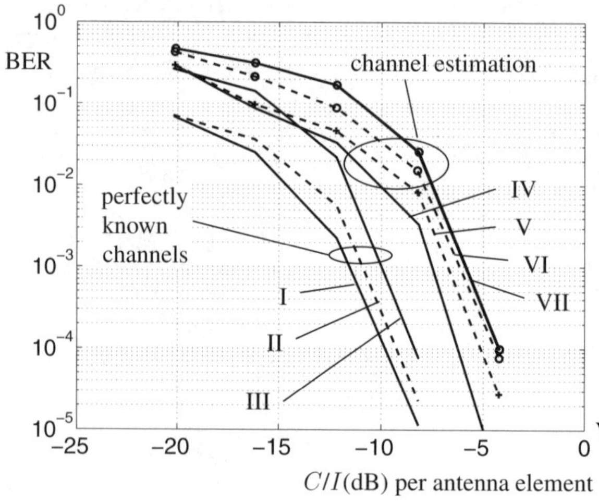

I: $\underline{\mathbf{R}}_s$ in data detection

II: $\widehat{\underline{\mathbf{R}}}_s$ in data detection

III: $\underline{\mathbf{R}}_s = \mathbf{I}^{(K_a)}$ in data detection

IV: $\underline{\mathbf{R}}_s$ in data detection and channel estimation

V: $\widehat{\underline{\mathbf{R}}}_s$ in data detection and channel estimation

VI: $\widehat{\underline{\mathbf{R}}}_s$ in data detection and $\underline{\mathbf{R}}_s = \mathbf{I}^{(K_a)}$ in the channel estimation

VII: $\underline{\mathbf{R}}_s = \mathbf{I}^{(K_a)}$ in data detection and channel estimation

Fig. 5.8. Coded BER versus carrier–to–interference ratio C/I per antenna element obtained by simulations of the BS receiver concept of Fig. 5.6

responses $\underline{\mathbf{h}}^{(k_s,k_a)}$ of (5.45) by $K_s K_d$ directional channel impulse responses $\underline{\mathbf{h}}_d^{(k_s,k_d)}$, which are defined for a reference point located somewhere within the array of the antenna elements and K_d discrete directions of arrival (DOA) [24]. The properties of the multielement antenna can be described by the steering factors $\underline{a}^{(k_s,k_a,k_d)}$, which are valid for burst signal k_s, antenna element k_a and DOA k_d [13]. Then,

$$\underline{\mathbf{h}}^{(k_s,k_a)} = \sum_{k_d=1}^{K_d} \underline{\mathbf{h}}_d^{(k_s,k_d)} \, \underline{a}^{(k_s,k_a,k_d)} , \tag{5.55}$$

holds. If K_d is smaller than K_a, then the effective number of unknowns is no longer $K_a K_s W$, but $K_d K_s W$, whereas the number of equations remains the same. For more details of the above approach for improving channel estimation the reader is referred to [13].

5.4 Downlink with Adaptive BS Antennas

5.4.1 Transmission Model

Whereas in the UL spatial-temporal signal processing can be jointly performed at the BS (Fig. 5.5c) such joint processing is not possible in the DL. Rather, the spatial processing has to be performed at the BS, whereas the temporal processing has to take place at the MS. This leads to the DL transmission model shown in Fig. 5.9 [23]. The DL transmitter generates the K_s burst signals. These signals are fed to an antenna-weighting network, which

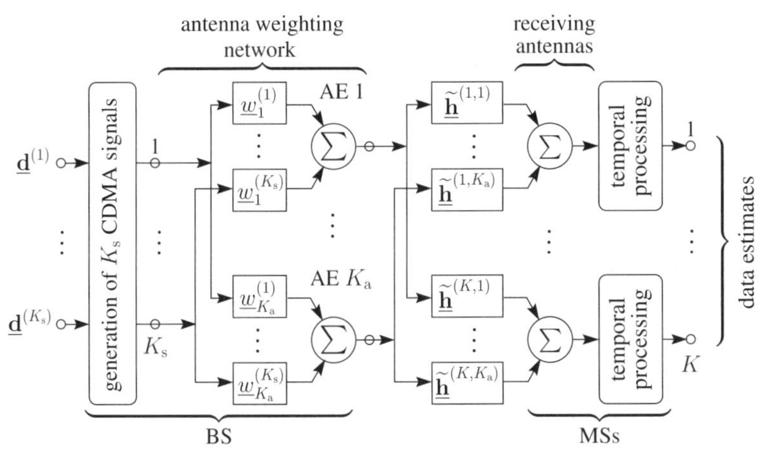

Fig. 5.9. DL transmission model with K_a antenna elements (AE), valid for the support of the MS μ_k, $k=1,\ldots,K$ [23]

generates the signals to be fed into the K_a antenna elements by utilizing the weight vectors

$$\underline{\mathbf{w}}^{(k_s)} = \left(\underline{w}_1^{(k_s)} \ldots \underline{w}_{K_a}^{(k_s)}\right)^{\mathrm{T}}, \; k_s = 1, \ldots, K_s. \tag{5.56}$$

The weighting network can be considered to perform the spatial signal processing at the BS. The signals radiated by the K_a antenna elements are transmitted to each MS μ_k via K_a radio channels with the impulse responses $\underline{\tilde{\mathbf{h}}}^{(k,k_a)}$ of (5.44). At each MS the sum of the K_a signals originating in the K_a transmit signals appears at the output of the receive antenna. This output signal is then temporally processed in the individual MS in order to obtain at each MS an estimate of some or all of the partial data vectors $\underline{\mathbf{d}}^{(k_s)}$ of (2.1), or of the total data vector $\underline{\mathbf{d}}$ of (5.9).

The channel impulse responses that is valid for the link between the input port k_s of the weighting network and the output of the receive antenna of MS μ_k, see Fig. 5.9, is termed $\underline{\mathbf{h}}^{(k,k_s)}$. In what follows $\underline{\mathbf{h}}^{(k,k_s)}$ is expressed by the properties of the weighting network and the radio channels. The transmission from the input of the BS antenna element k_a, $k_a = 1, \ldots, K_a$ to the output of the antenna of MS μ_k, $k = 1, \ldots, K$ can be described by the channel impulse response $\underline{\tilde{\mathbf{h}}}^{(k,k_a)}$ of (5.44). The channel impulse responses $\underline{\tilde{\mathbf{h}}}^{(k,k_a)}$, $k_a = 1, \ldots, K_a$ valid for MS μ_k can be arranged in the matrix

$$\underline{\mathbf{H}}^{(k)} = \left(\underline{\tilde{\mathbf{h}}}^{(k,1)} \ldots \underline{\tilde{\mathbf{h}}}^{(k,K_a)}\right). \tag{5.57}$$

With this matrix and $\underline{\mathbf{w}}^{(k_s)}$ of (5.56) we obtain

$$\underline{\mathbf{h}}^{(k,k_s)} = \underline{\mathbf{H}}^{(k)} \underline{\mathbf{w}}^{(k_s)}. \tag{5.58}$$

The response $\underline{\mathbf{h}}^{(k,k_\mathrm{s})}$ of (5.58) and the CDMA code $\underline{\mathbf{c}}^{(k_\mathrm{s})}$ yield the composite channel impulse response

$$\underline{\mathbf{b}}^{(k,k_\mathrm{s})} = \underline{\mathbf{c}}^{(k_\mathrm{s})} * \underline{\mathbf{h}}^{(k,k_\mathrm{s})}, \tag{5.59}$$

valid for the transmission of the burst signal k_s, $k_\mathrm{s} = 1, \ldots, K_\mathrm{s}$ to MS μ_k. Then, following the rationale already introduced in Sect. 5.2.2, see (5.7) and (5.8), the total system Matrix $\underline{\mathbf{A}}^{(k)}$ valid for MS μ_k can be generated. With this matrix, and with the total data vector $\underline{\mathbf{d}}$ of (5.9) and the noise $\underline{\mathbf{n}}^{(k)}$ received at MS μ_k, the signal received at MS μ_k takes the form

$$\underline{\mathbf{e}}^{(k)} = \underline{\mathbf{A}}^{(k)}\underline{\mathbf{d}} + \underline{\mathbf{n}}^{(k)}. \tag{5.60}$$

Again, we obtain an expression analogous to (5.10). This means that data detection at the MS can be performed by the algorithms explained in Sect. 5.2.3.

5.4.2 Determination of Weight Vectors

Up to now the question how to choose the weight vectors $\underline{\mathbf{w}}^{(k_\mathrm{s})}$ of (5.56) has not been addressed. Generally speaking, these vectors should be chosen in such a way that the system performance becomes as high as possible. One rationale to make this choice is outlined in what follows [23]. This rationale relies on the knowledge of the channel impulse responses $\underline{\widetilde{\mathbf{h}}}^{(k,k_\mathrm{a})}$ of (5.44) at the BS. In the case of TDD systems, thanks to the reciprocity theorem, this knowledge is at least approximately available at the BS from the UL channel estimation. Let us assume that the K_a BS antenna elements are decoupled. Then the total energy radiated by the K_a antenna elements in order to transmit one symbol of magnitude 1 of the burst signal k_s becomes

$$T^{(k_\mathrm{s})} = \underline{\mathbf{c}}^{(k_\mathrm{s})*\mathrm{T}}\underline{\mathbf{c}}^{(k_\mathrm{s})}\underline{\mathbf{w}}^{(k_\mathrm{s})*\mathrm{T}}\underline{\mathbf{w}}^{(k_\mathrm{s})}. \tag{5.61}$$

With $\underline{\mathbf{b}}^{(k,k_\mathrm{s})}$ of (5.59) and with (5.58), the corresponding energy received at MS μ_k can be written as

$$R^{(k,k_\mathrm{s})} = \underline{\mathbf{b}}^{(k,k_\mathrm{s})*\mathrm{T}}\underline{\mathbf{b}}^{(k,k_\mathrm{s})} = \left(\underline{\mathbf{c}}^{(k_\mathrm{s})} * \underline{\mathbf{H}}^{(k)}\underline{\mathbf{w}}^{(k_\mathrm{s})}\right)^{*\mathrm{T}} \left(\underline{\mathbf{c}}^{(k_\mathrm{s})} * \underline{\mathbf{H}}^{(k)}\underline{\mathbf{w}}^{(k_\mathrm{s})}\right). \tag{5.62}$$

Now, with the assignment relation $z(k_\mathrm{s})$ of (5.42), let us assume that it is intended to serve MS $\mu_{z(k_\mathrm{s})}$ by the burst signal k_s. Then the desired energy received at MS $\mu_{z(k_\mathrm{s})}$ and generated by sending the burst signal k_s follows from setting k equal to $z(k_\mathrm{s})$ in (5.62), which results in $R^{(z(k_\mathrm{s}),k_\mathrm{s})}$. The rationale proposed by the authors to determine the weight vector $\underline{\mathbf{w}}^{(k_\mathrm{s})}$ of (5.56) consists in maximizing the ratio

$$\frac{R^{(z(k_\mathrm{s}),k_\mathrm{s})}}{T^{(k_\mathrm{s})}} = \frac{\left(\underline{\mathbf{c}}^{(k_\mathrm{s})} * \underline{\mathbf{H}}^{(z(k_\mathrm{s}))}\underline{\mathbf{w}}^{(k_\mathrm{s})}\right)^{*\mathrm{T}} \left(\underline{\mathbf{c}}^{(k_\mathrm{s})} * \underline{\mathbf{H}}^{(z(k_\mathrm{s}))}\underline{\mathbf{w}}^{(k_\mathrm{s})}\right)}{\underline{\mathbf{c}}^{(k_\mathrm{s})*\mathrm{T}}\underline{\mathbf{c}}^{(k_\mathrm{s})}\underline{\mathbf{w}}^{(k_\mathrm{s})*\mathrm{T}}\underline{\mathbf{w}}^{(k_\mathrm{s})}} \tag{5.63}$$

of the desired energy $R^{(z(k_s),k_s)}$ and the transmitted energy $T^{(k_s)}$ from (5.61). For a given CDMA code $\underline{\mathbf{c}}^{(k_s)}$ and a given channel impulse response, the weight vector $\underline{\mathbf{w}}^{(k_s)}$, which maximizes the ratio formulated in (5.63), is the eigenvector corresponding to the dominant eigenvalue of the matrix $[\underline{\mathbf{c}}^{(k_s)} * \underline{\mathbf{H}}^{(z(k_s))}]^{*\mathrm{T}}[\underline{\mathbf{c}}^{(k_s)} * \underline{\mathbf{H}}^{(z(k_s))}]$. As the crux of the proposed choice of the weight vectors $\underline{\mathbf{w}}^{(k_s)}$ from (5.56), the multielement transmit antenna is utilized to minimize the transmitted energy necessary to generate a certain desired energy at the MS. This approach reduces the generated intercell MAI per served MS and, thus, ultimately allows an increase in the total system load in an interference-limited system.

5.4.3 Partial JD

As a side effect of maximizing the energies of the desired burst signals at the MS, the energies of the burst signals that are not of interest at the MS go down. In [23] this effect could be verified by simulations. If undesired burst signals arrive with low energies at the MS, an obvious idea would be to exclude all or some of these undesired signals from JD, an approach termed partial JD (PJD) by the authors. PJD has two opposite outcomes:

- The undesired burst signals excluded from JD now emerge as an additional intracell interference.
- Because in the case of PJD the number of jointly detected burst signals is smaller than in the case of conventional JD, PJD has a SNR degradation, see (2.24) and (2.25), smaller than that of conventional JD.

Simulations showed that the beneficial effect of reduced SNR degradation increasingly outweighs the detrimental effect of an additional intracell interference with an increasing number K_{a} of BS antenna elements [23]. Therefore, especially for larger values of K_{a}, the preferred detection scheme at the MS should be PJD, which is also computationally less expensive than conventional JD.

5.5 Multistep JD

5.5.1 Basic Principle

For time-slotted CDMA approaches to optimum data detection, which unfortunately are prohibitively expensive, as well as suboptimum less-expensive linear data detection schemes are presented in Sect. 5.2.3. In this section we address improved suboptimum detection schemes for time-slotted CDMA, which take, with respect to complexity and performance, a favorable position between the optimum detectors and the linear detectors described in Sect. 5.2.3. Promising detectors of this kind resort to the iterative (turbo) principle [7–9]. In the following, we present such a scheme, which is termed multistep JD (MSJD) by the authors.

In MSJD the K_s received burst signals are partitioned into two groups $g=1$ and $g=2$, which, with

$$1 \leq K_G \leq K_s - 1, \tag{5.64}$$

consist of the burst signals $k_s = 1, \ldots, K_G$ and $k_s = K_G+1, \ldots, K_s$, respectively. With $\underline{\mathbf{d}}^{(k_s)}$ of (5.1) the data vectors of groups 1 and 2 become

$$\underline{\mathbf{d}}_G^{(1)} = \left(\underline{\mathbf{d}}^{(1)^T} \ldots \underline{\mathbf{d}}^{(K_G)^T}\right)^T \tag{5.65}$$

and

$$\underline{\mathbf{d}}_G^{(2)} = \left(\underline{\mathbf{d}}^{(K_G+1)^T} \ldots \underline{\mathbf{d}}^{(K_s)^T}\right)^T, \tag{5.66}$$

respectively. With $\underline{\mathbf{A}}^{(k_s)}$ of (5.7) the system matrices

$$\underline{\mathbf{A}}_G^{(1)} = \left(\underline{\mathbf{A}}^{(1)} \ldots \underline{\mathbf{A}}^{(K_g)}\right) \tag{5.67}$$

and

$$\underline{\mathbf{A}}_G^{(2)} = \left(\underline{\mathbf{A}}^{(K_G+1)} \ldots \underline{\mathbf{A}}^{(K_s)}\right)^T \tag{5.68}$$

of groups 1 and 2, respectively, can be formed. Equations (5.65)–(5.68) can be used to express the received signal originating in the transmitted burst signals of the two groups as

$$\underline{\mathbf{r}}_G^{(g)} = \underline{\mathbf{A}}_G^{(g)} \underline{\mathbf{d}}_G^{(g)}, \; g = 1, 2. \tag{5.69}$$

With $\underline{\mathbf{r}}_G^{(g)}$ of (5.69) the received signal $\underline{\mathbf{e}}$ of (2.10) can be represented as

$$\underline{\mathbf{e}} = \underline{\mathbf{r}}_G^{(1)} + \underline{\mathbf{r}}_G^{(2)} + \underline{\mathbf{n}}. \tag{5.70}$$

Let us now consider the detection scheme shown in Fig. 5.10, which consists of the two branches $g = 1, 2$. In branch g, with $\underline{\mathbf{e}}$ of (5.70), first the signal

$$\underline{\mathbf{e}}_G^{(g)} = \underline{\mathbf{e}} - \underline{\mathbf{r}}_G^{(3-g)} = \underline{\mathbf{r}}_G^{(g)} + \underline{\mathbf{n}}, \; g = 1, 2, \tag{5.71}$$

is formed, where the subtraction of $\underline{\mathbf{r}}_G^{(3-g)}$ is a hypothetical step because $\underline{\mathbf{r}}_G^{(g)}$ is not known in the receiver. The signals $\underline{\mathbf{e}}_G^{(g)}$ of (5.71) are processed in the blocks "JD & FEC decoding", which leads to the estimates $\hat{\underline{\mathbf{u}}}_G^{(g)}$, $g=1, 2$ of the uncoded data vectors corresponding to the coded data vectors $\underline{\mathbf{d}}_G^{(g)}$ of (5.65) and (5.66), respectively. From (5.64) the numbers of burst signals contained in the two signals $\underline{\mathbf{e}}_G^{(g)}$, $g=1, 2$ of (5.71) are K_G and K_s–K_G, respectively, and, therefore, are smaller than the number K_s of burst signals constituting the total received signal $\underline{\mathbf{e}}$ of (2.10). Therefore, the SNR degradation, see (2.24) and (2.25), observed when performing JD in each of the two branches of the

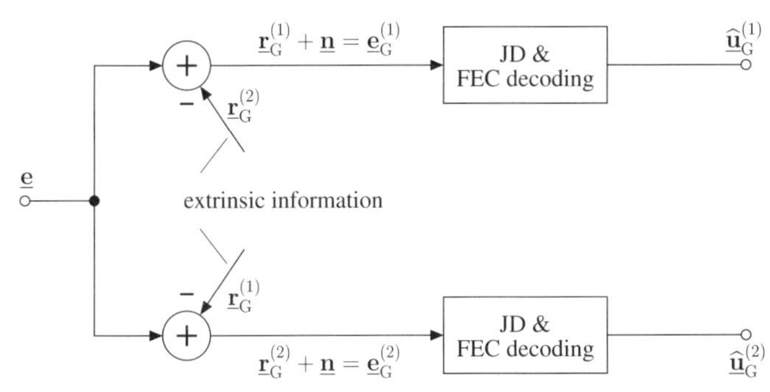

Fig. 5.10. Reduction of the number of burst signals subject to JD and FEC decoding

scheme shown in Fig. 5.10 is smaller than the SNR degradation that would be valid if all K_{s} burst signals were simultaneously evaluated in one single JD process. Consequently, the estimates of the uncoded data contained in the vectors $\widehat{\underline{\mathbf{u}}}_{\mathrm{G}}^{(1)}$ and $\widehat{\underline{\mathbf{u}}}_{\mathrm{G}}^{(2)}$ at the outputs of the scheme in Fig. 5.10 have a higher quality than the corresponding data estimates obtained by conventional JD.

The signal $\underline{\mathbf{r}}_{\mathrm{G}}^{(3-g)}$ fed to the adder in branch g of the scheme of Fig. 5.10 can be considered as extrinsic information [27] for the detection process to be performed in this branch. The question is how this extrinsic information can be obtained in real-world systems. To answer this question the scheme of Fig. 5.10 is extended as displayed in Fig. 5.11. In the scheme shown in Fig. 5.11 the detection is performed in a series of consecutive steps – therefore the designation MSJD – with the step number being termed i. Instead of subtracting the exact signals $\underline{\mathbf{r}}_{\mathrm{G}}^{(g)}$, $g=1,\,2$ of (5.69) from the received signal $\underline{\mathbf{e}}$ of (5.10), in MSJD estimates $\widehat{\underline{\mathbf{r}}}_{\mathrm{G}}^{(g)}$, $g=1,\,2$ of these signals are subtracted. Therefore, instead of (5.71) we now have

$$\underline{\mathbf{e}}_{\mathrm{G}}^{(g)} = \underline{\mathbf{e}} - \widehat{\underline{\mathbf{r}}}_{\mathrm{G}}^{(3-g)},\ g = 1,\,2. \tag{5.72}$$

As shown in Fig. 5.11, the estimates $\widehat{\underline{\mathbf{r}}}_{\mathrm{G}}^{(g)}$, $g=1,\,2$ of $\underline{\mathbf{r}}_{\mathrm{G}}^{(g)}$ from (5.69) to be subtracted from $\underline{\mathbf{e}}$ prior to performing the detection step i are termed $\widehat{\underline{\mathbf{r}}}_{\mathrm{G}}^{(g)}(i-1)$ and are obtained by reconstructing $\underline{\mathbf{r}}_{\mathrm{G}}^{(g)}$ based on the estimate $\widehat{\underline{\mathbf{u}}}_{\mathrm{G}}^{(g)}(i-1)$ of the uncoded data vector $\underline{\mathbf{u}}_{\mathrm{G}}^{(g)}$, obtained in the previous step $i-1$.

In order to assess MSJD, simulations were performed for the following situation:

- Data symbol set $\mathbb{V}_{\mathrm{d}}$ of (5.2) according to the modulation scheme QPSK, that is

$$\mathbb{V}_{\mathrm{d}} = \left\{ \frac{1+\mathrm{j}}{\sqrt{2}},\ \frac{-1+\mathrm{j}}{\sqrt{2}},\ \frac{-1-\mathrm{j}}{\sqrt{2}},\ \frac{1-\mathrm{j}}{\sqrt{2}} \right\}; \tag{5.73}$$

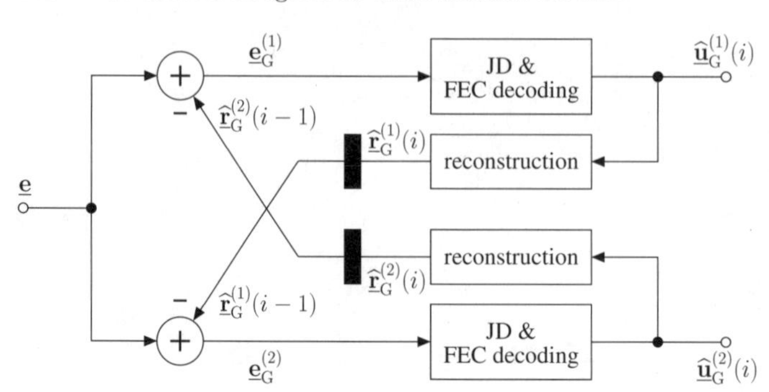

Fig. 5.11. Multistep joint detection (MSJD)

- System parameters

$$Q = 16, \quad K_s = 16, \quad K_G = 8, \tag{5.74}$$

  Q equal to K_s means full system load;
- Orthogonal CDMA codes $\underline{\mathbf{c}}^{(k_s)}$, $k_s = 1, \dots, K_s$ of (5.3);
- FEC coding by a convolutional code with the following values [27] for constraint length R_c, free distance d_f and number N_a of connections of the shift register in the FEC encoder:

$$R_c = 1/2, \quad d_f = 7, \quad N_a = 7; \tag{5.75}$$

- JD by ZF–BLE, see (5.23);
- Hard bits utilized for signal reconstruction [27];
- No interleaving;
- Channel model according to COST 207 BU [28].

Figure 5.12 shows the coded bit error rate (BER) of MSJD versus E_b/N_0, with the number i of iterations as the parameter. For the sake of comparison the BER curves for the following cases are also included in Fig. 5.12:

- Conventional JD with K_s and $Q = 16$, that is, full system load;
- Conventional JD with $Q = 16$ and $K_s = 8$, that is, 50% system load;
- Conventional JD with $K_s = Q = 16$ and transmission over AWGN channels.

The curves of Fig. 5.12 for full system load show that after only three iterations MSJD achieves a performance much better than that of conventional JD. One also recognizes that MSJD with full system load comes close to conventional JD with only 50% system load. While conventional JD with full system load shows a dramatic performance degradation when considering COST 207 channels instead of AWGN channels, this degradation is much smaller in the case of MSJD.

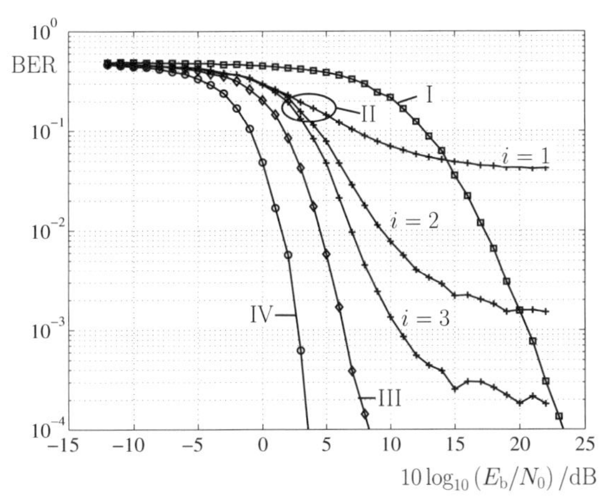

Fig. 5.12. Coded BER versus E_b/N_0; system values see (5.73)–(5.75); I–III: COST 207 BU channel model; IV: AWGN channels; I: Conventional JD at full system load; II: MSJD at full system load; III: Conventional JD at 50% system load; IV: Conventional JD at full system load

5.5.2 Approximate Closed Form Analysis

In this section the performance of MSJD achieved after infinitely many iterations, that is, for $i \to \infty$, will be compared with the performance of conventional JD. This comparison follows the approach developed in [27] and is performed under the following conditions:

- Full system load, that is,

$$K_s = Q; \tag{5.76}$$

- QPSK data symbol set, see Sect. 5.5.1;
- JD by ZF–BLE, see (5.23).
- Same size of the two groups $g = 1, 2$, that is,

$$K_G = K_s/2; \tag{5.77}$$

- All K_s burst signals arrive with the same average power C at the receiver input, and the power of the noise at the receiver input is σ^2. Consequently, the SNR at the receiver input of each of the K_s burst signals becomes

$$\gamma = C/\sigma^2. \tag{5.78}$$

The mean power of the reconstruction error is defined as

$$\sigma_{rec}^2 = \frac{1}{2}\frac{1}{NQ+W-1}\mathrm{E}\left\{\left|\hat{\underline{\mathbf{r}}}_G^{(g)} - \underline{\mathbf{r}}_G^{(g)}\right|^2\right\}. \tag{5.79}$$

Then, the average SNR at the inputs of the two blocks “JD & FEC decoding” in the system of Fig. 5.11 becomes

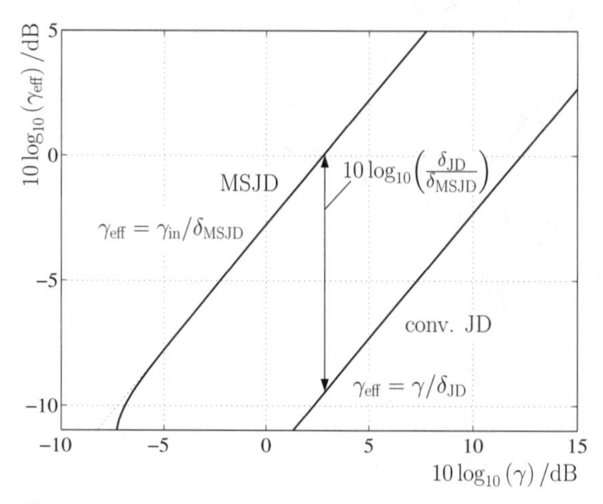

Fig. 5.13. Effective SNR γ_{eff} versus SNR γ for conventional JD and MSJD

$$\gamma_{\text{in}} = \frac{C}{\sigma^2 + \sigma_{\text{rec}}^2(\gamma_{\text{in}})} = \frac{\gamma}{1 + \sigma_{\text{rec}}^2(\gamma_{\text{in}})/\sigma^2}. \tag{5.80}$$

As indicated in (5.80), σ_{rec}^2 of (5.79) itself depends on γ_{in}. The value γ_{in} achieved after infinitely many iterations, that is, for $i \to \infty$, is the solution of the implicit equation (5.80) for γ_{in}. In order to solve this equation, the function $\sigma_{\text{rec}}^2(\gamma_{\text{in}})$ has to be determined. To this purpose, let us assume that the SNR γ of (5.78) at the receiver input is so large that the error probabilities at the outputs of the two blocks "JD & FEC decoding", see Fig. 5.11, are much smaller than 1. Let us further assume that the applied FEC code is a convolutional code with rate R_{c}, free distance d_{f} and a number of N_{a} connections of the shift register of the FEC encoder. Then, as shown in [27], the approximation

$$\sigma_{\text{rec}}^2(\gamma_{\text{in}}) = 2CN_{\text{a}}R_{\text{c}}Q\exp\left[-\frac{Q(Q/2+1)}{2(Q+1)}\gamma_{\text{in}}d_{\text{f}}\right] \tag{5.81}$$

holds. Substitution of (5.81) in (5.80) yields

$$\gamma_{\text{in}} = \frac{\gamma}{1 + 2N_{\text{a}}R_{\text{c}}Q\gamma\exp\left[-\frac{Q(Q/2+1)}{2(Q+1)}\gamma_{\text{in}}d_{\text{f}}\right]}. \tag{5.82}$$

This equation determines, depending on γ, the value $\gamma_{\text{in}}(\gamma)$ obtained for $i \to \infty$. The effective SNR [29] results from γ_{in} through division by the SNR degradation. If we set $Q = K_{\text{s}}/2$ in (5.25), we obtain the SNR degradation δ_{MSJD} of MSJD. If we set $Q = K_{\text{s}}$ in (5.25), we obtain the SNR degradation of conventional JD. These SNR degradations lead to the effective SNR

$$\gamma_{\text{eff,MSJD}} = \frac{\gamma_{\text{in}}(\gamma)}{\delta_{\text{MSJD}}} = \frac{\gamma_{\text{in}}(\gamma)(Q/2+1)}{Q+1} \tag{5.83}$$

and

$$\gamma_{\mathrm{eff,JD}} = \frac{\gamma}{Q+1} \tag{5.84}$$

for MSJD and conventional JD, respectively. In Fig. 5.13 $\gamma_{\mathrm{eff,MSJD}}$ and $\gamma_{\mathrm{eff,JD}}$ are depicted versus γ of (5.78) for the values Q, R_{c}, d_{f} and N_{a} of (5.74) and (5.75), respectively. The curves of Fig. 5.13 impressively show the superiority of MSJD over conventional JD.

5.6 Intercell Interference Reduction

5.6.1 Motivation

Mobile radio systems usually operate in the interference-limited mode, that is, the receiver noise is of minor importance. In the case of time-slotted CDMA utilizing JD, the only relevant type of interference is intercell MAI, because intracell MAI can be eliminated by JD in the form of ZF–BLE or MMSE–BLE (Sect. 5.2.3). Basically, in a cellular scenario the number of potential sources of intercell MAI is equal to the total number of cochannel transmitters outside the considered cell, which is termed the reference cell, and therefore is very large in extended cell nets. This tends to make intercell MAI reduction extremely complicated. However, especially in the case of time-slotted CDMA, the number of those intercell MAI sources, which in reality significantly contribute to intercell MAI, is usually rather low. This is due to three effects:

- Thanks to the TDMA component of time-slotted CDMA, see Fig. 5.1, only $1/N_{\mathrm{fr}}$ of the total number of potential intercell MAI sources is simultaneously active.
- Sources of intercell MAI located in cells far away from the reference cell have only a marginal impact on the intercell MAI in the reference cell because of the large path losses.
- Of the intercell MAI sources located in cells closer to the reference cell some are always heavily shadowed with respect to the perturbed receivers in the reference cell.

Simulations of time-slotted CDMA scenarios on the system level showed that, on the average, as much as about 20% of the total intercell MAI power is caused by the strongest intercell interferer, and that the strongest ten intercell interferers generate about 80% of the total intercell MAI power [31]. This situation is illustrated by Fig. 5.14, where the average contributions of the strongest intercell interferers to the total intercell MAI power are shown versus the number K_{s} of CDMA signals per cell for the UL [30, 31]. As a consequence of this composition of the intercell MAI power, the system performance could be dramatically improved if only a few of the strongest intercell MAI signals could be eliminated in the reference cell. This finding

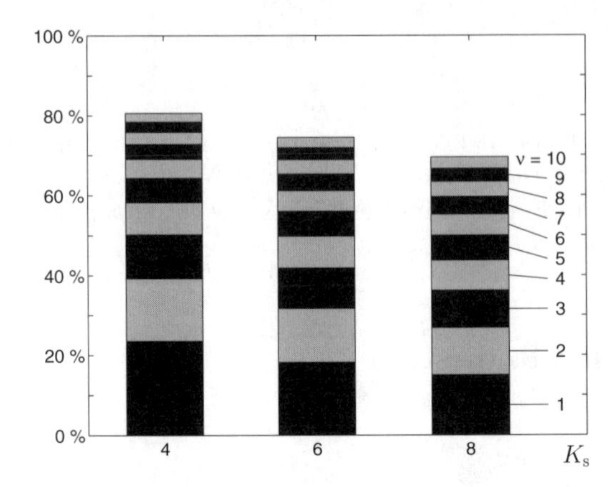

Fig. 5.14. Average contributions of the strongest intercell interferers $\nu=1,\dots,10$ to the total intercell MAI power versus the number K_s of burst signals per cell [30, 31]; cluster size $r=3$, attenuation exponent $\alpha=3.6$, standard deviation of lognormal fading $\sigma=8$dB

opens the way for designing schemes of intercell MAI reduction. To be able to perform such a reduction, two problems have to be solved in the reference cell:

- Identification of the strongest, that is, relevant, intercell MAI signals.
- Inclusion of these signals into the process of JCE and JD in the reference cell, and in this way eliminating them (Fig. 5.15).

These two problems are briefly addressed in Sects. 5.6.2 and 5.6.3, respectively. A more detailed investigation of intercell MAI reduction, including simulation results, is reported in [32]. As an important prerequisite of intercell MAI reduction, in each cell both the CDMA and midamble codes utilized by potential intercell interferers must be known. Also, the system should be synchronized, which means that the frame and burst patterns of all cells should be synchronous, which, because of the path delays, cannot be achieved if the cell sizes exceed certain limits.

5.6.2 Identification of Relevant Intercell Interferers

The identification of the strong intercell MAI signals in the reference cell must use specific characteristics of these signals. Two approaches are described in [32]. An obvious approach to identification consists in exploiting a priori given characteristics of the intercell MAI signals as, for instance, their CDMA or midamble codes. As an example, matched filters for the midambles of the intercell interferers could be provided in the receivers of the reference cell. Then, based on the levels of the output signals of these matched filters, decisions could be taken concerning the relevance of the intercell interferers.

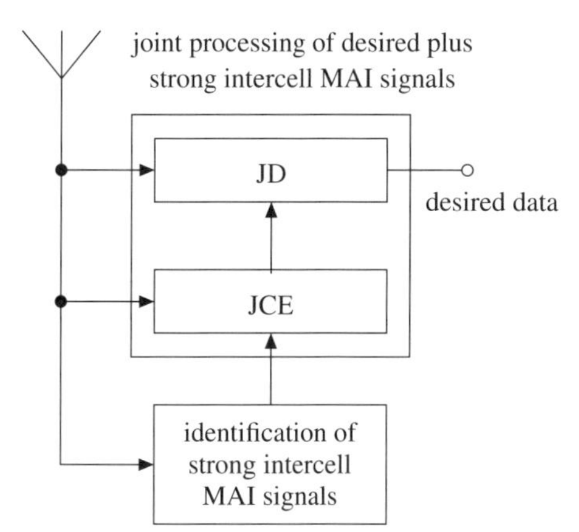

Fig. 5.15. Intercell MAI reduction by identifying strong intercell MAI signals and including these signals in the process of JCE and JD

Another possibility for enabling identification consists in supplementing the burst signals transmitted in time-slotted CDMA with signal-specific signature signals of low power. Unmodulated sine signals, which could be detected by FFT-based signature detectors in the receivers of the reference cell, would be well suited as such signature signals. In contrast to exploiting a priori given signal characteristics as, for instance, CDMA or midamble codes, the introduction and utilization of additional signature signals may lead to more reliable identification. On the other hand, additional interference would be generated in the system by such signals.

5.6.3 Elimination of Strong Intercell Interferers

Once the relevant intercell MAI signals are identified in the reference cell, the further aim is to eliminate them. To this purpose, first the channel impulse responses of the radio channels between the relevant sources of intercell MAI and the receivers in the reference cell have to be determined together with the channel impulse responses valid for the burst signals of the reference cell by JCE. After that, one can try to include the relevant intercell MAI signals into the JD process in the reference cell. The number of intercell MAI signals that can be included into the JD process in the reference cell is limited by two factors:

- For a given midamble dimension L_{m}, see (5.29), the total number of channel taps including those of the interferer channels, which can be estimated by a Steiner estimator [21, 22], is limited.
- The number of CDMA signals, including those of the intercell interferers, that can be jointly detected for a given dimension Q of the CDMA codes is limited.

In [32] the issue of eliminating strong intercell MAI signals is treated in detail.

5.7 Transmit Signal Optimization by Joint Transmission

5.7.1 Introduction

In the case of TDD the same channel impulse responses are valid for both the UL and the DL, if – and this is assumed in what follows – the time elapsing between UL and DL transmission is sufficiently small compared to the coherence time of the mobile radio channels. This identity of UL and DL channel impulse responses is not exploited in state-of-the-art time-slotted CDMA, which requires channel estimators at the receivers of both the BS and the MS. This is true because the knowledge of the channel impulse responses is a prerequisite for performing JD (Sect. 5.2.3). In what follows a novel DL transmission technique for time-slotted CDMA is presented, which needs channel estimators only in the BS receivers, and which, for DL transmission, utilizes the knowledge of the channel impulse responses at the BS transmitter in such a way that channel estimators are no longer required at the receivers of the MS. In addition, compared to conventional time-slotted CDMA characterized by using JD, the computational expense of data detection in the MS is dramatically reduced, and system performance, especially capacity, is simultaneously enhanced by this transmission scheme. The crux of the proposed scheme consists in determining one common transmit signal at each BS for the service of all K MS to be supported by this BS, which, after having passed the mobile radio channels, at each MS μ_k yields the data $\underline{\mathbf{d}}^{(k)}$ sent for this MS by simple linear filtering with a time-invariant filter. The authors term this approach joint transmission (JT). Its rationale is the inverse of the one followed in JD, where a received signal common to all users is jointly processed at the receiver in order to obtain the data sent by the individual transmitters. In a way, JT is related to known CDMA concepts using transmit signals which originate by predistortion of CDMA coded signals under consideration of the known channel impulse responses [33]. However, JT is a more general concept, because the transmit signals are directly generated and do not result by first generating and then distorting CDMA coded signals. Therefore, the application of the proposed JT scheme is not restricted to time-slotted CDMA systems, but is also applicable to other systems having a TDMA component.

Section 5.7 is structured as follows. In Sect. 5.7.2 the signal transmission model of JT is developed. Section 5.7.3 deals with data transmission and detection in a general DL system utilizing JT. In Sect. 5.7.4 the application of JT in the DL of time-slotted CDMA is considered. The topic of Sect. 5.7.5 is a brief performance assessment of JT in time-slotted CDMA.

5.7.2 Signal Transmission Model

As in the considerations of Sects. 5.4 and 5.4, at the BS an array of K_{a} transmit antenna elements is utilized, whereas each MS μ_k, $k = 1, \ldots, K$

is equipped with a single receiving antenna element. This situation can be described by the KK_{a} channel impulse responses $\underline{\tilde{\mathbf{h}}}^{(k,k_{\mathrm{a}})}$ of (5.44). Into each of the K_{a} transmit antenna elements a signal

$$\underline{\mathbf{s}}^{(k_{\mathrm{a}})} = \left(\underline{s}_1^{(k_{\mathrm{a}})} \dots \underline{s}_S^{(k_{\mathrm{a}})}\right)^{\mathrm{T}}, \; k_{\mathrm{a}}=1,\dots,K_{\mathrm{a}}, \tag{5.85}$$

of dimension S is fed. The K_{a} signals $\underline{\mathbf{s}}^{(k_{\mathrm{a}})}$ of (5.85) can be compiled to form the total transmit signal

$$\underline{\mathbf{s}} = \left(\underline{\mathbf{s}}^{(1)^{\mathrm{T}}} \dots \underline{\mathbf{s}}^{(K_{\mathrm{a}})^{\mathrm{T}}}\right)^{\mathrm{T}} \tag{5.86}$$

of dimension $K_{\mathrm{a}}S$. Signal $\underline{\mathbf{s}}$ of (5.86) represents the transmit signal of one TDMA burst. With the KK_{a} channel impulse responses $\underline{\tilde{\mathbf{h}}}^{(k,k_{\mathrm{a}})}$ of (5.44), the $(S+W-1)\times S$ matrices

$$\begin{aligned}
\underline{\tilde{\mathbf{H}}}^{(k,k_{\mathrm{a}})} &= \left(\underline{\tilde{H}}_{i,j}^{(k,k_{\mathrm{a}})}\right), i=1,\dots,S+W-1, \\
&\quad j=1,\dots,S, \\
\underline{\tilde{H}}_{i,j}^{(k,k_{\mathrm{a}})} &= \begin{cases} \underline{\tilde{h}}_{i-j+1}^{(k,k_{\mathrm{a}})} & \text{for } 1 \le i-j+1 \le W, \\ 0 & \text{otherwise}, \end{cases} \qquad (5.87) \\
&\quad k=1,\dots,K,\; k_{\mathrm{a}}=1,\dots,K_{\mathrm{a}}, \qquad (5.88)
\end{aligned}$$

can be formed. The K_{a} matrices $\underline{\tilde{\mathbf{H}}}^{(k,k_{\mathrm{a}})}$, $k=1,\dots,K$, valid for a given value of k, that is, for a given MS μ_k, can be arranged in an $(S+W-1)\times(K_{\mathrm{a}}S)$ matrix

$$\underline{\tilde{\mathbf{H}}}^{(k)} = \left(\underline{\tilde{\mathbf{H}}}^{(k,1)} \dots \underline{\tilde{\mathbf{H}}}^{(k,K_{\mathrm{a}})}\right). \tag{5.89}$$

Now, the signal originating in $\underline{\mathbf{s}}$ of (5.86) and received at MS μ_k can be represented by a column vector

$$\underline{\mathbf{e}}^{(k)} = \underline{\tilde{\mathbf{H}}}^{(k)}\underline{\mathbf{s}},\; k=1,\dots,K, \tag{5.90}$$

of dimension $S+W-1$. Then $\underline{\mathbf{e}}^{(k)}$ is termed the partial received signal. The K signals $\underline{\mathbf{e}}^{(k)}$ of (5.90) can be arranged in a vector

$$\underline{\mathbf{e}} = \left(\underline{\mathbf{e}}^{(1)^{\mathrm{T}}} \dots \underline{\mathbf{e}}^{(K)^{\mathrm{T}}}\right)^{\mathrm{T}} \tag{5.91}$$

of dimension $K(S+W-1)$, which is termed the total received signal and corresponds to the reception of one TDMA burst. The K matrices $\underline{\tilde{\mathbf{H}}}^{(k)}$ of (5.89) can be arranged in the $[K(S+W-1)]\times(K_{\mathrm{a}}S)$ matrix

$$\underline{\tilde{\mathbf{H}}} = \left(\underline{\tilde{\mathbf{H}}}^{(1)^{\mathrm{T}}} \dots \underline{\tilde{\mathbf{H}}}^{(K)^{\mathrm{T}}}\right)^{\mathrm{T}}, \tag{5.92}$$

which is termed the channel matrix. In the case of time-variant channels, the elements of $\underline{\widetilde{\mathbf{H}}}$ of (5.92) are nonconstant. When designing a system care should be taken that the coherence time of the channels is sufficiently larger than the duration T_{bu} of the TDMA bursts. With $\underline{\widetilde{\mathbf{H}}}$ of (5.92) the total received signal $\underline{\mathbf{e}}$ of (5.91) can be, in the noise-free case, concisely expressed as

$$\underline{\mathbf{e}} = \underline{\widetilde{\mathbf{H}}}\,\underline{\mathbf{s}}\,. \tag{5.93}$$

Note that the entire signal $\underline{\mathbf{e}}$ of (5.93) cannot be observed at a single MS μ_k. Signal $\underline{\mathbf{e}}$ consists of K received partial signals $\underline{\mathbf{e}}^{(k)}$, see (5.91), with each of these signals being available only at the corresponding MS μ_k.

5.7.3 Data Transmission and Detection, General Case

In Sect. 5.7.2 the component values $\underline{s}_s^{(k_\mathrm{a})}$, $s=1,\ldots,S$, $k_\mathrm{a}=1,\ldots,K_\mathrm{a}$ of the transmit signals $\underline{\mathbf{s}}^{(k_\mathrm{a})}$ of (5.85), and therefore the component values of the total transmit signal $\underline{\mathbf{s}}$ of (5.86) have not yet been specified. In this section these values are determined in such a way that data transmission from the BS to the MS μ_k, $k=1,\ldots,K$ can be performed. It is assumed that, per burst signal, N_k data symbols have to be transmitted from the BS to the MS μ_k, $k=1,\ldots,K$, where, in contrast to the considerations in the previous sections, the number of data symbols may differ from MS to MS. The N_k data symbols sent for MS μ_k are arranged in the partial data vector

$$\underline{\mathbf{d}}^{(k)} = \left(\underline{d}_1^{(k)} \ldots \underline{d}_{N_k}^{(k)}\right)^{\mathrm{T}}, k=1,\ldots,K. \tag{5.94}$$

The K partial data vectors $\underline{\mathbf{d}}^{(k)}$ $k=1,\ldots,K$, of (5.94) are stacked to form the total data vector

$$\underline{\mathbf{d}} = \left(\underline{\mathbf{d}}^{(1)^{\mathrm{T}}} \ldots \underline{\mathbf{d}}^{(K)^{\mathrm{T}}}\right)^{\mathrm{T}} = \left(\underline{d}_1 \ldots \underline{d}_{N_\mathrm{t}}\right)^{\mathrm{T}} \tag{5.95}$$

of dimension

$$N_\mathrm{t} = \sum_{k=1}^{K} N_k. \tag{5.96}$$

With $\underline{\mathbf{s}}$ of (5.86), $\underline{\mathbf{d}}$ of (5.95) and a $(K_\mathrm{a}S)\times N_\mathrm{t}$ matrix $\underline{\mathbf{M}}$ to be determined in what follows, the linear modulation process to be performed at the BS transmitter can be represented by a matrix–vector operation

$$\underline{\mathbf{s}} = \underline{\mathbf{M}}\,\underline{\mathbf{d}}. \tag{5.97}$$

The matrix $\underline{\mathbf{M}}$ is termed the modulator matrix and has to be known at the BS transmitter as a prerequisite for modulation. In order to perform data demodulation at the MS μ_k, an $N_k\times(S+W-1)$ matrix $\underline{\mathbf{D}}^{(k)}$ is multiplied by the partial received signal $\underline{\mathbf{e}}^{(k)}$, see (5.90), received at this MS. The matrix $\underline{\mathbf{D}}^{(k)}$ has to be a priori agreed upon by the BS and MS μ_k. Therefore, $\underline{\mathbf{D}}^{(k)}$ can

$$\underline{\mathbf{D}} = \begin{pmatrix} \underline{D}_{1,1}^{(1)} & \underline{D}_{N_1,1}^{(1)} & 0 & 0 & 0 & 0 \\ \underline{D}_{1,2}^{(1)} & \underline{D}_{N_1,2}^{(1)} & 0 & 0 & 0 & 0 \\ \underline{D}_{1,3}^{(1)} & \underline{D}_{N_1,3}^{(1)} & 0 & 0 & 0 & 0 \\ \underline{D}_{1,S+W-1}^{(1)} & \underline{D}_{N_1,S+W-1}^{(1)} & 0 & 0 & 0 & 0 \\ 0 & 0 & \underline{D}_{N_2,1}^{(2)} & 0 & 0 & 0 \\ 0 & 0 & \underline{D}_{N_2,2}^{(2)} & 0 & 0 & 0 \\ 0 & 0 & \underline{D}_{N_2,3}^{(2)} & 0 & 0 & 0 \\ 0 & 0 & \underline{D}_{N_2,S+W-1}^{(2)} & 0 & 0 & 0 \\ 0 & 0 & 0 & \underline{D}_{1,2}^{(K)} & \underline{D}_{2,2}^{(K)} & \underline{D}_{N_3,2}^{(K)} \\ 0 & 0 & 0 & \underline{D}_{1,3}^{(K)} & \underline{D}_{2,3}^{(K)} & \underline{D}_{N_3,3}^{(K)} \\ 0 & 0 & 0 & \underline{D}_{1,S+W-1}^{(K)} & \underline{D}_{2,S+W-1}^{(K)} & \underline{D}_{N_3,S+W-1}^{(K)} \end{pmatrix}^{\mathrm{T}}$$

Fig. 5.16. Structure of the demodulator matrix $\underline{\mathbf{D}}$ for the example given by (5.101)

be assumed to be known both at the BS transmitter and the MS receivers. However, there exists a great number of ways to choose this matrix. This implies a high degree of freedom for the system designer, and finding optimum matrices $\underline{\mathbf{D}}^{(k)}$ is a challenge for further research. One of the ways to chose the matrix $\underline{\mathbf{D}}^{(k)}$ is considered in Sect. 5.7.4. The total transmit signal $\underline{\mathbf{s}}$ of (5.86) should be chosen such that, in the noise-free case, the matrix–vector multiplication

$$\underline{\mathbf{d}}^{(k)} = \underline{\mathbf{D}}^{(k)}\underline{\mathbf{e}}^{(k)},\ k=1,\ldots,K, \tag{5.98}$$

yields the desired partial data vector $\underline{\mathbf{d}}^{(k)}$ free from ISI and intracell MAI. Substituting (5.90) in (5.98) yields

$$\underline{\mathbf{d}}^{(k)} = \underline{\mathbf{D}}^{(k)}\underline{\widetilde{\mathbf{H}}}^{(k)}\underline{\mathbf{s}},\ k=1,\ldots,K. \tag{5.99}$$

The K matrices $\underline{\mathbf{D}}^{(k)}$, $k=1,\ldots,K$ can be combined to form the total demodulator matrix

$$\underline{\mathbf{D}} = \mathrm{blockdiag}\left[\underline{\mathbf{D}}^{(1)}\ldots\underline{\mathbf{D}}^{(K)}\right] \tag{5.100}$$

of the dimension $N_{\mathrm{t}}\times[K(S+W-1)]$. As an example, the special case

$$S+W-1=4,\quad K=3,\quad N_1=2,\quad N_2=1,\quad N_3=3 \tag{5.101}$$

is briefly considered. In this case the total demodulator matrix $\underline{\mathbf{D}}$ takes the form illustrated in Fig. 5.16.

With the matrix $\underline{\mathbf{D}}$ of (5.100) and the matrix $\underline{\widetilde{\mathbf{H}}}$ of (5.92), the matrix

$$\underline{\mathbf{B}} = \underline{\mathbf{D}}\,\underline{\widetilde{\mathbf{H}}} \tag{5.102}$$

can be established, which is termed the system matrix. Because both $\underline{\mathbf{D}}$ and $\underline{\widetilde{\mathbf{H}}}$ are known at the BS transmitter, this is also true for $\underline{\mathbf{B}}$ from (5.102). With

$\underline{\mathbf{B}}$ the K matrix–vector equations given by (5.99) can be combined to give the single matrix–vector equation

$$\underline{\mathbf{d}} = \underline{\mathbf{B}}\ \underline{\mathbf{s}}. \tag{5.103}$$

In (5.103), $\underline{\mathbf{B}}$ and $\underline{\mathbf{d}}$ are known at the BS transmitter. Therefore (5.103) can be considered as a system of equations that can be solved to determine the modulator matrix $\underline{\mathbf{M}}$ appearing in (5.97). In addition, based on $\underline{\mathbf{d}}$ of (5.95), the component values of the total transmit signal $\underline{\mathbf{s}}$ from (5.86) can be determined.

In (5.103) the total number of equations is N_{t}, and the total number of unknowns is $K_{\mathrm{a}}S$. When determining the modulator matrix $\underline{\mathbf{M}}$ and the component values of $\underline{\mathbf{s}}$, the three cases

$$N_{\mathrm{t}} \gtreqless K_{\mathrm{a}}S, \tag{5.104}$$

have to be distinguished. If "<" holds in (5.104), then the system of equations described by (5.103) is underdetermined and has infinitely many solutions $\underline{\mathbf{s}}$. One possible solution in this case would be

$$\underline{\mathbf{s}} = \underbrace{\underline{\mathbf{B}}^{*\mathrm{T}} \left(\underline{\mathbf{B}}\ \underline{\mathbf{B}}^{*\mathrm{T}}\right)^{-1}}_{\underline{\mathbf{M}}} \underline{\mathbf{d}}. \tag{5.105}$$

This is the solution of minimum energy $\|\underline{\mathbf{s}}\|^2/2$ [34], which is optimum in the sense that the transmit power and, consequently, the MAI caused by each BS in the cellular system are minimized. If "=" holds in (5.104), then the system of equations (5.103) is determined and has the unique solution

$$\underline{\mathbf{s}} = \underbrace{\underline{\mathbf{B}}^{-1}}_{\underline{\mathbf{M}}} \underline{\mathbf{d}}. \tag{5.106}$$

Finally, if ">" holds in (5.104), then the system of equations (5.103) is overdetermined and can be, in the Gaussian sense, approximately solved by

$$\underline{\mathbf{s}} = \underbrace{\left(\underline{\mathbf{B}}^{*\mathrm{T}}\underline{\mathbf{B}}\right)^{-1} \underline{\mathbf{B}}^{*\mathrm{T}}}_{\underline{\mathbf{M}}} \underline{\mathbf{d}}. \tag{5.107}$$

The problem posed by (5.97) concerning the determination of the modulator matrix $\underline{\mathbf{M}}$ has been solved by (5.104)–(5.107).

In Fig. 5.17 the conventional transmission scheme, which, for instance, is applied in conventional time-slotted CDMA, and the novel transmission scheme JT are compared with each other. As shown in Fig. 5.17, in both cases the transmission is characterized by the three processes: modulation, transmission over the mobile radio channel and demodulation, which are described by the matrices $\underline{\mathbf{M}}$, $\underline{\widetilde{\mathbf{H}}}$ and $\underline{\mathbf{D}}$, respectively. Prior to demodulation

a

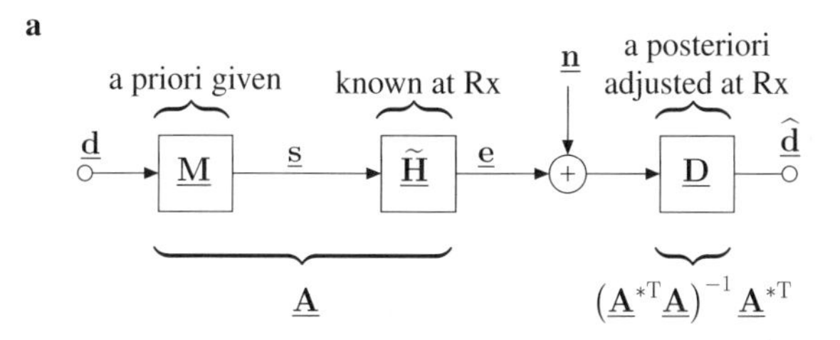

b

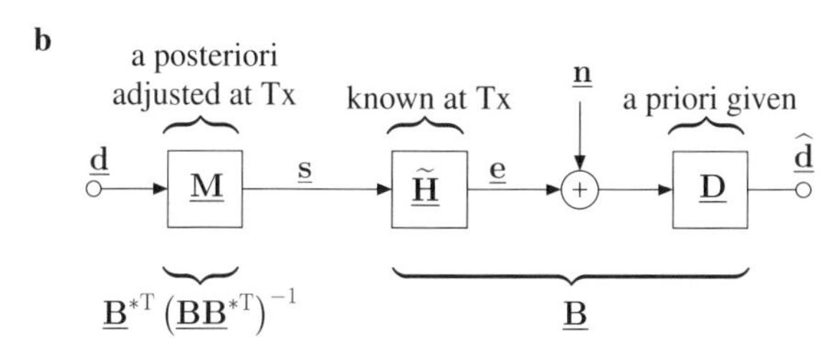

Fig. 5.17. Transmission schemes: **a** conventional JD, used in state-of-the-art time-slotted CDMA; **b** novel scheme Joint Transmission (JT)

a noise signal given by a vector $\underline{\mathbf{n}}$ of length $K(S+W-1)$ representing thermal noise plus intercell MAI is fed into the system. At the detector output an estimate $\hat{\underline{\mathbf{d}}}$ of the total data vector $\underline{\mathbf{d}}$ of (5.95) is obtained.

In the well-known case of conventional transmission considered in the previous sections (Fig. 5.17a) the modulator matrix $\underline{\mathbf{M}}$ is a priori given and determines, together with the channel matrix $\underline{\widetilde{\mathbf{H}}}$, the system matrix [6]

$$\underline{\mathbf{A}} = \underline{\widetilde{\mathbf{H}}}\,\underline{\mathbf{M}}. \tag{5.108}$$

The demodulator matrix is a posteriori derived from $\underline{\mathbf{A}}$ of (5.108) and takes, for instance, the form [6]

$$\underline{\mathbf{D}} = \left(\underline{\mathbf{A}}^{*\mathrm{T}}\underline{\mathbf{A}}\right)^{-1}\underline{\mathbf{A}}^{*\mathrm{T}}, \tag{5.109}$$

if the ZF–BLE is utilized and if the noise covariance matrix $\underline{\mathbf{R}}_{\mathrm{n}}$, see (5.23), is proportional to the unit matrix. If a multielement antenna is applied at the BS, that is, $K_{\mathrm{a}} > 1$, the antenna weights have to be included in the channel matrix, see e.g. [23]. In the considered DL situation, only a section $\underline{\mathbf{A}}^{(k)}$ of the matrix $\underline{\mathbf{A}}$ of (5.108) and only a section $\underline{\mathbf{e}}^{(k)}$ of the total received vector $\underline{\mathbf{e}}$ of (5.91) are available at each MS μ_k, $k=1,\ldots,K$. This has to be considered if the system is to be modeled in a realistic way. In the case of JT (Fig. 5.17) the demodulator matrix $\underline{\mathbf{D}}$ is a priori given and determines, together with the channel matrix $\underline{\widetilde{\mathbf{H}}}$, the system matrix $\underline{\mathbf{B}}$ of (5.102). From $\underline{\mathbf{B}}$ follows a posteriori the modulator matrix $\underline{\mathbf{M}}$.

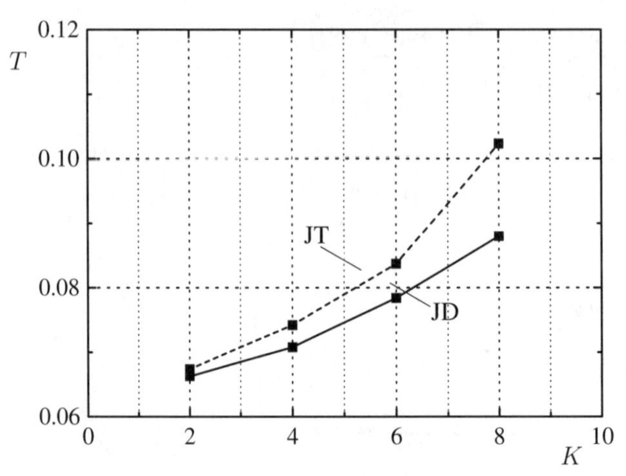

Fig. 5.18. Normalized average transmit energy T per symbol for JD and JT; $N=10$, $Q=16$; channel model COST 207 RA [28]

5.7.4 Data Transmission and Detection, Special Case Related to Time-Slotted CDMA

Now

$$N_k = N \;\forall\; k=1,\ldots,K \tag{5.110}$$

is assumed. In the DL of conventional time-slotted CDMA not utilizing CDMA code pooling, one of the K_s CDMA codes $\underline{\mathbf{c}}^{(k_\mathrm{s})}$ of (5.3) is uniquely assigned to each MS μ_k, $k=1,\ldots,K$. With (5.42) an obvious assignment would be

$$z(k_\mathrm{s})=k_\mathrm{s}, \text{ i.e. } k=k_\mathrm{s}. \tag{5.111}$$

Then, in the case of AWGN channels, the optimum linear demodulator at each MS μ_k would consist of a filter matched to the CDMA code $\underline{\mathbf{c}}^{(k)}$. As a main feature of the proposed JT scheme in connection with time-slotted CDMA, this simple and inexpensive demodulator will also be maintained if the radio channels are no longer AWGN channels, but show multipath behavior. Nevertheless, it will be required, like in the case of employing JD, that ISI and intracell MAI do not occur at the demodulator outputs. Like in conventional time-slotted CDMA, the dimension of the K_a signals $\underline{\mathbf{s}}^{(k_\mathrm{a})}$ of (5.85) is assumed to be

$$S = QN. \tag{5.112}$$

Consequently, the dimension of $\underline{\mathbf{s}}$ of (5.86) becomes

$$K_\mathrm{a}S = K_\mathrm{a}QN\,. \tag{5.113}$$

Now the matrix $\underline{\mathbf{D}}$ of (5.100), which appears in (5.102), takes the form

$$\underline{\mathbf{D}} = \text{blockdiag}\left[\underline{\mathbf{C}}^{(1)} \dots \underline{\mathbf{C}}^{(K)}\right]^{*\mathrm{T}} \tag{5.114}$$

with

$$\underline{\mathbf{C}}^{(k)} = \underline{\mathbf{D}}^{(k)*\mathrm{T}} \tag{5.115}$$

and

$$\begin{aligned}\underline{\mathbf{C}}^{(k)} &= \left(\underline{C}_{i,j}^{(k)}\right),\ i=1,\dots,S+W-1,\\ &\quad j=1,\dots,N,\ k=1,\dots,K,\\ \underline{C}_{i,j}^{(k)} &= \begin{cases}\underline{c}_{i-Q(j-1)}^{(k)}, & \text{for } 1 \leq i-Q(j-1) \leq Q,\\ 0, & \text{else.}\end{cases}\end{aligned} \tag{5.116}$$

According to (5.113) the number of unknowns in the system of equations (5.103) is equal to $K_{\mathrm{a}}QN$. On the other hand, the number of equations is equal to the number KN of elements of $\underline{\mathbf{d}}$ of (5.95). In what follows we assume

$$K_{\mathrm{a}}QN > KN, \tag{5.117}$$

which is also a reasonable requirement in conventional time-slotted CDMA employing JD in order to keep the SNR degradation sufficiently low, see (5.24) and (5.25). Equation (5.117) means that the system of equations given by (5.103) is underdetermined.

5.7.5 Performance Assessment of Joint Transmission in Time-Slotted CDMA

As shown in Sect. 5.7.4, the complexity of the time-slotted CDMA receivers in the MS is greatly reduced if JT is applied instead of JD. However, besides the receiver complexity, the system performance is also an important issue when comparing JT and JD in an adequate manner. One such comparison relies on the average transmit energy T required per symbol in each case for obtaining the same SNR at the demodulator outputs. In this comparison the following assumptions are made:

- For the elements of the total data vector $\underline{\mathbf{d}}$ from (5.95)

$$|\underline{d}_i| = 1,\ \forall\, i=1,\dots,N_{\mathrm{t}}. \tag{5.118}$$

- For the elements of the CDMA codes $\underline{\mathbf{c}}^{(k_{\mathrm{s}})}$ from (5.3)

$$|\underline{c}_q^{(k_{\mathrm{s}})}| = 1,\ \forall\, q=1,\dots,Q,\ \forall\, k_{\mathrm{s}}=1,\dots,K_{\mathrm{s}}. \tag{5.119}$$

- From TDMA burst to TDMA burst the total data vector $\underline{\mathbf{d}}$ of (5.95) fluctuates in such a way that the data symbols $\underline{d}_i$ are all statistically independent of each other.
- The channel impulse responses $\underline{\tilde{\mathbf{h}}}^{(k,k_{\mathrm{a}})}$ of (5.44) are subject to stationary fast fading, whereas slow fading is eliminated by transmit power control.

– Intercell MAI at the receiver inputs of the MS is modelled as uncorrelated complex Gaussian noise with the variance σ^2 of the real and imaginary part at all MS μ_k, $k=1,\ldots,K$.

The average transmit energies T for JT and JD cannot be determined in closed form, but have to be determined by simulations. To date, simulation results for the following case are available:

– Only one antenna element is utilized at the BS, that is, $K_\mathrm{a}=1$.
– The well-known COST 207 rural area channel model proposed in [28] is chosen.
– The dimension Q of the CDMA codes $\underline{\mathbf{c}}^{(k)}$ is 16, and $N=10$ symbols are transmitted per burst and MS.

In Fig. 5.18 the normalized average required transmit energy T is depicted versus the number K of MS. JT needs a slightly larger T than JD. However, the receiver complexity of JT is significantly smaller than that of JD.

5.8 Summary

Time-slotted CDMA is a promising multiple-access scheme for third-generation (3G) mobile radio systems and, under the acronyms TD–CDMA and TD–SCDMA, forms part of the 3G standards. CDMA concepts with a TDMA component are superior to CDMA concepts without a TDMA component. This is especially true when time-slotted CDMA is combined with the TDD duplexing scheme, which offers an inherent gain over systems utilizing the FDD duplexing scheme in terms of asymmetric capacity. While the basics of time-slotted CDMA have been studied since the early 1990s, a number of ideas to enhance the performance of this concept have emerged in recent years. These ideas concern joint temporal–spatial signal processing in the uplink receivers, efficient downlink schemes of BS transmit antenna beamforming and temporal processing at the MS, including the new concept of partial joint detection (PJD), advanced joint detection algorithms utilizing the turbo principle, intercell interference reduction and transmit signal optimization by a scheme termed joint transmission (JT). The background and functional principles of these schemes are explained in this chapter, and the achievable benefits are illustrated. Hopefully, the presented ideas increase the understanding of time-slotted CDMA and offer encouragement for further research in this field.

References

1. P. W. Baier (1996) A critical review of CDMA. In: Proc. IEEE 46th Vehicular Technology Conference (VTC'96), Atlanta, pp. 6–10
2. R. Prasad, W. Mohr, W. Konhäuser (2000) Third-generation mobile communication systems. Artech House, Boston
3. A. Klein and P. W. Baier (1992) Simultaneous cancellation of cross interference and ISI in CDMA mobile radio communications. In: Proc. IEEE 3rd International Symposium on Personal, Indoor and Mobile Radio Communications (PIMRC'92), Boston, pp. 118–122
4. R. Padovani (1994) Reverse link performance of IS-95 based cellular systems. IEEE Personal Communications, 1:28–34
5. S. Verdú (1998) Multiuser detection. Cambridge University Press, Cambridge
6. A. Klein (1996) Multi-user detection of CDMA signals – algorithms and their application to cellular mobile radio. Fortschrittberichte VDI, Reihe 10, Nummer 423, VDI-Verlag, Düsseldorf
7. H. V. Poor (2000) Turbo multiuser detection: An overview. In: Proc. IEEE 6th International Symposium on Spread Spectrum Techniques & Applications (ISSSTA'00), Parsippany, pp. 583–587
8. J. Hagenauer (1996) Forward error correcting for CDMA systems. In: Proc. IEEE 4th International Symposium on Spread Spectrum Techniques & Applications (ISSSTA'96), Mainz, pp. 566–569
9. T. Weber, J. Oster (2001) Hard– and soft–descision multi–step joint detection for TD–CDMA. In: Proc. URSI International Symposium on Signals, Systems & Electronics (ISSSE'01), Tokyo, pp. 347–350
10. J. J. Blanz, A. Klein, M. M. Naßhan, A. Steil (1994) Performance of a cellular hybrid C/TDMA mobile radio system applying joint detection and coherent receiver antenna diversity. IEEE Journal on Selected Areas in Communications, 12:568–579
11. J. Mayer, J. Schlee, T. Weber (1996) Realtime feasibility of joint detection CDMA. In: Proc. 2nd European Personal Mobile Communications Conference (EPMCC'97), Bonn, pp. 245–252
12. P. W. Baier (1994) CDMA or TDMA? CDMA for GSM? In: Proc. IEEE 5th International Symposium on Personal, Indoor and Mobile Radio Communications (PIMRC'94), Den Haag, pp. 1280–1284
13. A. Papathanassiou (2000) Adaptive antennas for mobile radio systems using time division CDMA and joint detection. Forschungsberichte Mobilkommunikation, Band 6, Universität Kaiserslautern, Kaiserslautern
14. Third Generation Partnerschip Project (2000) 3GPP specifications home page, http://www.3gpp.org/specs/specs.htm
15. A. Klein, G. K. Kaleh, P. W. Baier (1996) Zero forcing and minimum mean-square-error equalization for multiuser detection in code-division multiple-access channels. IEEE Transactions on Vehicular Technology, 45:276–287
16. J. J. Blanz (1998) Empfangsantennendiversität in CDMA-Mobilfunksystemen mit gemeinsamer Detektion der Teilnehmersignale. Fortschrittberichte VDI, Reihe 10, Nummer 535, VDI-Verlag, Düsseldorf
17. J. G. Proakis (2001) Digital communications. McGraw-Hill, New York
18. B. Friedrichs (1996) Kanalcodierung: Grundlagen und Anwendungen in modernen Kommunikationssystemen. Springer, Berlin
19. R. Lupas, S. Verdú (1989) Linear multiuser detectors for synchronous code-division multiple-access channels. IEEE Transactions on Information Theory, 35:123–136

20. M. Weckerle (2002) Adaptive array processing exploiting covariance matrices for the TD–CDMA uplink. Forschungsberichte Mobilkommunikation, Universität Kaiserslautern, Kaiserslautern
21. B. Steiner, P. Jung (1994) Optimum and suboptimum channel estimation for the uplink of CDMA mobile radio systems with joint detection. European Transactions on Telecommunications and Related Technologies (ETT), 5:39–50
22. B. Steiner (1995) Ein Beitrag zur Mobilfunkkanalschätzung unter besonderer Berücksichtigung synchroner CDMA–Mobilfunksysteme mit Joint Detection. Fortschrittberichte VDI, Reihe 10, Nummer 337, VDI-Verlag, Düsseldorf
23. Y. Lu, P. W. Baier (2000) Performance of adaptive antennas for the TD–CDMA downlink under special consideration of multi–directional channels and CDMA code pooling. International Journal of Electronics and Communications (AEÜ), 54:249–258
24. M. Weckerle, A. Papathanassiou (1999) Performance analysis of multi-antenna TD–CDMA receivers with estimation and consideration of the interference covariance matrix. International Journal on Wireless Information Networks, 6:157–170
25. M. Weckerle (2001) Turbo methods for exploiting spatial and temporal covariance matrices in the TD-CDMA uplink with multi-element antennas. In: Proc. 4th European Personal Mobile Communications Conference (EPMCC'01), Vienna, session 29
26. International Telecommunications Union (1997) Requirements for the radio interface(s) for international mobile telecommunications–2000 (IMT–2000). Recommendation ITU-R M.1034-1
27. T. Weber, J. Oster, M. Weckerle, P. W. Baier (2002) Turbo multiuser detection for TD–CDMA. International Journal of Electronics and Communications (AEÜ), 56:120–130
28. COST - European Cooperation in the Field of Scientific and Technical Research (1989) COST 207: Digital land mobile radio communications. Final report, Office for Official Publications of the European Communities, Luxemburg
29. P. W. Baier, C. A. Jötten, T. Weber (2001) Review of TD–CDMA. In: Proc. 3rd International Workshop on Commercial Radio Sensors and Communication Techniques, Linz, pp. 11–20
30. A. Steil (1997) Statistics of the carrier–to–interference ratio in C/TDMA cellular mobile radio systems applying multi–user detection. Wireless Personal Communications, 5:259–277
31. A. Steil (1996) Spektrale Effizienz digitaler zellularer CDMA–Mobilfunksysteme mit gemeinsamer Detektion. Fortschrittberichte VDI, Reihe 10, Nummer 437, VDI-Verlag, Düsseldorf
32. J. Oster (2001) Ein Beitrag zur Interzellinterferenzreduktion in zeitgeschlitzten CDMA–Systemen. Forschungsberichte Mobilkommunikation, Universität Kaiserslautern, Kaiserslautern
33. H. Matsutani, Y. Sanada, M. Nakagawa (1997) A forward link intracell orthogonalization technique using multicarrier pre-decorrelation for CDMA wireless local communication system. In: Proc. IEEE 8th International Symposium on Personal, Indoor and Mobile Radio Communications (PIMRC'97), Helsinki, pp. 125–129
34. A. Ben-Israel, T. N. E. Greville (1974) Generalized inverses, theory and applications. Wiley, New York

6 Future Applications and Fundamental Problems of Microwave Photonics

Abstract Microwave photonics integrates both the merits of radio wave propagation and of optical-fiber cable transmission. In other words, it is an integration of the optical-fiber trunk/backbone network and the versatile radio-zone network. Microwave photonics is expected to have future broadband and multimedia communications applications. In this chapter the characteristic features, applications, and fundamental problems/technologies of microwave photonics are discussed. The material discussed here is gathered from recent international conferences and workshops.

6.1 Introduction

In recent years, there have been a wide variety of needs for communications. To meet such needs, a complex and flexible network topology is required. The technology of microwave photonics gives us a promising solution to this issue. Microwave photonics integrates both the merits of microwaves as a medium propagating in the atmosphere and of lightwaves as a medium transmitting in flexible optical-fiber cables. Microwaves give us a movable wireless link with very low cost, whereas fiber optics gives us a low-loss and broadband link free from electrical interference and fading.

Radio waves have been widely used for terrestrial communications and broadcasts. Recent wide availability of mobile/portable communications has made radio waves indispensable for modern communications and human life. Since the amount of information that can be conveyed is basically proportional to frequency bandwidth, the most essential direction of exploiting radio wave technology is extending the upper limit of available frequencies. Microwaves up to 15 GHz have relatively low loss in atmospheric propagation and have been used for various terrestrial communication and broadcasting purposes. Frequencies beyond 15 GHz, especially millimeter waves, suffer from relatively large atmospheric propagation losses. If, therefore, we can develop a method to transmit millimeter waves with low loss for long distances, we can extend the versatility of radio waves up to millimeter-waves.

Since the first success in realizing optical fiber communications in the early 1970s, ever higher-speed baseband transmission has been pursued through optical fibers. Let us call this baseband transmission by an optical fiber baseband-on-fiber transmission. To date baseband-on-fiber transmission of several tens gigabits per second or more has been realized. This means that an optical fiber can carry very high frequencies up to several tens of gigahertz and more, and that we can transmit radio waves up to several tens of gigahertz free from fading and interference by means of an optical fiber.

The frequency response of baseband transmission has low-pass characteristics and is flat from the baseband frequency down to DC. In the case of transmitting a radio wave modulated by baseband signals, however, the frequency response has bandpass characteristics and does not contain DC components. Let us call this transmission radio-on-fiber transmission. This radio wave is called a subcarrier because the radio wave itself modulates an optical frequency and is carried in an optical fiber by a light wave. The subcarrier is in the microwave or millimeter-wave band in many cases. In radio-on-fiber transmission, we can obtain a broader frequency bandwidth than in baseband transmission. This is because the frequency bandwidth of a microwave device is basically characterized by its fractional bandwidth, and it is easier to obtain a broader bandwidth for bandpass characteristics than for low-pass characteristics. The upper-limit frequency of signals is mainly determined not by the fiber optics but by the frequency responses of the microwave/lightwave conversion devices or microwave electronic devices. In the case of bandpass characteristics, the upper-limit frequency will extend by 100 GHz or more, which means that we can transmit ever higher-speed (or more broadband) signals than baseband transmission by introducing radio-on-fiber systems.

Until recently, microwaves and fiber optics developed separately. Although the modulation speed of an optical modulator has increased up to the order of several tens gigabits per second, it was intended solely for baseband-on-fiber transmission. Microwave photonics, especially radio-on-fiber systems, is intended to transmit microwaves (or millimeter waves) not necessarily containing low frequencies or DC. Microwave photonics will give us new communications applications that are different from the baseband transmission developed so far. Design consideration for obtaining broader bandpass characteristics, i.e. a higher center frequency and broader bandwidth, will impose different challenges than those in the case of baseband transmission.

This chapter first presents the characteristic features of microwave photonics and then discusses applications in communication systems. Fundamental problems/technologies concerning the devices in microwave photonics for realizing these applications will then be discussed. Discussion here is limited to communication applications, and therefore, radio-on-fiber technology will mainly be dealt with. Information introduced here has been gathered from recent international conferences and workshops dealing with these themes.

6.2 Attenuation of Transmission Media

Let us first briefly review the loss of transmission media. Figure 6.1 shows the transmission attenuation of metallic waveguides, radio-wave propagation, and optical fibers, as a function of frequency. There are two kinds of metallic

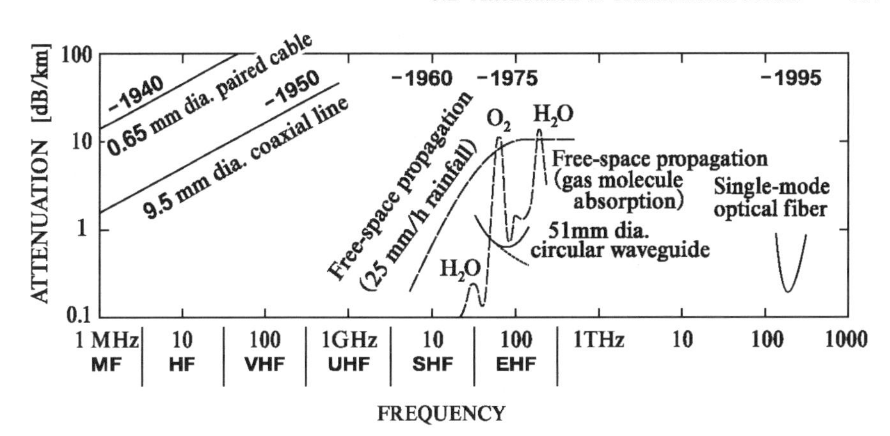

VLF: Very Low Frequency LF: Low Frequency MF: Medium Frequency HF: High Frequency
VLF: Very High Frequency UHF: Ultra High Frequency SHF: Super High Frequency
EHF: Extremely High Frequency

Fig. 6.1 Specific attenuation as a function of frequency. Paired cable, coaxial line, circular waveguide, optical fiber, and atmospheric propagation are shown

weveguides, i.e. two-wire waveguides and hollow waveguides. A two-wire waveguide, which is composed of two electrically insulated metallic lines, has no cutoff characteristics in the fundamental (the TEM) mode. That is the fundamental mode can propagate to frequencies as low as DC. The attenuation from the resistive loss of the two-wire TEM (Transversal Electromagnetic) mode increases by the square root of frequency.

Hollow waveguides, which are made of one conductor, cannot transmit DC and the fundamental mode is not the TEM mode. The fundamental mode of the hollow waveguide is not suitable for long-distance transmission. The transmission bandwidths of the fundamental modes of hollow waveguides are generally not so broad, and the dimension of the hollow inevitably determines the order of the wavelength of the fundamental mode. The loss of the fundamental mode of the hollow waveguide, therefore, is more than one order of magnitude larger than that of the two-wire TEM mode.

Millimeter-wave transmission [1] guided by a circular waveguide was widely investigated in the 1960s and 1970s when optical-fiber technology was still in its infancy. The TE_{01} mode (Transverse Electrical 01 mode, one of the higher-order modes of a circular waveguide) transmission through an oversized circular waveguide was quite a unique and sophisticated idea. The loss of the TE_{01} mode of circular waveguides theoretically decreases with frequency, but practically, mode-conversion loss due from mechanical imperfections gradually increases for higher frequencies. The transmission of the TE_{01} circular waveguide, therefore, shows a bandpass frequency response. The inside of the waveguide is filled with dry nitrogen to avoid oxygen absorption around 60 GHz. The TE_{01} mode of an oversized circular waveguide can provide a transmission medium for a frequency band of 30-100 GHz with very low attenuation, comparable to microwave (e.g. 4–6 GHz region) atmospheric propagation.

The attenuation of radio wave atmospheric transmission is sufficiently low roughly up to 15 GHz. Beyond 15 GHz, however, the attenuation caused by absorption by rain and fog gradually increases. Molecular gas absorption, e.g. by oxygen and water vapor/molecules, increases for frequencies higher than several tens of gigahertz. Therefore millimeter waves are not adequate for long-distance atmospheric transmission.

The loss of optical fibers also has a bandpass frequency response, whose lower and upper frequency limits are determined by mechanical dispersion and infrared absorption, respectively [2]. The minimum loss now practically available is 0.2 dB/km at the 1.5-μm wavelength band. When we consider a loss up to 1 dB/km for long-distance transmission, optical fiber has a bandwidth of 40,000 GHz.

6.3 Advantageous Characteristics of Radio Waves and Optical Fibers

The advantage of radio waves and optical fibers are well known. The merit of radio-wave propagation exists in its wireless connection, but the propagation characteristics are inevitably affected by atmospheric conditions and therefore may suffer from fading and mutual interference. The merit of optical fiber is in its low loss and broad bandwidth, but the connection of many moving users is basically impossible. Microwave photonics is intended to utilize the advantageous characteristics of both radio waves and optical fibers.

The advantageous characteristics of radio waves are summarized as follows:
(1) Capability to transmit in any direction;
(2) Easy construction and reconstruction of equipment;
(3) Capability to connect moving or portable terminals;
(4) Low cost of transmission (atmospheric propagation);
(5) High frequency utilization efficiency by use of cellular systems.

The advantageous characteristics of optical fibers are:
(1) Thin, light weight, and mechanical flexibility;
(2) Ultra-broad bandwidth and very low loss;
(3) Capability of multiplexing in wavelength (or frequency) and in space;
(4) Easy multiplexing and demultiplexing and flexible network architecture;
(5) Immunity to electrical interference.

6.4 Characteristic Features of Radio-on-Fiber Network Systems

The radio-on-fiber network system has three major characteristic features: transparent connection of radio waves and lightwaves, skillful utilization of the broad bandwidth of optical fibers for millimeter-wave transmission, and radio wave generation by means of optical techniques.

Conversion between radio waves and lightwaves should be transparent, that is, the spectrum of radio waves should be unchanged when they propagate in the air or in an optical fiber. The radio waves are coupled into an optical fiber and guided anywhere and for any distance without suffering from serious attenuation, fading or inter-radio wave interference, which always exists for open-space radio wave propagation. In this case, the optical fiber functions as a low-loss radio transmission medium that is free from transmission instabilities. Conversion between radio waves and light waves should be performed easily, as required. The whole network appears to be composed of a big radio wave network, accessed from radio wave terminals. The end radio terminals do not recognize optical fibers. By incorporating an optical-fiber network into the radio wave network, we will be able to design and synthesize a radio network with high frequency-reuse efficiency and a high degree of flexibility. Although fading and interference have been the biggest restricting factors for radio wave systems, by introducing optical-fiber radio-wave waveguides, we can realize a radio wave system with better stability and higher density.

A single-mode optical fiber has a sufficiently broad bandwidth to transmit even higher frequencies than millimeter waves, which means that an optical fiber can function as a low-loss long-distance transmission waveguide for millimeter waves and above. Circular waveguides, which were developed in the 1970s for millimeter-wave transmission, indeed were excellent waveguides from the perspective of their broad bandwidth and low losses. Nevertheless, they possessed a serious flaw that they had no mechanical flexibility. Their lack of mechanical flexibility was the biggest impediment to introducing the guided millimeter-wave system into wide-range practical applications. By utilizing optical fibers for long-distance transmission waveguides, we can construct trunk backbone networks as well as cellular communication/distribution networks using millimeter-waves and even higher frequencies.

Another noteworthy capability of optical technology is the generation of frequencies even higher than millimeter waves by means of optical devices. A distributed-feedback (DFB) laser is a single-frequency laser oscillator whose frequency spectrum is sufficiently pure and stable. We modulate an optical carrier fed from a DFB laser with an external modulator driven by a radio-frequency oscillator. Then we obtain on the optical carrier upper and lower sidebands, whose frequency difference is twice the radio modulation frequency. If we inject the modulated optical carrier to a two-mode (two-frequency) laser oscillator, e.g. a Fabry-Perot laser whose oscillation frequencies are nearly equal to the injected sidebands, the two modes are phase-locked by the two injected sidebands. As a result, we obtain two amplified coherent optical spectra whose frequencies are separated by twice the modulating radio frequency. Heterodyne detection of these two coherent optical spectra gives us a radio-frequency carrier whose frequency and phase are controlled by a radio-frequency oscillator with one-half frequency. Injection-locking is also useful for modulating a frequency at a higher-order subharmonic of the difference frequency of the two modes. Thus we obtain a radiofrequency carrier whose frequency is a multiple of the source frequency. The stability and purity of the frequency and phase are controlled by a subharmonic source oscillator. By this method we can send radio-frequency carrier through an optical fiber with a small loss, which we utilize for sending a radio-frequency local-oscillator power to a remote location.

There are two types of radio-on-fiber networks, i.e. a long-distance trunk/backbone communication network and a subscriber communication/distribution network, for example, a cellular radio and a radio LAN. In the first type, the radio-frequency (microwave or millimeter-wave) subcarrier modulated by a hierarchically multiplexed baseband signals is carried in an optical fiber. The radio-on-fiber of this type is intended to operate in parallel with existing terrestrial radio-relay networks. The only different feature, compared with the terrestrial radio-relay systems equipped so far, is that the radio-frequency signals are transmitted in the optical fibers instead of propagating in the air. This radio-on-fiber will be introduced as a part of the terrestrial radio-relay systems for avoiding fading/interferences or for making some detours. In the second type, end mobile users/terminals are accommodated in a radio zone (a cell) and are connected to a radio base station by radio waves. The network inside the base station will typically be built by radio-on-fibers. When the frequency of radio waves is high, e.g. millimeter waves or beyond, the radio zones will become very small, that is, microcells or picocells, and a great number of radio base stations will be required. In this case, realizing a radio base station with simple construction and low cost is of primary importance. The functions in a radio base station should be as simple and as few as possible, and the only function left will be radio-wave and light-wave (or electrical and optical) conversion.

6.5 Comparison of Baseband-on-Fiber and Radio-on-Fiber Systems

6.5.1 Frequency Response of Electrical Devices

In a baseband-on-fiber system, baseband signals are converted to light waves by an optical modulator and are carried in an optical fiber. The frequency response is the Fourier transform of the time waveform of the baseband signals. The frequency response, therefore, is of a low-pass nature and extends from the baseband signal frequency down to DC. In order to transmit the baseband signals without distortion, a flat low-pass frequency response is required of the electrical input circuit of the optical modulators and the electrical output circuit of the optical detectors.

In a radio-on-fiber system, on the other hand, baseband signals are first converted to radio frequencies and then are further converted to light waves. Since the baseband signals are converted to sidebands of a carrier (a subcarrier from the light wave) radio frequency, signals of a bandpass nature are carried in the optical fiber. In order to transmit the signals without distortion, a flat bandpass frequency response is required of the electrical input circuit of the optical modulators and the electrical output circuit of the optical detectors.

6.5.2 Electrical Circuit Requirement on Modulators/Demodulators

Modulation/demodulation means baseband and light wave conversion in baseband-on-fiber systems, and radio wave and light wave conversion in radio-on-fiber systems. Since the frequency of the light waves is more than three orders of magnitude higher than that of the radio waves or the baseband frequency, the frequency bandwidth of modulation/demodulation devices is determined mainly by the frequency bandwidth of the radio-wave or baseband circuits. A bandpass characteristic, which contains no DC component, gives us greater circuit design flexibility than a low-pass characteristic from three points of view, that is, transmission line, impedance matching, and frequency bandwidth, each of which is closely related to the others.

Let us first discuss the transmission lines that we use for electrical circuits in modulators/demodulators. The transmission line for baseband signals is limited to a two-wire transmission line (for example, a stripline, a coaxial line, and a coplanar line) and to its TEM mode due to a DC component contained in the baseband signal. For transmission of signals with a bandpass characteristic, on the other hand, we can freely select appropriate transmission lines and their modes because the transmission lines used are not required to support DC frequencies. We can adopt three-dimensional waveguides, such as rectangular waveguides, as well as planar waveguides. Furthermore, devices for waveguide conversion and mode conversion are also freely selectable for bandpass transmission.

Design flexibility in impedance-matching circuits is another important issue. Lossless impedance transformation for a broadband signal with low-pass characteristics is difficult to make. In particular lossless transformation using inductances and capacitances is impossible in DC circuits. A device for modulation and demodulation principally consists of a resistance, which represents its active function, and a parallel capacitance, which represents the junction and/or electrode-plate capacitance. Since the parallel capacitance works as a low-pass filter, no inductances, in parallel or in series, can be used for an impedance-matching circuit in transmission of signals containing DC components. Compensating for the parallel capacitance is not possible and therefore the frequency response of these circuits is solely determined by this parallel capacitance.

The effect of the parallel capacitance can be successfully eliminated by a parallel inductance for circuits with bandpass characteristics. We can enhance the center frequency of the pass band by inserting an inductance into the matching circuit to compensate for the capacitance. An impedance transformer further broadens the bandwidth. For bandpass characteristics an impedance transformer is easily designed by using a transmission-line transformer and a mode transformer.

A modulator/demodulator inevitably has a capacitance from its semiconductor junction and electrode metals. If we can ideally compensate for such reactance, regardless of the operating frequency, the absolute value of the bandwidth is not a function of the frequency. In practical cases, however, because the reactance is determined by its size and shape, the reactance becomes a function of the frequency in distributed-constant circuits. It is more difficult, therefore, to have a broader bandwidth for lower frequencies. For that reason, the frequency bandwidth of a semiconductor device is generally characterized by its fractional bandwidth, i.e. the bandwidth normalized by its center frequency, which means that a device working at a higher center frequency will yield a broader bandwidth.

For radio-on-fiber systems, the electrical circuits for modulation/detection should have a bandpass frequency response. If we roughly assume that the transmission capacity is proportional to the bandwidth and that the fractional bandwidth is constant for different frequencies, we can send a higher capacity of information with a higher carrier frequency. In practice, a bandwidth-limiting factor caused by the parasitic resistance becoming relatively larger for higher frequencies must be carefully minimized.

6.6 Prospective Applications

6.6.1 Subcarrier Multiplex System

In subcarrier multiplex (SCM) systems, multiplexed microwave frequency bands are carried by the light waves in a single optical fiber. Each microwave frequency band carries one or a group of multiplexed baseband signals, that is, baseband signals are piled up by multistage multiplexing and then are converted to microwaves. Since the microwave carriers are further multiplexed on optical frequencies, this system is called a microwave subcarrier multiplex system (SCM). The final carrier is the light wave in an optical fiber. The baseband signals are multiplexed and embedded into microwaves by the standard multiplex and modulation technique by digital gate circuits and analog filters. In some cases, optical carriers are further multiplexed by wavelength-division multiplexing onto a single optical fiber, in which multiple optical carriers with different wavelengths are put together and transmitted in a single optical fiber. Microwave-to-light wave conversion is performed with a bandpass-type optical modulator. At optical reception, optical signals are converted to microwave signals with an optical detector. Demultiplexing in the microwave and baseband stages is done by the standard techniques.

An SCM system for cellular digital mobile telephone systems has been proposed [3, 4]. This SCM is used for offering services in radio-blind areas where the radio waves do not directly reach mobile terminals from the base station, such as the inside of tunnels, underground shopping malls, and buildings, which are electrically shielded by conducting walls. Figure 6.2 shows the system configuration of an SCM system. The link consists of one base unit (BU) and multiple access units (AU). One BU and two sets of AU (three AU for one set in Fig. 6.2) are connected by an uplink and a downlink of a 1.3-μm single-mode optical fiber by a bus-type topology. The BU is connected to the base station by coaxial cables. Each AU has a Fabry-Perot laser diode and a photodiode. To reduce the optical

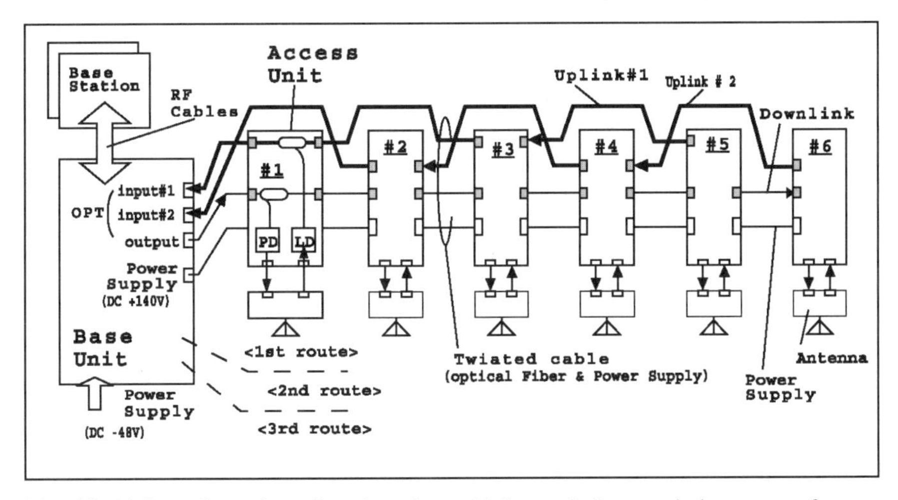

Fig. 6.2 Link configuration of a subcarrier multiplex optical transmission system for personal digital cellular telephones (Table 6.1). Conversion between microwaves and lightwaves is made by a laser diode and a PIN photodiode [3]

beat interference, spectrum broadening with low-frequency intensity-modulation is adopted. Table 6.1 shows the main system parameters. The subcarrier frequency ranges are 800 MHz and 1.5 GHz. The output power, carrier-to-noise ratio, and third-order intermodulation are so determined as to secure sufficient transmission quality within a specified 20-m-radius service area.

Table 6.1 A subcarrier multiplex optical transmission system for personal digital cellular telephones. This system was developed for providing service in radio-blind areas [3]

	Downlink	Uplink
Frequency Range [MHz]	810 – 818	940 – 948
	1487 – 1491	1439 – 1443
Number of carriers	24	24
RF input power level	– 3 dBm/carrier	–27 dBm/carrier
RF output power level	0 dBm/carrier	–27 dBm/carrier
Carrier to noise ratio (CNR) [dB/21kHz]	> 45	> 55
Composite triple-beat (CTB) [– dBc]	> 45	–
3rd order IM (IM3) [– dBc]	–	> 55

In the SCM system optical beat interference (OBI) is the major source of degradation of the transmission quality of the uplink. The OBI is caused by simultaneous detection of multiple optical carriers whose wavelengths are different but very closely located. To reduce the OBI, increasing the optical modulation index (OMI) is discussed for transmission of PSK (phase-shift keying)-modulated microwave signals [5]. Larger OMI spreads the optical spectrum and reduces the OBI peak power, and as a result reduces the OBI. This method is effective for modulation formats without amplitude variation, but is not suitable for modulation formats that include amplitude variation such as quadrature amplitude modulation (QAM). This is because nonlinear distortion is generated in the amplitude of microwave signals as the OMI assumes larger values, especially near 100%.

A method in which multiple optical carriers are multiplexed in the time domain rather than in the frequency domain has been proposed [6], which uses an optical switch with multiple input ports and one output port. Figure 6.3 shows the proposed system configuration. Multiple channels of microwave signals are converted to the respective frequencies of light waves in remote access units (RAU) and transmitted through different optical fibers to a multiplexer. The incoming optical carriers with different frequencies are multiplexed and are sequentially allocated by an optical sampling multiplexer in the transmitter. In this system, no simultaneous detection of different optical frequencies is done, resulting in the elimination of OBI. The optical switch (multiplexer) is operated with a multiplexing clock frequency in the transmitter. In the receiver, detection is performed in the frequency domain, thus no precise synchronization between the transmitter and receiver is required.

For future systems where base stations are more densely installed, systems with very small cells (called microcells or picocells), millimeter-wave transmission through optical fibers, and optical fiber links are of interest. In microcellular or picocellular systems, there will be many base stations connected to a central station, and therefore realizing simple, small, and economical base stations is of primary importance. As many functions as possible should be centralized to the central station. The base stations should be as simply constructed as possible. We discuss

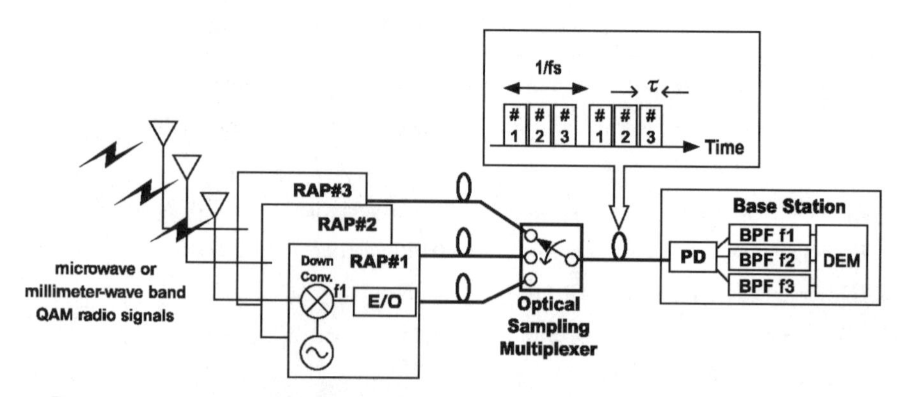

Fig. 6.3 A subcarrier multiplex system using time-sampling multiplexing in order to suppress optical beat interference [6] RAP: remote access point

this issue more specifically in Sect. 6.7.4. In these systems, millimeter-wave subcarriers are preferred, because higher frequency subcarriers have greater transmission capacity and because lossy characteristics of millimeter-waves in atmospheric propagation are suited to picocells.

6.6.2 Intelligent Transport System

In recent years the role of automobiles has increased significantly. The role of automobiles includes not only the transportation of people and goods, but also the provision of moving living-spaces for everyday human life. In this sense communications between automobiles and the outside world has become substantially important. The communication signals range from necessary traffic (or traffic control) information to leisure entertainment materials. The intelligent transport system (ITS) offers integrated multimedia information for automobile traffic and transport systems. Because of the broadband multimedia characteristics of the information and the network complexity/flexibility, the distribution network should be constructed from optical fibers and radio waves (microwaves or millimeter waves).

The network is composed of a fixed optical-fiber backbone network and a cellular radio network. The cellular radio zones are provided along the routes of mobile terminals (automobiles), and each cellular zone has one base station. The role of the base station is the conversion of optical and radio signals and the delivery/collection of information to and from mobile terminals. Since there are a number of base stations to be constructed, the function of the base station should be minimized and they should be constructed as simply as possible. The network topology of the center and base stations is basically a star configuration.

There are two kinds of communications, i.e. center-to-mobile and mobile-to-mobile communications. The former communication is between the central station and the mobile terminals and includes traffic control information from the central station to mobile terminals, and the information from mobile terminals to the central station, which is necessary for the knowledge of the present traffic conditions at the central station. There are two modes of communications, i.e. one-way and two-way (or interactive) communications. Most of the traffic control signals and entertainment materials need not be interactive.

There are two modes of communications between two mobile terminals, i.e. communications between automobile drivers and those between automobiles themselves. The former is basically ordinary communications between two persons. The latter is necessary for securing more efficient and safer transportation, for example, automated driving, convoy transportation, and collision surveillance. The communication between automobiles is not always necessarily recognized by drivers. Since the ITS has the capability to completely change the infrastructure of not only traffic/transport systems but also the overall lifestyle of human society, it is investigated as a large nationwide project, and systems are now being proposed.

A conceptual ITS system is shown in Fig. 6.4 [7]. In this system, the ITS network will be overlaid onto the existing communication network and will interface

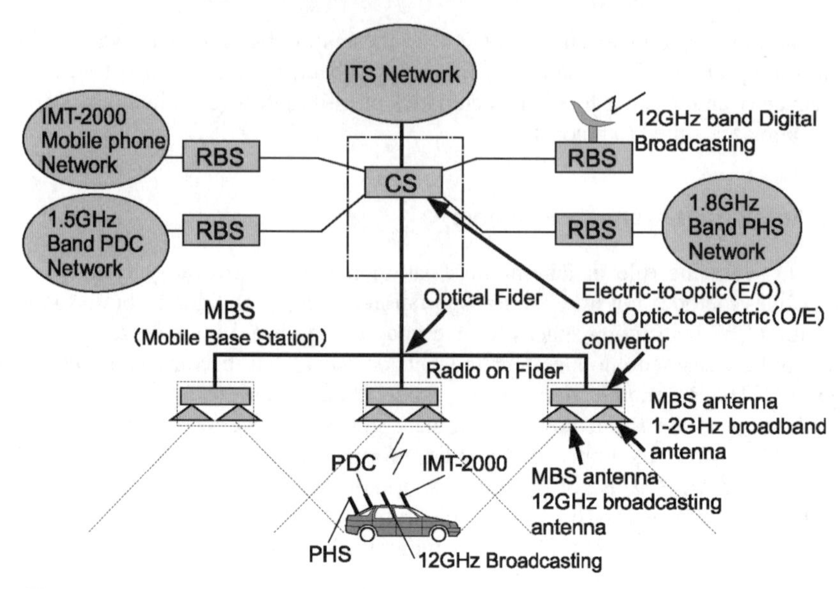

Fig. 6.4 Intelligent transport system network architecture by a radio-on-fiber system planned by Communication Research Laboratories. Mobile telephone signals, broadcasting, and other control signals are transmitted [7]

with the existing network through a gateway. All information from automobiles is collected by mobile base stations (MBS), converted to optical signals, and then transmitted to a control station (CS). The existing wireless networks include the 1.5-GHz-band personal digital communication (PDC), IMT-2000 mobile communication, 1.5-GHz-band personal handy-phone system (PHS), satellite broadcasting, and so on, which are connected to the ITS network via remote base stations (RBS). The main functions of a CS are electrical (radio wave)-to-optical (E/O) conversion and optical-to-electrical (O/E) conversion. Connection between CS and RBS will be either by radio, metallic cables, or optical fibers.

There will be many radio-frequency bands employed in the system. Simplification of the MBS configuration is an important problem. In particular, accommodating many antennas for different frequency bands is not practical from the points of view of economy and equipment size. Furthermore, an automobile must have many antennas, each of which corresponds to its own frequency and system. In order to simplify the antenna system of automobiles and MBS, adopting a common frequency band covered by a single antenna will be acceptable [7]. In this case, a CS is required to also provide frequency conversion to the common frequency band in addition to E/O and O/E conversions.

6.6.3 Radio LAN

The combination of millimeter waves and fiber optics is applied for small-scale radio local-area networks (LAN). For LAN in limited areas, i.e. the inside of

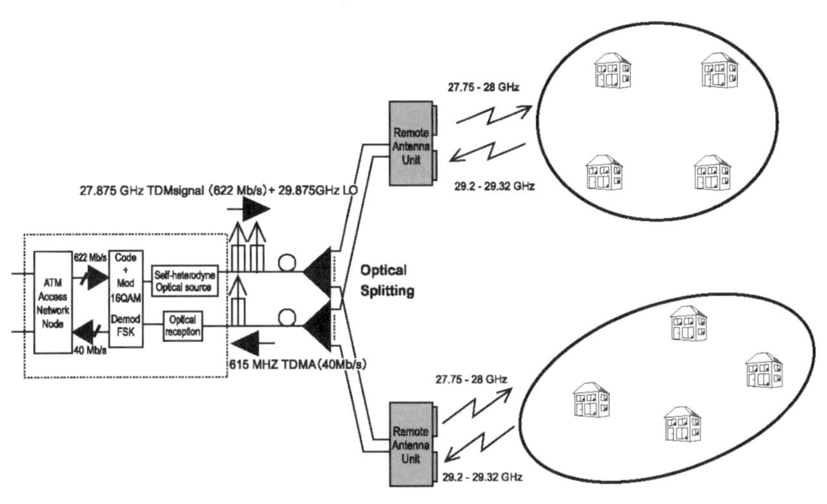

Fig. 6.5 Network for fiber radio ATM network and services (ACTS FRANS). Multiple signals, i.e. video on demand, high-speed Internet, visiophony, etc. are carried by an optical fiber and delivered to customers by millimeter waves. An optically generated millimeter wave source is carried to the remote antenna unit [8]

premises and buildings, high-attenuation characteristics of quasi-millimeter waves or millimeter waves are utilized. Figure 6.5 shows the link configuration of Fibre radio ATM network and services (ACTS FRANS) project [8]. The aim of the ACTS FRANS project is to demonstrate the feasibility of radio-over-fiber services of delivery for video on demand, high-speed Internet, visiophony, and so on. The test configuration is composed of an optical transmitter/receiver, optical-fiber cables, and remote antenna units (RAU). The downlink frequencies from RAU to customer premises equipment (CPE) are 27.75–28 GHz and the uplink frequencies (from CPE to RAU) are 29.2–29.32 GHz.

Two frequencies, one modulated signals (27.75–28 GHz) and the other the local-oscillator frequency (29.875 GHz), are created in the optical transmitter in the central station (CS) on an optical carrier frequency, and they are sent to the RAU through an optical fiber. How two frequencies are generated on an optical carrier frequency is shown in Fig. 6.6. The output optical power from a 1.55-μm distributed feedback (DFB) laser is fed to a 3-dB optical coupler. In one path, the lightwave is modulated by a $LiNbO_3$ Mach-Zehnder modulator (MZM1). Two optical spectra are generated by a suppressed-carrier double-sideband modulation scheme. The frequency difference of the two spectra is 59.75 GHz, which is twice the driving frequency of 29.875 GHz. One of these frequencies is selected by an optical filter following the modulator. In the other path, the optical carrier is modulated at a microwave subcarrier frequency (2 GHz) also by double-sideband modulation (MZM2). The optical powers of two paths are combined and amplified, and are then sent through the optical fiber. Thus we obtain two microwave frequencies on the optical frequency domain.

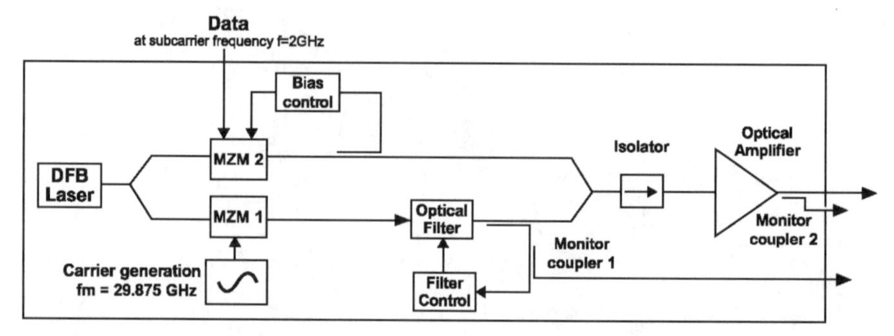

Fig. 6.6 An optically generated of millimeter-wave source. The millimeter-wave source is remotely generated by two modulators in the central station and heterodyne-detected in a remote antenna unit [8]

The RAU demodulates coherent optical sources through a high-speed high-responsivity photodiode. The resulting mixture signals in the range 27.75–28 GHz are transmitted to CPE. Unwanted sidebands produced by mixing are suppressed. The local-oscillator power of 29.875 GHz also acts as the local-oscillator power for the uplink signals in the RAU. Downconverted signals (615-MHz IF) from 29.2–29.32-GHz uplink signals are transmitted to the CS.

The transmission over fiber of wideband-code division multiple access (W-CDMA) signals with a carrier frequency of 2 GHz for indoor communications application is also discussed [9].

6.6.4 Broadband Radio Highway

Let us consider a network that consists of radio wave networks placed around end terminals and an optical-fiber network placed inside the radio wave networks. The radio wave networks and the optical-fiber network are connected by transparent radio wave/light-wave converters. Conversion between radio waves and light waves is freely done as required. Since the optical-fiber network is transparent to the radio wave networks, the network viewed from the end terminals appears to be a big radio wave network, in which the optical-fiber network looks like a virtual radio wave network, and which is completely free from fading and interference. As a result, we can build a comprehensive radio wave network with a very broad bandwidth, high efficiency in frequency reuse, and a high degree of freedom in network design.

A conceptual network configuration [10, 11] of a radio highway is shown in Fig. 6.7, which consists of control stations (CS), radio base stations (RBS), and connecting optical fibers. The radio highway incorporates different kinds of wireless systems and offers universal capability and flexibility with as many air interfaces as possible. A key point of this network is that a radio base station is equipped with only E/O and O/E converters, and other complex functions, i.e. modulation/demodulation, multiplexing/demultiplexing, and so on, are left to a CS.

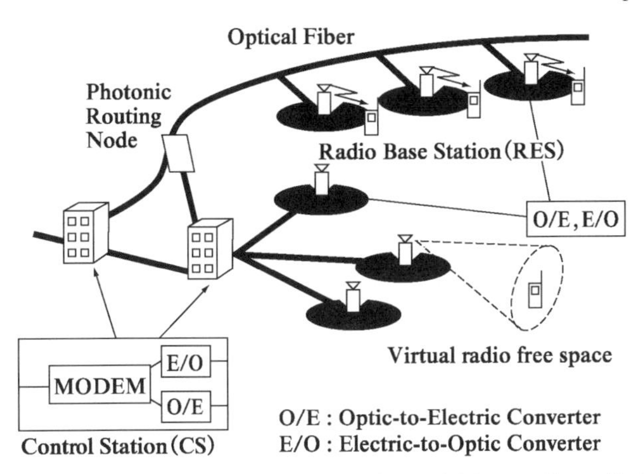

Fig. 6.7 Conceptual configuration of radio highway. The radio highway provides universal and flexible applicability of all different air interfaces. Each radio base station has only O/E and E/O converters and is made as simply as possible. The main important functions are concentrated at the control station [10]

Various types of future networks with a combination of optical fibers and micro- or picocells are discussed in [12–14]. Figure 6.8 shows a broadband wireless access system architecture [13]. The radio wave service area consists of microcells with diameters less than 500 m, in which millimeter waves are used for transmitting broadband signals. A centralized network operation center (NOC) integrates the complex functions of network operation, e.g. radio/network switching, routing, and service-mixing functions. The NOC is connected to the external networks and the backbone network.

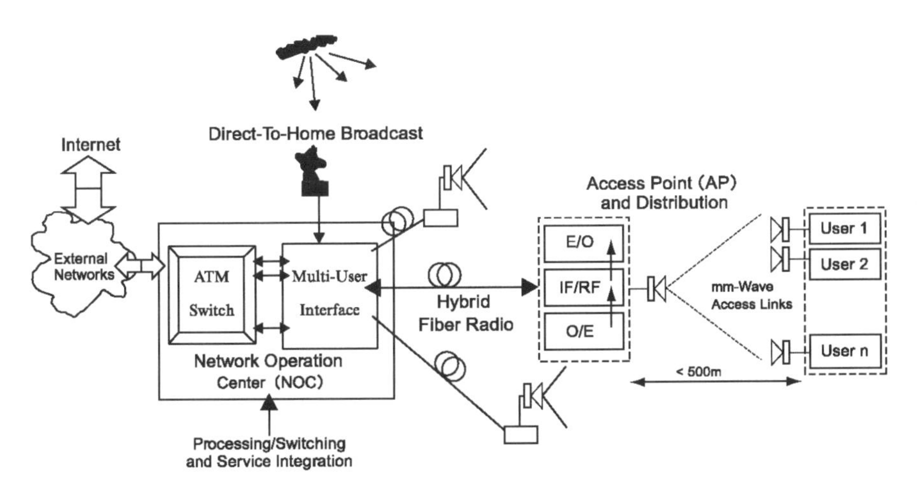

Fig. 6.8 A radio-on-fiber system for the next-generation Internet network. A short-range (less than 500 m) access with millimeter waves satisfies various broadband needs of inner urban users [13]

6.7 Fundamental Device Problems and Technologies in Microwave Photonics

In this section, we discuss important problems and technologies in microwave photonics. Here we take up transparency in microwave/light wave and light wave/microwave conversion, simplification of base-station functions, and optical generation of microwaves. Transparency is discussed from the points of view of linearity, efficiency/bandwidth, and impedance matching.

6.7.1 Linearity and Dynamic Range

In radio wave and light wave conversion, a linear relationship of input and output is of primary importance. Since the information is carried in the amplitude, frequency, or phase of the carrier wave, linearity in these three quantities is required. Higher-order distortion and mutual modulation in modulators and detectors (or demodulators) should be as low as possible. In radio-on-fiber systems, IM (intensity modulation) of light waves is commonly used. In order to obtain a broader dynamic range of linearity, AM/PM (amplitude/phase) and AM/FM (amplitude/frequency) conversion should be minimized [15–18].

In light-wave and microwave converters, saturation is caused by the excess space charge excited by light illumination. To decrease the excess space charge, both quickly sweeping light-excited charges out of the junction and diluting the density of light-excited charges are effective. Compensating for nonlinear distortion by use of two elements is another effective method. The dynamic range of linearity relates to the efficiency and bandwidth, which will be discussed in the next section.

Figure 6.9 shows a photodetector composed of distributed balanced photodiodes [15]. In this detector, a segment (the lower photograph in the figure) of the detector is made of multiple metal-semiconductor-metal diodes located in an array, and segments are further distributed in an array along the optical waveguide. The diode is operated in balanced mode. The diode array creates periodic capacitance loading to slow the microwave velocity down to the light-wave velocity in the guide. This technique is called velocity matching. Excess space-charge density is suppressed by distributing the light illumination, and the coupling between light waves and microwaves is enhanced by velocity matching. Because of this distributed loading, good linearity between the detector current and the input optical power, and between the microwave response and the input optical power have been obtained.

The linearization method of modulation by distortion compensation [16] is shown in Fig. 6.10. This modulator is composed of two 3-dB optical couplers and two identical multiquantum-well (MQW) electroabsorption modulators (EAM). An incident optical signal is divided by a 3-dB coupler into two paths, and the signals on two paths are separately modulated and then again combined by another 3-dB coupler. The two modulators are identical except that the bias voltage and

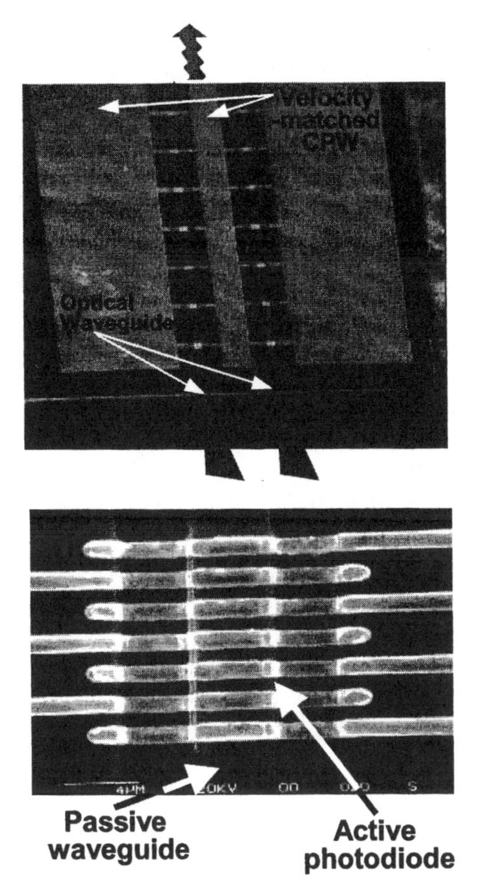

Fig. 6.9 Metal-semiconductor-metal (MSM) photodiodes are distributed along the optical waveguide. By balancing these diodes, very broad linearity has been obtained [15]

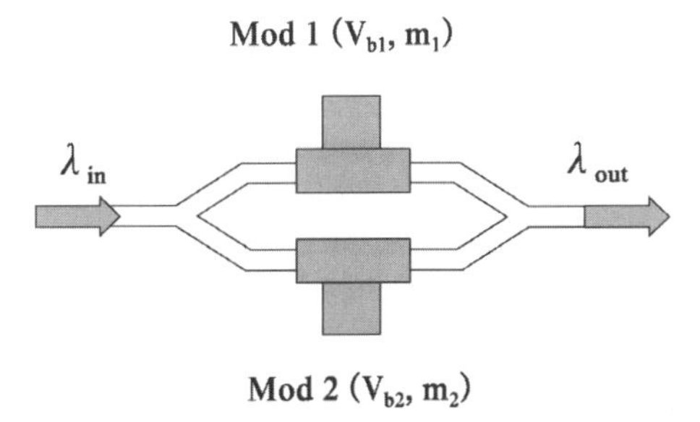

Fig. 6.10 A high-speed linear analog MQW electroabsorption modulator. Two modulators are biased so that the distortion components are of opposite sign and compensate for each other; V_b and m express the bias point of the modulator and amplitude of the signal, respectively [16]

modulation amplitude are applied in such a way that distorted signal terms caused by the nonlinear light-to-voltage characteristic are of opposite sign and compensate for each other.

The transfer function (the output as a function of input) of an EAM is expressed as a function of bias voltage, absorption coefficient, optical confinement factor, and device length. Band-gap wavelength and device length are other controllable parameters for distortion compensation. Figure 6.11 shows an EAM with two modulators with different band-gap wavelengths and modulator lengths, coupled by an optical coupler [17]. The two parameters along with bias voltage are adjusted so that the total value of nonlinearity of the transfer function is minimized.

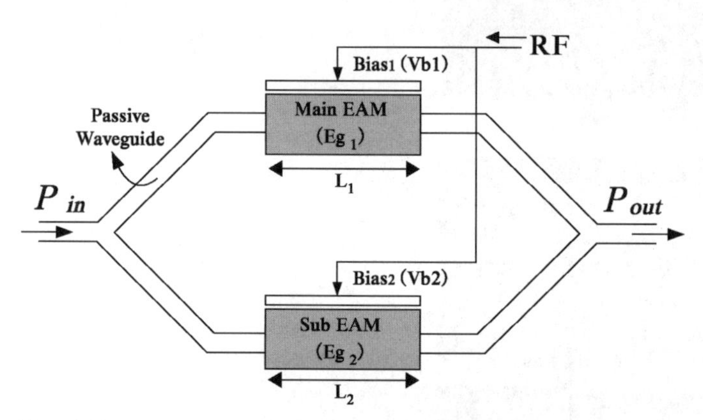

Fig. 6.11 A dual electroabsorption modulator, which is composed of a main and submodulators. The band gap and modulator length of the two modulators are different. The relative transmission characteristics of the submodulator are controlled so as to minimize the distortion of output power P_{out} [17]

6.7.2 Efficiency and Bandwidth

The efficiency and bandwidth of devices can be discussed together, because the product of efficiency and bandwidth of a device may be estimated as a quality factor of the device. Higher efficiency at a higher operating frequency gives a higher quality factor. There are three factors that determine the efficiency and bandwidth, i.e. the intrinsic, extrinsic, and external factors. These categories, however, have no definite distinction, so there is some overlap between them.

The "intrinsic factors" means the factors relating to the semiconductor junctions. The intrinsic factors of a semiconductor junction include the drift velocity of carriers, the junction structure, the materials/impurities, and the built-in potential. The carrier with the higher drift velocity and the lower excess space charge density at the light-absorption layer gives a higher value of the quality factor. The effect of low drift-velocity carriers should be minimized. The saturation effect is caused by excess space charge due to high-density photocarrier generation. Three techniques are important to decreasing the space charge density, that is, (1) pulling out photocarriers as quickly as possible outside the absorption layer, which directly relates to the built-in potential and therefore to the junction structure; (2) selecting the electrode configuration (distributing/travelling-wave electrodes) so as to dilute photocarrier generation; and (3) matching the velocities of the carriers for the light waves and the modulating microwaves. The factors related to the electrode are classified into the "extrinsic factors," which are discussed further later in this section.

In photodiodes, the transport of holes, whose drift velocity is much lower than that of electrons, restricts the bandwidth, output power, and efficiency. Successful utilization of electrons while controlling hole transport is a key issue. Figure 6.12 shows the junction structure of a unitravelling carrier photodiode (UTC-PD) made by an InP/InGaAs junction [19]. Photocarriers (electrons and holes) are generated in the p-type light-absorption layer. The electrons are injected into the depleted carrier-collection layer. In the light-absorption layer, the hole current actually exists just to maintain the current continuity relation. The quasi-neutral polarity of the light-absorption layer is maintained, and the hole density is much higher than the electron density. As a result, the hole drift current quickly responds to the electron drift current with dielectric relaxation time. InP/InGaAs UTC-PD with a 220 μm thick InGsAs absorption layer gives a 3 dB bandwidth of 94 GHz, a peak photocurrent of 184 mA (4.6 V output voltage for a load resistance of 25 Ω).

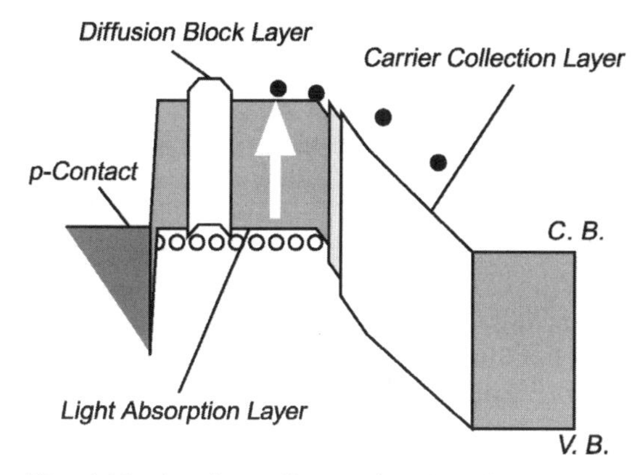

Fig. 6.12 A uni-traveling-carrier photodiode. Unlike the conventional PIN photodiode, light is absorbed at a p-type neutral layer adjacent to the carrier collection layer (depletion layer). This junction structure and the uniqueness of the polarity (i.e. only electrons are used) produce broad bandwidth (high speed), high efficiency, and high saturation power [19]

Figure 6.13 shows the cross-sectional view and junction diagram of an InP/InGaAs double-heterojunction phototransistor (HPT) [20]. The HPT is integrated with a heterojunction bipolar transistor (HBT). In this HPT, holes of low

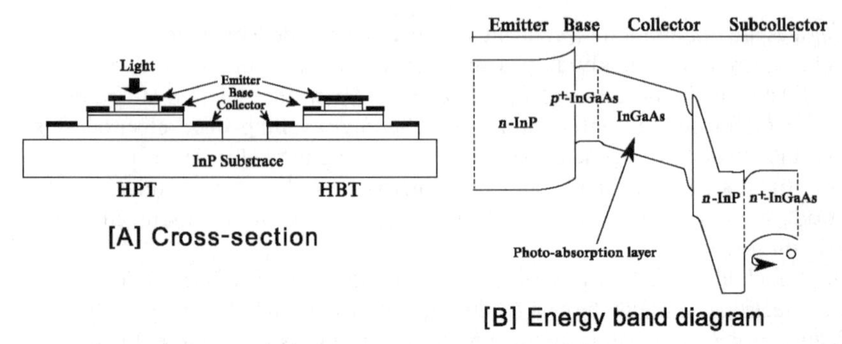

Fig. 6.13 An InP/InGaAs double-heterostructure phototransistor for millimeter-wave generation. The transistor features (1) elimination of low-speed holes by introducing a double-heterostructure, (2) high unity-gain cutoff frequency, and (3) high f_{max} obtained from reduction of size by self-aligned process [20]

conducting speed across the junction are eliminated by the double-heterostructure, and high values of unity-current-gain cutoff frequency (f_T) and f_{max} have been obtained by a self-aligned process, resulting an optical gain cutoff frequency of 82 GHz.

"Extrinsic factors" means the factors relating to the electrode and its parasitics, for example, series resistance, inductance, and capacitance, and other parasitic elements around the junction and electrode. To lower the electrode width across which the carriers drift without changing the cross-sectional area of the electrode, an electrode, whose cross section is similar to that of a mushroom, has been devised for high electron-mobility transistors (HEMT). By this shape the active length of the electrode can be shortened without increasing the series resistance (ohmic loss) of the electrode.

To obtain stronger and distributed coupling between microwaves and light waves, a distributed/travelling-wave type electrode has been proposed for a Mach-Zehnder optical modulator. In this modulator, the microwave electrode is extended to the direction of the optical guide along which the light waves propagate. The microwaves couple to the light waves while propagating along the electrode. By distributing the light-wave absorption, the peak space charge density can be decreased. The velocity of the microwaves is designed to match to the velocity of the light waves (velocity matching). The velocity-matching technique also distributes the optical and microwave interaction in the direction of the wave propagation. Various devices for distributing and velocity matching in photodiodes have been reported [15, 21–25]. Figure 6.14 shows the configuration of a waveguide-integrated photodiode [21], in which the waveguides of the light waves and the absorbing layer are separated but are placed in parallel, and the input light waves interact with the absorber while propagating along the waveguide. Unlike the conventional photodiodes, in which the waveguide and absorber are identical and therefore the light-generated current density is a steep decreasing function of waveguide length, a flat current density with respect to the waveguide length has been obtained. A velocity-matched PIN photodiode is shown in Fig. 6.15 [22].

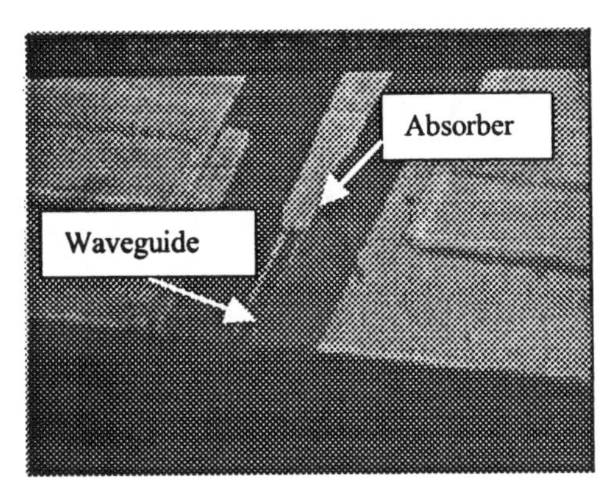

Fig. 6.14 A waveguide-integrated photodide with distributed absorption. The coupling between the lightwave and diode is distributed and, as a result, high saturation current, high responsivity, and high speed are achieved [21]

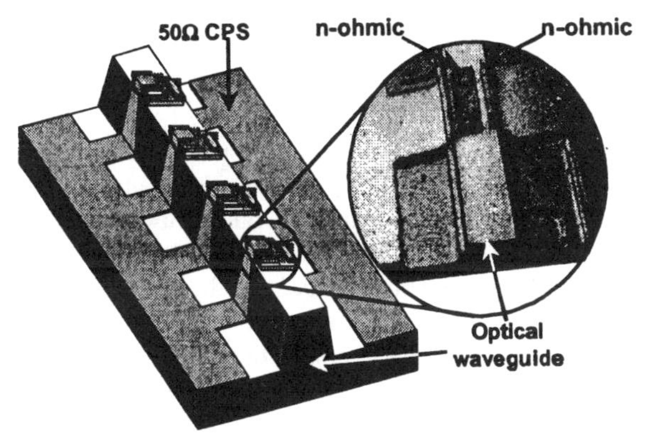

Fig. 6.15 A velocity-matched distributed photodetector with PIN photodiodes. It consists of an array of PIN diodes along which the lightwaves propagate and couple to microwaves. Since this detector sustains the merit of a PIN diode, it is advantageous for linear operation under high-power illumination [22]

The most efficient technique to minimize the degrading effects of extrinsic parasitics is to integrate devices monolithically. Monolithically integrated photonics and microwave circuits are successfully fabricated for the millimeter-wave region [26–28]. A photoreceiver designed for the 60 GHz band is shown in Fig. 6.16 [26], in which a photodiode, three HEMT and coupling circuits are integrated. The microfabricated electromechanical system (MEMS) provides a three-dimensional structure with lower parasitics in microwave circuits and optical waveguides [29].

The "external" factors are those factors that relate to an interface circuit between a microwave photonic device and its external microwave circuit. The exter-

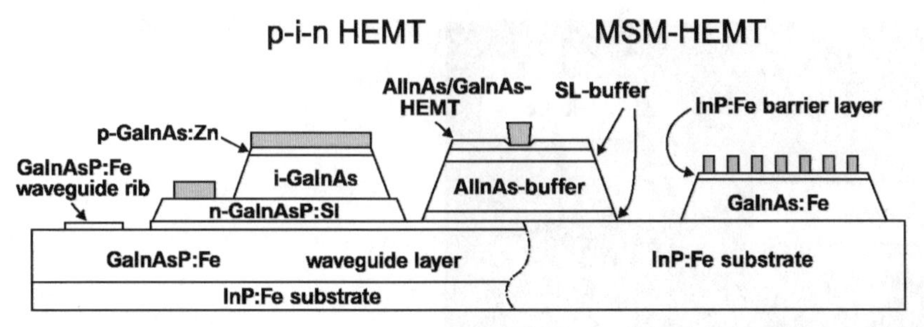

Fig. 6.16 An optoelectronic integrated photoreceiver. In this circuit, lightwave circuits and millimeter-wave circuits are integrated on a plane. The effect of parasitics is decreased by the integration [26]

nal factors include impedance matching, load impedance optimization, and reactive termination of idler frequencies. Detailed and specific discussion on these points is expected to become more active. We discuss impedance matching in the following section.

6.7.3 Impedance Matching

In electrical-optical interaction semiconductor devices, electrons excited by light illumination are pulled out to the electrical circuit by the built-in potential of the junction and an externally applied voltage. The generated microwave source by illumination is expressed in the microwave equivalent circuit as a current source whose generator impedance is very high, or ideally infinity. On the other hand, the impedance of microwave devices and the characteristic impedance of transmission lines are rather low. Electrical-optical interaction semiconductor devices, therefore, are used in fairly bad impedance-matching conditions if we directly connect them. From the point of view of power efficiency, the load impedance should be somewhat higher than the characteristic impedance of microwave transmission lines. An impedance transformer inserted between the diode and load improves the efficiency and frequency response.

Figure 6.17 shows the equivalent circuit of a PIN photodetector and amplifier [30], in which an impedance transformer is employed between the diode and amplifier. The amplifier is represented by a resistive load following the step transformer. The capacitance, inductance, and resistance in the resonant circuit represent the junction capacitance, inductance of the bonding wire, and series resistance, respectively. In this circuit, the parasitics of the diode and connecting circuit are successfully utilized as an impedance-matching circuit. The load impedance and the diode are well impedance-matched and a bandpass characteristic with a flat frequency response from 27–31 GHz has been obtained.

In microwave semiconductor diodes, the current through the junction is expressed by a nonlinear function of voltage, i.e. an exponential function of voltage. In exponential-function devices, the device impedance is low and constant-voltage source excitation (with a relatively low generator impedance) is used. The design

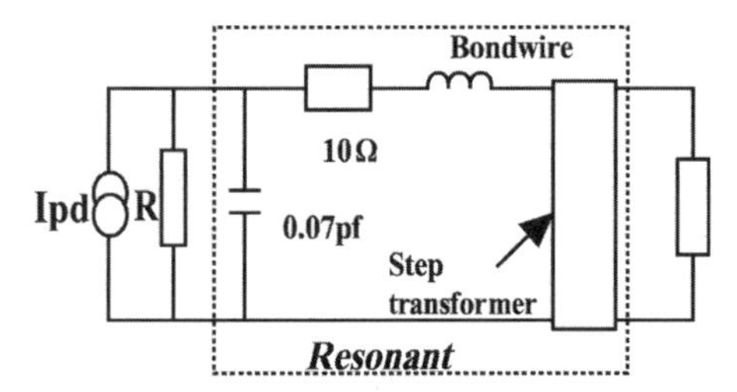

Fig. 6.17 A matching network between a PIN photodiode (the current source, resistance, and capacitance of the left side) and a millimeter-wave amplifier (the load resistance of the right side). The matching network is composed of a resonant circuit (a junction-capacitance compensating inductance) and an impedance transformer. The bond-wire between the PIN diode and millimeter-wave circuit is utilized as the resonant inductance. We obtain band-pass characteristics by adjusting the resonant inductance and step-up ratio of the transformer [30]

theories based upon exponential functions are well discussed and are relatively well established. For optically illuminated diodes, however, since illumination is expressed by a current source, the nonlinearity is expressed as a logarithmic function, i.e. the voltage is expressed by a logarithmic function of current. In logarithmic-function devices, the device impedance is fairly high compared with the characteristic impedance of the transmission line, and a constant-current source excitation (with a relatively high generator impedance) is used. Although a logarithmic function is the reverse function of an exponential function and both expressions are equivalent in the equivalent circuit expression, the logarithmic-function expression has more realistic physical meaning [31].

Figure 6.18 shows the equivalent circuit and theoretically calculated conversion efficiency of a microwave photonics mixer [31], in which an optical signal and microwave local-oscillator (LO, with frequency f_{LOC}) power are applied to a PIN diode and are mixed in the diode. The optical signal conveys a microwave subcarrier (with frequency f_S) and is fed to the diode through an optical fiber. The LO power is applied through a microwave transmission line. The optical signal is detected by the diode and converted to a microwave current source whose frequency

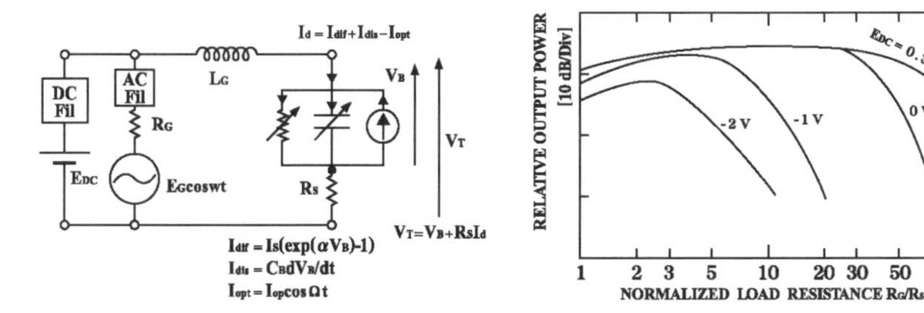

Fig. 6.18 Output power (relative value) of subcarrier downconverter is calculated as a function of local-oscillator generator resistance (R_G). The generator resistance is normalized by the series resistance (R_S) of the PIN diode [31]

is f_S, as shown in Fig. 6.18a. Two frequencies, f_S and f_{LOC}, are mixed by the voltage-current and voltage-capacitance nonlinearities of the diode, and an intermediate frequency (IF), f_{IF}, where $f_{IF} = f_S - f_{LOC}$, is produced. Figure 6.18b shows the conversion output power (relative value) from f_S to f_{IF} as a function of the IF load resistance, where the LO generator impedance and IF load impedance are assumed to be equal. It shows that the optimum IF load resistance is higher than for the exponential-function case.

6.7.4 Simplification of Base Station

In radio-on-fiber systems, the microcellular or picocellular mobile/LAN is the most typical application. It is very important, therefore, to make radio wave base stations with simple configurations at low cost, and to concentrate complex network functions to the central station [32–42]. The base station should only have E/O and O/E converting functions and, therefore, have as few active microwave and/or optical devices as possible. There are various propositions for simplifying base-station configurations. Among them the optical generation of microwaves is a new technology and is very important in the sense that the radio wave signals are transmitted inside an optical fiber without suffering serious fading, interference, or transmission loss.

The heterojunction phototransistor (HPT) is a device developed for use in microwave photonic purposes. The basic junction structure is similar to that of a heterojunction bipolar transistor (HBT), which is commonly used in the microwave/millimeter-wave region. A trial experiment is shown [32], in which an InP/InGaAs HPT was used for two functions for light wave and radio wave conversion in the 30-GHz region, i.e. photodetection and upconversion mixing. HPT must be adopted to delivery of broadband multiservice data to professional and residential customers.

Image rejection in mixers by optical methods is one of the most important techniques for realizing simple, and high-performance base stations. In Fig. 6.19, a 60 GHz image-rejection mixing by use of optical heterodyne and optical filtering is shown [33]. Realizing image rejection in the millimeter-wave range is comparatively difficult due to difficulty in electric circuit implementation in these freque n-cies. Here a successful image rejection by use of optical heterodyne detection and optical filtering is proposed to simplify the construction of a base station, an electroabsorption (EA) device with multiple quantum-well (MQW) structure is expected to be a promising multifunction device [34–36]. An EA works as an optical detector and as an optical modulator; in other words, it works as a light-wave-to-radio wave converter and as a radio wave-to-light-wave converter. Its operation is based upon the fact that the responsivity of detection and the modulation efficiency of an MQW device have different responses with respect to optical wavelength and have their optima at different wavelengths. Therefore we can operate it as a detector and a modulator for two different optical frequencies. For InGaAsP MQW EA, responsivity and modulation efficiency become maximum around wavelengths of 1.52 and 1.56 μm, respectively. The EA device is capable of

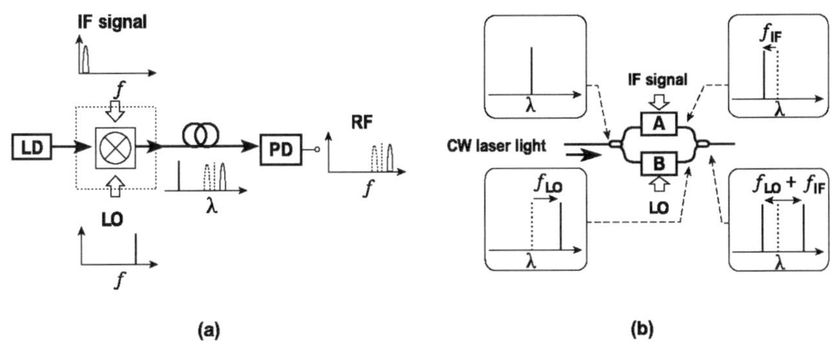

Fig. 6.19 Optoelectronic mixing at 60 GHz with high image rejection: **a** A schematic of optoelectronic image rejection mixing. A local oscillator and an IF signal modulate a laser optical source (in the central station), and the modulated optical signal is transmitted to a remotely placed photodiode. Two optical frequencies $f_{LO}+f_{IF}$ apart are sent to the photodiode. Image-rejected millimeter-wave signals are output from the photodiode. **b** Another type of image rejection by two modulators [33]

operating as a modulator and a detector at these wavelengths, so we can construct an E/O and O/E transmitter/receiver (that is, an electroabsorption transceiver, EAT) with a single MQW device. Figure 6.20 is a radio-on-fiber link utilizing an EAT (EAM in the figure) in a base station [34]. The EAT works as a detector (O/E converter) in the downlink and a modulator (E/O converter) in the uplink. The link is fully duplexed by a single optical fiber.

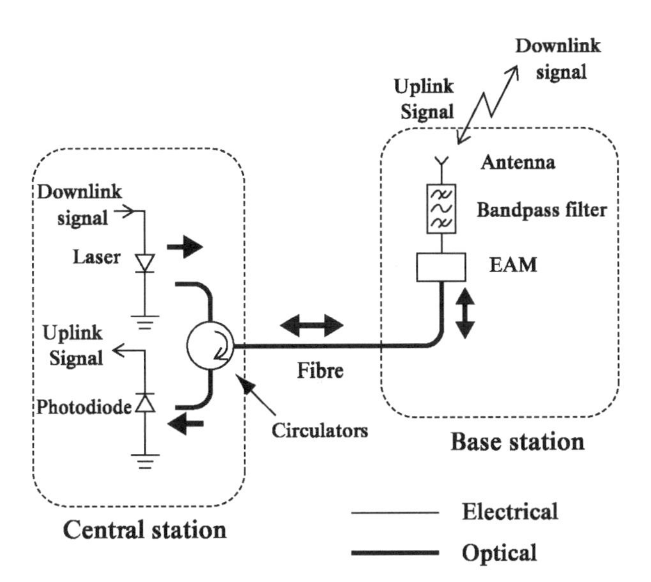

Fig. 6.20 An electroabsorption modulator (EAM) works as a modulator (for the uplink) and detector (for the downlink) in a radio base station, which leads to crucial simplification and reduces the cost of the radio base station [34]

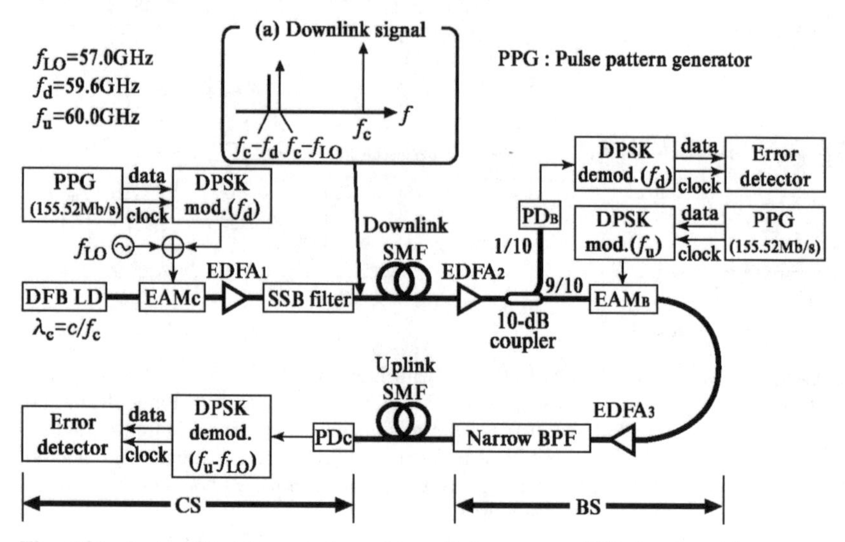

Fig. 6.21 A simplified base station without light sources. With local-oscillator power remotely sent to the base station and an EAM in the base station, full duplex millimeter-wave transmission is realized [37]

Another type of a base station is shown in Fig. 6.21 [37], in which the base station is only equipped with a photodetector and an external modulator. A laser oscillator and a microwave local oscillator are centralized to the central station. Full-duplex connection is realized by an optical fiber loop around the central station and base station. An optical downlink signal is transmitted from the central station conveying downlink signals and a microwave local-oscillator frequency. A portion of the transmitted optical signal is tapped, detected, and converted to downlink signal at the base station. The rest of the optical carrier is looped back to the central station with modulated uplink signals and the local-oscillator frequency that is sent from the central station. The uplink signals and local-oscillator frequency are detected and mixed to yield the intermediate frequency. The local-oscillator power sent from the central station is utilized for detection in both the central station and base station.

6.7.5 Generation of Radio Waves by Optical Methods

Generation of microwaves by means of optical methods have been investigated [43–47]. The basic principle of microwave generation is heterodyne detection of two different optical frequencies whose frequency difference falls into the microwave range and whose frequency and phase differences are kept constant with time, which is called optical phase locking. When the optical power of a laser oscillating with one frequency (or multiple frequencies) is injected into another oscillating laser, the frequency and phase (or frequencies and phases) of the second laser are pulled and locked by the injected source(s). If the injected frequencies have a frequency difference falling into the microwave range and the phases are kept constant, a microwave frequency source with low phase noise is obtained by heterodyne detection of the output power of the injection-locked (the second) laser. For the source (master) laser, an intensity-modulated single-mode distributed feedback (DFB) laser is commonly used, and for injection-locked (slave) laser, a multimode Fabry-Perot laser is commonly used. When the microwave frequency modulating the master laser is a subharmonic of the difference frequency of the slave laser, the locking phenomenon is also observed; this is called subharmonic locking. Subharmonic locking is used to obtain even higher frequencies by harmonic multiplication of injected source frequencies. Optical generation of microwaves makes it possible to send microwave/millimeter-wave power remotely from central stations to base stations. The injection-locking and heterodyne detection are realized by various circuit configurations.

Figure 6.22 shows a 60 GHz source created by use of a master laser, a slave laser, and a subharmonic (30 GHz) oscillator [43]. The master oscillator is a DFB laser and is oscillating at a single frequency. The output of the master laser is modulated at 30 GHz, generated by a frequency-stabilized 30 GHz signal generator, by a Mach-Zehnder (MZ) external modulator. The output of the modulator contains three frequencies, an optical carrier and two sideband frequencies, which are 60 GHz apart. The output of the modulator is injected through an optical circulator to a Fabry-Perot (FP) slave laser oscillating with two modes whose frequency difference is about 60 GHz. In order for the two frequencies of the slave laser to be injection locked, the oscillating frequency of the master laser is set around the center of the two frequencies of the slave laser. The output from the FP laser then is led through the optical circulator to a photodetector, where the two frequencies are mixed and the difference frequency, 60 GHz, is generated. The phase stability of the 60 GHz output is determined by the 30 GHz signal generator. A satisfactory low phase noise (-93 dBc/Hz at 100 kHz offset from 60 GHz) has been measured. In such a process, we obtain low-noise harmonic millimeter-wave carriers with optical modulation and the heterodyne detection technique.

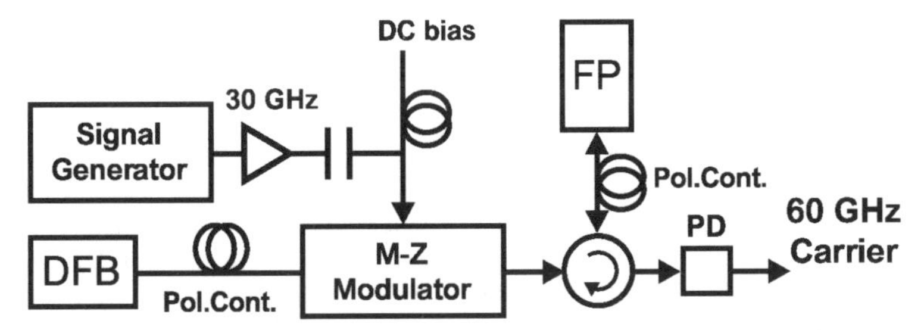

Fig. 6.22 Millimeter-wave generation by its subharmonic. The frequency of the master oscillator (DFB) is set near the center frequency of the two frequencies of the slave oscillator (FP). The master oscillator is modulated by a frequency of one-half of the difference (60 GHz) of the two frequencies of the slave oscillator. Then 60 GHz with low phase noise is obtained by heterodyning the output of the slave oscillator [43]

By amplitude or phase modulation of an optical modulator, we obtain a microwave generator that produces an array of frequencies, called comb frequencies, each of which is allocated by an identical frequency difference. A fiber-based optical comb frequency generator is shown in Fig. 6.23 [44], and basically consists of a reference laser oscillator, an optical modulator, two optical couplers, and a loop of optical fiber. An optical reference laser signal from an external cavity laser is injected into the optical-fiber loop and goes through a $LiNbO_3$ phase modulator, which is modulated by a microwave RF source. The output of the three optical frequencies, an optical carrier and upper and lower sidebands, are sent into the optical fiber. By undergoing multiple round trips along the optical loop, a series of optical sidebands are generated, which are allocated exactly by the modulation frequency of the phase modulator. Modulation-by-modulation, the number of sidebands consecutively increases, and as a result, we obtain an array of frequencies (comb frequencies) on the optical frequency axis. More than one hundred sideband frequencies can be generated. By detecting this optical comb with an optical detector, we obtain microwave comb frequencies. Typical spectra of the optical comb are shown in Fig. 6.24 [44].

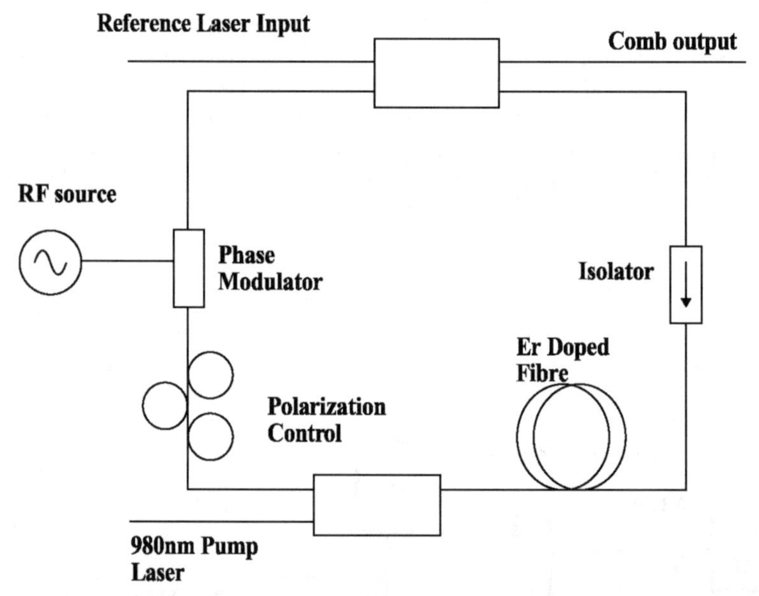

Fig. 6.23 Microwave comb frequency generation. An optical phase modulator is modulated by an RF source, and the output of the modulator is connected to its input by an optical-fiber loop with a certain length. The number of modulation sidebands increases, while the light wave travels around the optical loop. Thus an optical comb is obtained in which an array of many sidebands with the RF modulation frequency spacing are formed on the optical frequency axis. By detecting this optical comb by an optical detector, we obtain the microwave comb frequencies [44]

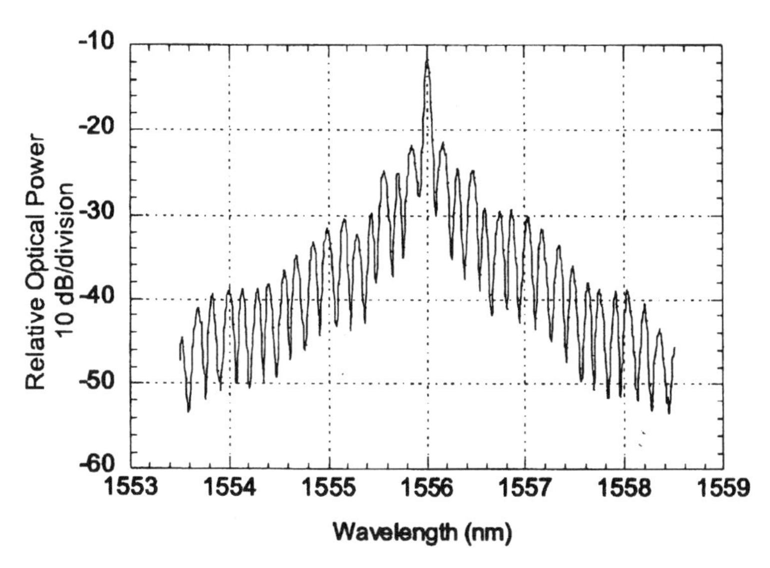

Fig. 6.24 Typical spectrum of an optical comb. Each line has a spacing of 18 GHz [44]

6.8 Conclusion and Acknowledgements

Microwave photonics is a new technical field in which the two technologies of microwaves and photonics are integrated. Because radio waves with a certain bandwidth are carried by an optical fiber as subcarriers in microwave photonics, it differs from baseband transmission that includes DC components, and therefore, it presents possibilities of new and more versatile applications in trunk and subscriber communications fields. Since the bandwidth of optical fibers is quite broad, the frequencies of the subcarrier profitably fall into the millimeter-wave region. Millimeter-wave applications for communications will be enhanced. Microwave photonics may also give a practical solution to realizing broadband and low-cost subscriber networks.

Here expected and/or envisaged applications have been presented along with the fundamental technical issues we are now facing and will encounter henceforth. Since the technology of this field is still in its infancy, we may still encounter many new problems to be solved. A number of papers have been presented in international conferences and workshops, and among them practical systems will be created. The material for this chapter is extracted mainly from the recent international conferences. In the figures referenced here, are indicated their original papers. The author would like to express his sincere thanks to the authors of those papers for allowing the use of their figures in this chapter.

References

[1] W. D. Warters (1977), WT4 Millimeter Waveguide System, The Bell System Technical Journal, Vol. 56, No.10, December

[2] C. K. Kao (1988), Optical Fibre, Peter Preregrinus, London

[3] Y. Fuke and Y. Ebine (1999), RF sub-carrier transmission using a bus type topology for personal digital cellular systems, 1999 International Topical Meeting on Microwave Photonics (MWP'99) Digest, T-8.19

[4] Y. Ebine (1999), Development of fiber-radio systems for cellular mobile communications, 1999 International Topical Meeting on Microwave Photonics (MWP'99) Digest, F-10.1

[5] I. Seo, T. Yomioka, and S. Ohshima (2000), Error-free transmission of radio QPSK signals in an optical subcarrier multiple access system suppressing optical beat interference with over-modulation, 2000 International Topical Meeting on Microwave Photonics (MWP 2000) Digest, TU1.4

[6] Y. Shoji and H. Ogawa (2000), Optical subcarrier multiplexing system using time sampling multiplexer to suppress optical beat interference, 2000 International Topical Meeting on Microwave Photonics (MWP 2000) Digest, TU1.5

[7] M. Fujise and H. Harada (1999), An experimental study on multi-service radio on fiber transmission system for ITS road-vehicle communications, 1999 International Topical Meeting on Microwave Photonics (MWP'99) Digest, F-10.4

[8] F. Deborgies, M. Mittrich, H. Schmuck, P. Jaffré, and Pescod, "Progress in ACTS FRANS project," 1999 International Topical Meeting on Microwave Photonics (MWP'99) Digest, T-7.1, pp. 115-118

[9] R. E. Schuh, D. Wake, B. Verri, and E. Sundberg(2000), Penalty free W-CDMA radio signal transmission over fiber, 2000 International Topical Meeting on Microwave Photonics (MWP 2000) Digest, TU1.3

[10] K. Kumamoto, K. Tsukamoto, and S. Komaki (2000), Theoretical consideration on transferring transparency for RF signal bandwidth on direct optical switching CDMA radio-on-fiber networks, 2000 International Topical Meeting on Microwave Photonics (MWP 2000) Digest, WE2.3

[11] K. Kitayama (1999), Optical DSB signal based millimeter-wave fiber-radio system using external modulation technique, 1999 International Topical Meeting on Microwave Photonics (MWP'99) Digest, W-4.1

[12] R. E. Schuh (1999), Hybrid fiber radio for second and third generation wireless systems, 1999 International Topical Meeting on Microwave Photonics (MWP'99) Digest, T-8.20

[13] H. Izadpanah, D. J. Gregoire, F. A. Dolezal, W. Ng, D. Yap, and G. Tangonan (2000), An integrated fiber optics/broadband wireless access demonstrator for the next generation Internet network extension, 2000 International Topical Meeting on Microwave Photonics (MWP 2000) Digest, WE2.6

[14] S. H. Seo, J. M. Cheong, Y. S. Son, J. S. Kim, and S. Park (1999), Reference frequency signal transmission over optical fiber for CDMA-based microcel-

lular system, 1999 International Topical Meeting on Microwave Photonics (MWP'99) Digest, T-8.15

[15] M. S. Islam, T. Jung, S. Mathai, T. Chau, A. Rollinger, T. Itoh, M. C. Wu, D. L. Sivco, and A. Y. Cho (1999), High power distributed balanced photodetectors with high linearity, 1999 International Topical Meeting on Microwave Photonics (MWP'99) Digest, T-5.2

[16] M. Shin, J. Lim, C. Y. Park, J. Kim, J. S. Kim, K. E. Pyun, and S. Hong (1999), High-speed linear analog multiple quantum well electroabsorption modulator integrated with MMI couplers, 1999 International Topical Meeting on Microwave Photonics (MWP'99) Digest, W-2.5

[17] G.-W. Lee and S.-K. Han (1999), Linearization of a narrowband analog optical link using integrated dual electroabsorption modulator, 1999 International Topical Meeting on Microwave Photonics (MWP'99) Digest, W-2.4

[18] C. P. Liu, A. J. Seeds, Y. Bester, V. Sidorov, D. Ritter, and A. Madjar (1999), Two-tone third-order intermodulation distortion characteristics of an HBT optoelectronic mixer using a two-laser approach, 1999 International Topical Meeting on Microwave Photonics (MWP'99) Digest, T-5.4

[19] T. Ishibashi, H. Fushimi, H. Ito, and T. Furuta (1999), High power unitraveling-carrier photodiodes, 1999 International Topical Meeting on Microwave Photonics (MWP'99) Digest, T-5.1

[20] H. Kamitsuna, Y. Matsuoka, S. Yamahata, and N. Shigekawa (2000), A 82-GHz optical-gain-cutoff frequency InP/InGaAs double-heterostructure phototransistor and its application to a 40-GHz-band OEMMIC photoreceiver, 30th European Microwave Conference (EuMC'2000) Proceedings:388-391

[21] H. Jiang and P. K. L. Yu (2000), High-power waveguide integrated photodiode with distributed absorption," IEEE MTT-S 2000 International Microwave Symposium Digest, WE1D-1

[22] M. S. Islam, T. Jung, S. Murthy, T. Itoh, M. C. Wu, D. L. Sivco, and A. Y. Cho (2000), A velocity-matched distributed photodetectors with PIN photodiodes, 2000 International Topical Meeting on Microwave Photonics (MWP 2000) Digest, WE3.1

[23] W.-S. Cho, Y.-S. Lim, and Y.-W. Choi (1999), Bandwidth limitation factors and linear characteristics of traveling-wave CPW electroabsorption modulators, 1999 International Topical Meeting on Microwave Photonics (MWP'99) Digest, T-8.5

[24] S. R. Sakamoto, A. Jackson, and N. Dagli (1999), Substrate removed GaAs/AlGaAs Mach-Zehnder electro-optic modulators for ultra wide bandwidth operation, 1999 International Topical Meeting on Microwave Photonics (MWP'99) Digest, W-2.2

[25] A. Stöhr, R. Heizelmann, and D. Jäger (2000), Millimetre-wave bandwidth electroabsorption modulators and transceivers, 2000 International Topical Meeting on Microwave Photonics (MWP 2000) Digest, WE1.1

[26] A. Umbach, T. Engel, and G. Unterbrösch (1999), Optoelectronic integration of ultrafast photoreceivers, 1999 International Topical Meeting on Microwave Photonics (MWP'99) Digest, W-3.1

[27] T. Nagatsuma, Y. Royter, M. Shinagawa, T. Furuta, and H. Itoh (2000), A 120-GHz integrated photonic transmitter, 2000 International Topical Meeting on Microwave Photonics (MWP 2000) Digest, WE3.3

[28] K. Takahata, Y. Muramoto, and S. Fukushima (2000), Monolithically integrated millimeter-wave photonic emitter for 60-GHz fiber radio application, 2000 International Topical Meeting on Microwave Photonics (MWP 2000) Digest, WE3.4

[29] G. Tangonan, R. Loo, J. Schaffner, and JJ. Lee (1999), Microwave photonics applications of MEMS technology, 1999 International Topical Meeting on Microwave Photonics (MWP'99) Digest, T-6.3

[30] L. Gomez-Rojas, N. J. Gomes, X. Wang, P. A. Davies, and D. Wake (2000), High performance optical receiver using a PIN photodiode and amplifier for operation in the millimeter-wave region, 30th European Microwave Conference (EuMC'2000) Proceedings:392-394

[31] M. Akaike (2001), Analysis of conversion efficiency of microwave subcarrier mixers using a PIN photodiode," 2001 URSI International Symposium on Signals, Systems, and Electronics (ISSSE'01) Proceedings:374-377

[32] C. Gonzalez, J. Thuret, J. L. Benchimol, and M. Riet (1999), InP/InGaAs bipolar phototransistor as a front-end photoreceiver for HFR distribution network, 1999 International Topical Meeting on Microwave Photonics (MWP'99) Digest, W-3.2

[33] K. Nishikawa and M. Tsuchiya (1999), 60 GHz optoelectronic mixing with high image rejection ratio, 1999 International Topical Meeting on Microwave Photonics (MWP'99) Digest, F-9.5

[34] R. I. Killey, J. B. Song, C. P. Liu, A. J. Seeds, J. S. Chadha, and M. Whitehead (2000), Multiple quantum well asymmetric Fabry-Perot modulators for RF-over-fibre applications, 2000 International Topical Meeting on Microwave Photonics (MWP 2000) Digest, TU2.4

[35] K. Kitayama, T. Kuri, R. Heinzelmann, A. Stöhr, D. Jäger, and T. Takahashi (2000), A good prospect for broadband millimeter-wave fiber-radio access system, IEEE MTT-S 2000 International Microwave Symposium Digest:1745-1748

[36] D. Jäger, R. Heinzelmann, and A. Stöhr (2000), Microwave optical interaction devices: from concept to applications, 30th European Microwave Conference (EuMC'2000) Proceedings:384-387.

[37] T. Kuri, K. Kitayama, and Y. Takahashi (1999), Simplified BS without light source and RF local oscillator in full-duplex millimeter-wave radio-on-fiber system based upon external modulation technique, 1999 International Topical Meeting on Microwave Photonics (MWP'99) Digest, T-7.3

[38] T. Tomioka and S. Ohshima (1999), Polarization-free operation of external modulators in radio on fiber systems, 1999 International Topical Meeting on Microwave Photonics (MWP'99) Digest, T-8.17

[39] T. Ohno, S. Fukushima, Y. Doi, Y. Muramoto, and Y. Matsuoka (1999), Application of uni-traveling-carrier waveguide photodiodes in base station of a millimeter-wave fiber-radio system, 1999 International Topical Meeting on Microwave Photonics (MWP'99) Digest, F-10.2

[40] A. Nirmalathas, C. Lim, D. Novak, and R. Waterhouse (1999), Optical interfaces without light sources for base-station designs in fiber-wireless systems incorporating WDM, 1999 International Topical Meeting on Microwave Photonics (MWP'99) Digest, T-7.2

[41] D. S. Shin, G. L. Li, C. K. Sun, S. A. Pappert, W. S. C. Chang, and P. K. L. Yu (2000), Performance of electroabsorption modulator as integrated optoelectronic mixer for RF frequency conversion, 2000 International Topical Meeting on Microwave Photonics (MWP 2000) Digest, TU2.5

[42] S. J. Strutz and K. J. Williams (2000), A 6 to 11 GHz all-optical image rejection microwave downconverter, 2000 International Topical Meeting on Microwave Photonics (MWP 2000) Digest, TU2.6

[43] M. Ogusu, K. Inagaki, and Y. Mizuguchi (2000), 400 Mbit/s BPSK data transmission at 60 GHz-band millimeter-wave using a two-mode injection-locked Fabry-Perot slave laser, 2000 International Topical Meeting on Microwave Photonics (MWP 2000) Digest, TU1.1

[44] O. P. Gough, C. F. C. Silva, and A. J. Seeds (1999), Exact millimeter wave frequency synthesis by injection locked laser comb line selection, 1999 International Topical Meeting on Microwave Photonics (MWP'99) Digest, W-4.3

[45] L. A. Johansson and A. J. Seeds (2000), Fibre-integrated heterodyne optical injection phase-lock loop for optical generation of millimeter-wave carriers, IEEE MTT-S 2000 International Microwave Symposium Digest:1737-1740

[46] P. R. Herczfeld, A. J. C. Vieira, A. Rosen, and W. D. Jemison (2000), Generation and transmission of millimeter-wave signals using mode-locked microchip laser, 30th European Microwave Conference (EuMC'2000) Proceedings:395-398

[47] R. T. Logan Jr. (2000), All-optical heterodyne RF signal generation using a mode-locked-laser frequency comb: theory and experiments, IEEE MTT-S 2000 International Microwave Symposium Digest:1741-1744

7 Design of CMOS RF IC for Wireless Applications: System Level Compromised Considerations

7.1 Introduction

7.1.1 Wireless Market

The importance of wireless communication is undisputed, and it will surely continue to change and evolve because of the explosive growth of wireless personal mobile and cellular communications since the late 1980s. These growth areas, and they are enjoying the fastest growth rate in the telecommunications industry adding customers at a rate of 20–30% more each year as shown in Figs. 7.1 and 7.2 [1].

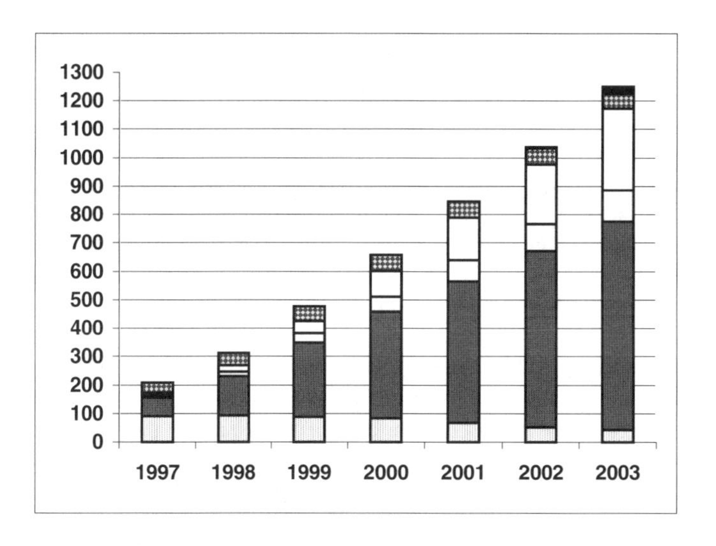

Fig. 7.1 Agilent Forecast: Subscribers (millions) by Technology: 1999 - 2003 CAGR = 27% [1]: Analog, GSM/EDGE, TDMA. CDMA2000/CDMAone, PDC/PHS, 3GPP respectively from the bottom to top

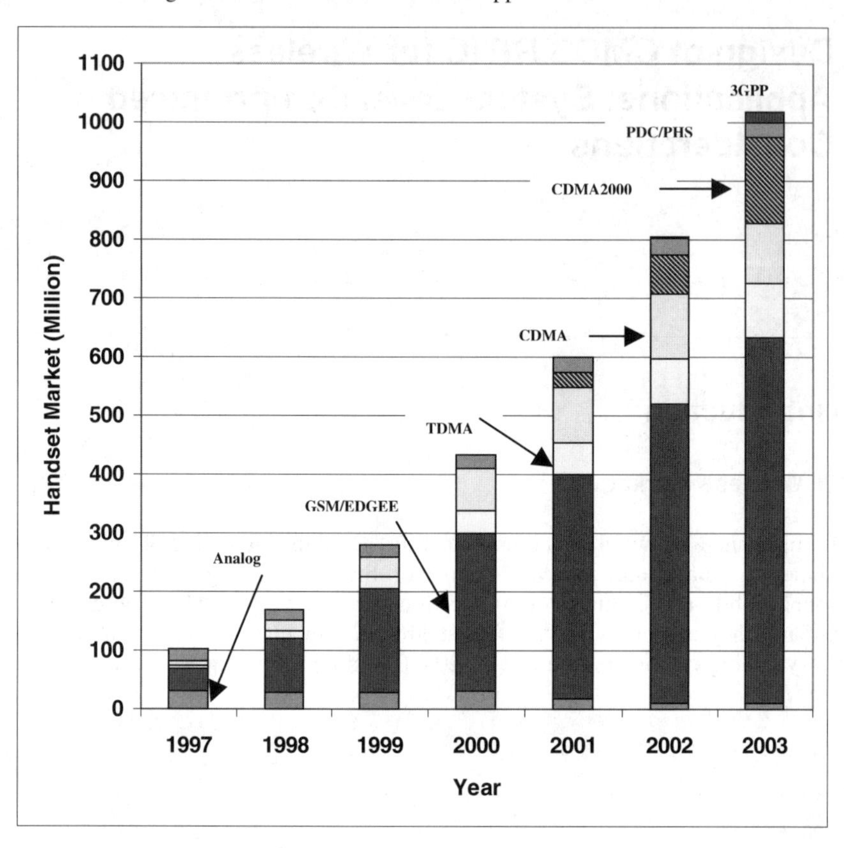

Fig. 7.2 Agilent Forecast: Handsets by Technology 1999–2003 CAGR = 38% [1]

Presently, in addition to mobile cellular communications, there are at least six satellite communication systems being developed so that wireless personal voice and data communications will be transmitted from any part of the earth to another part. There are also many other kinds of short-range wireless such as Bluetooth, HiperLAN, WLAN (Wireless Local Area Network), IEEE 802.11, and so on (see Fig 7.3).

In Fig 7.3 the radio-frequency (RF) spectrum is depicted. Using a combination of wireless telephones, wireless modem, terrestrial cellular telephones, satellites, and other short-range personal wireless networks, one can transmit data, voices, and other information without any physical limitations. The use of wireless remote sensing, RF identification, direct broadcasting, global positioning systems and navigation, smart highways and smart automobiles has become more popular in the past decade. Wireless communications, PDA (personal digital assistance), and sensors have become very popular consumer products in daily life. All of these wireless systems consist of a radio-frequency front-end, in which radio-frequency integrated circuits (RF ICs) are these applications are rapidly penetrating all aspects of our normal lives.

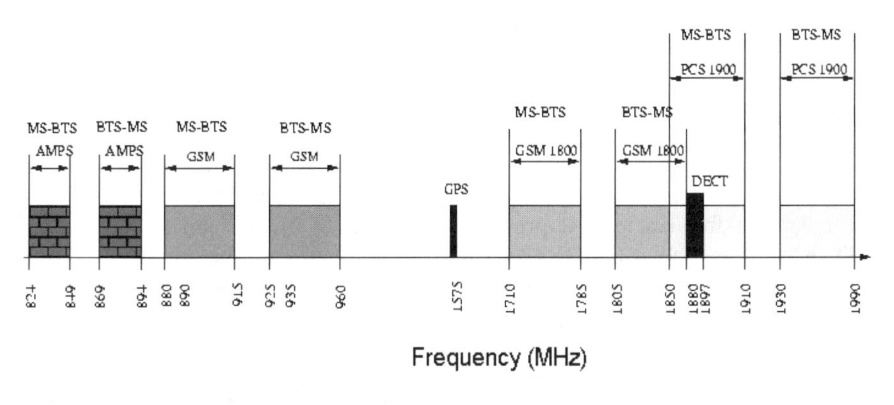

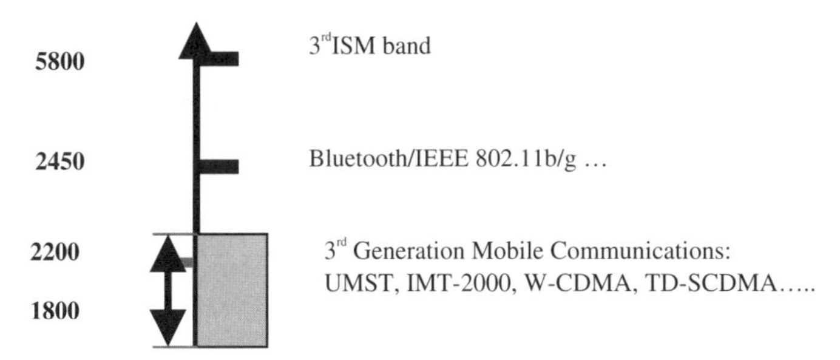

Fig. 7.3 Spectrum of the wireless applications

The market requirements for wide bandwidth and high speed also accelerate the development of RF products. The predicted frequency bandwidth demands are given in Fig 7.4. Because of the market, traditional semiconductor and integrated circuits and systems companies as well as new comers, small and large, analog and digital, have seen the statistics and predictions and are striving to capture the market share by introducing various RF products.

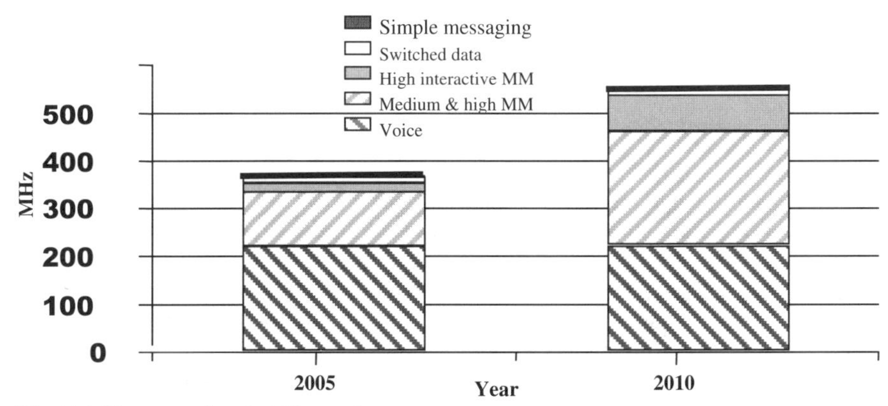

Fig. 7.4 Frequency bandwidth requirements

7.1.2 Technology Available

The rapid pace of development leads to a great demand on RF IC. For commercial applications the cost is always the first consideration. One decade ago RF circuits were mainly fabricated on GaAs, and it was impossible to design RF IC on complementary metal-oxide semiconductors (CMOS) because CMOS had very low operation frequencies. At present, GaAs and Si bipolar junction transistor (BJT) as well as bipolar CMOS (BiCMOS) technologies are the dominant section of the RF market, especially in power amplifiers and front-end RF switches. As the advanced CMOS technologies are introduced, the unit-gain frequency of f_T is increasing, and it is sufficient for RF applications. The comparisons are shown in Fig. 7.5, which shows that CMOS technologies are possible for RFIC. The minimum required f_T for wireless applications are around 20 GHz. Because CMOS RF ICs are low cost and easier to integrate with other parts of the entire integrated systems, they have become of great interest. Nevertheless, a number of practical issues surrounding RF CMOS technologies must be resolved: substrate isolation required by applications, temperature sensitivity of silicon, process variations, and RF device modelling for passive and transistors.

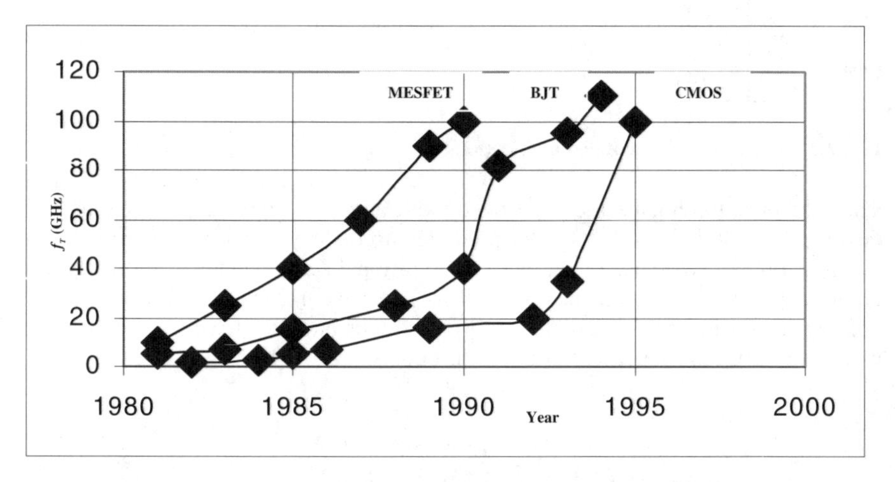

Fig. 7.5 Advanced CMOS technologies now become possible for RF IC [2] because of the increased f_T

As the gate length and the dimension of a CMOS are shrunk further, f_T increases as indicated in Table 7.1 as an example for the CSM processes.

It can be seen from Table 7.1 that based on 0.25 μm and 0.18 μm CMOS processes, most commercial wireless applications of interest can be realized by CMOS IC. Recent publications indicate that it is possible to realize a 10 GHz voltage-controlled oscillator (VCO) based on 0.25 μm CMOS technology [3].

As the gate length of a CMOS shrinks further, it will be no problem to realize all front-end blocks with the baseband together on a single chip using CMOS

Table 7.1 Technology trend

Device	Gate length (μm)		
	0.35	0.25	0.18
NMOS f_T (GHz)	27 GHz Vds = 3.0 V	39 GHz Vds = 2.5 V	50 GHz Vds = 1.8 V
NMOS f_{max} (GHz)	35 GHz Vds = 3.0 V	44 GHz Vds = 2.5 V	53 GHz Vds = 1.8V
NMOS NFmin (dB) @ 2.45GHz	1.9 dB @ Vds=2.0V, Vgs=1.0V	1.9 dB @ Vds=2.5V, Vgs=1.3V	1.9 dB @ Vds=250V, Vgs=1.3V
PMOS f_T (GHz)	15 GHz Vds = 3.0 V	20 GHz Vds = 2.5 V	23 GHz Vds = 1.8 V
PMOS f_{max} (GHz)	20 GHz Vds = 3.0 V	23 GHz Vds = 2.5 V	28 GHz Vds = 1.8V
Year available	1999	2000	2001

process for applications up to 10 GHz. Therefore CMOS RF ICs are of interest for all commercial wireless applications, and more and more companies are in this area. Market surveys show that in China alone more than 5 *million* new subscribers join the cellular phone system *every month*. This trend will continue to at least 2005, and more than *1 billion* handsets will be sold in 2003 as depicted in Fig. 7.1 [1], motivating competitive RF IC houses to provide chips with increasingly higher performance and lower prices. In fact, those chips also are the key for reducing the power consumption and price of cellular phones. Even though most commercial RF products for cellular phones are made by GaAs, Si BJT or BiCMOS processes due to the traditional developments, new products such as GPS, Bluetooth, and HiperLAN are mainly based on RF CMOS in the literature and in the markets, because of low costs, small sizes, and integration capabilities with the other parts of the systems.

7.2 RF IC Designs are Bottleneck

Compared to other parts of a wireless system, RF circuits contain very few transistors, but they still belong to one of the three key techniques of a wireless system, namely RF IC, base-band ICs the software. In contrast to the traditional IC designs like digital IC designs in terms of the numbers transistors, the RF IC is much simple. However, RF IC designs are more difficult than traditional IC designs. The main reasons are the required multidisciplinary knowledge for RF IC designs and the trade-offs among the contrasting conditions. The block diagram of a general wireless terminal, which is given in Fig. 7.6, consists of the RF section, including the receiving path and the transmitting path, the high-speed high-resolution analog digital converter (ADC) and digital analog converter (DAC), and the coder-decoder (CODEC) in the baseband.

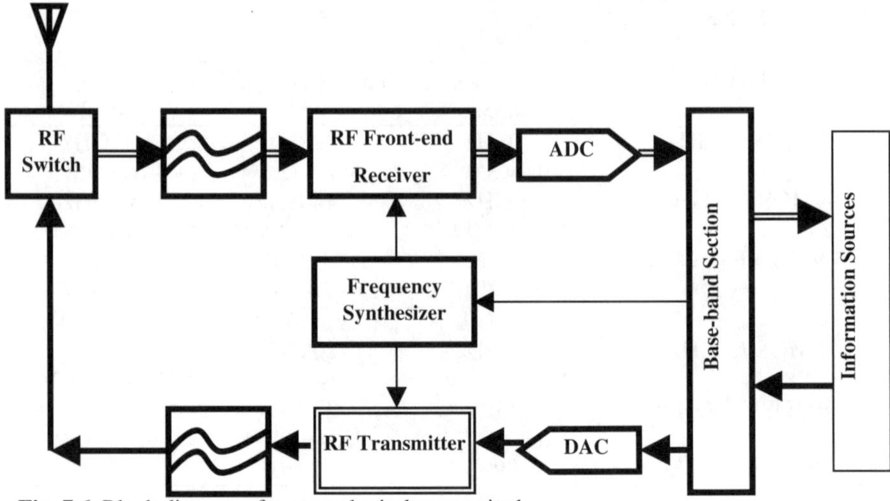

Fig. 7.6 Block diagram of a general wireless terminal

7.2.1 Multidisciplinary Knowledge

The RF section (receiver path and the transmitter path) is connected to the antenna, which is part of the field of microwave engineering. An RF section serves as the load for the antenna, which needs 50-Ω matching. However, the signal propagation and modulation are defined by the different communication standards, which are covered by communication theory. Generally, RF IC are also analog IC, therefore, all knowledge for analog IC designs is also required by RF IC designs.

RF designers must manufacture all passive components on chips such as inductors, capacitors, resistors and so on, which are generally external components for traditional analog IC designers. Thorough knowledge of all the disciplines shown in Fig. 7.7 is required of RF IC designer in order to design a good RF circuits.

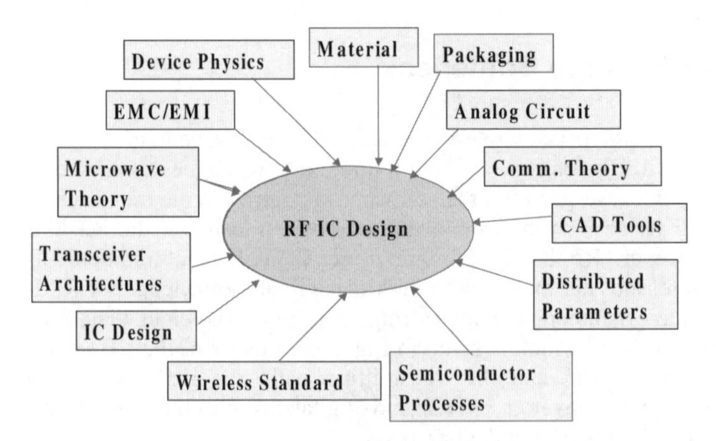

Fig. 7.7 Broad knowledge required for RF IC designs

7.2.2 Teamwork

The RF front-end section is only one of the subsystems defined by wireless applications. RF IC designers need to plan the RF system as system experts and to work with other system experts for other systems; they also need to define and develop individual subblocks in RF systems as IC designers. Meanwhile, they are also manufacturers for passive components according to the different semiconductor processes and must communicate with material as well process engineers. As indicated in Fig. 7.8 that an RF IC designer must work very closely with the people in other fields.

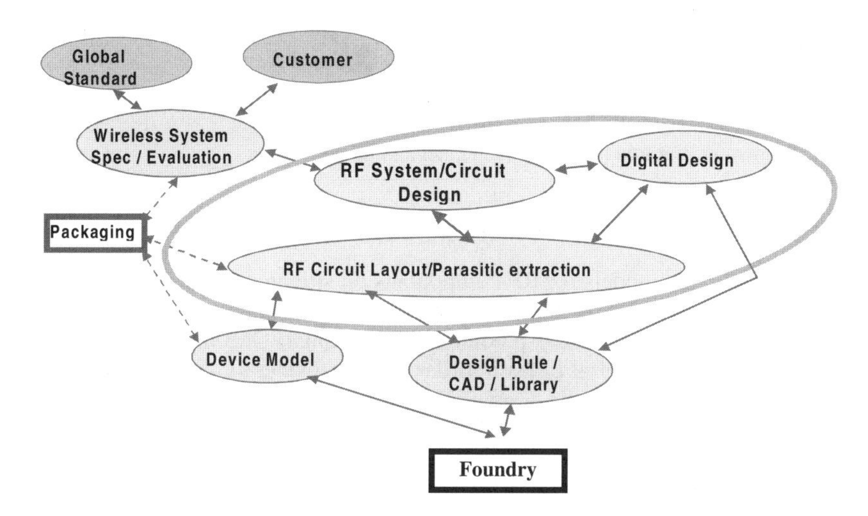

Fig. 7.8 RF design requirements

7.2.3 Trade-Off of Design Requirements

RF integrated circuits must process analog signals received by the antenna with a sufficient dynamic range in order to drive the subsequent ADC as a core requirement. Ideally, the wider the dynamic range is, the better. Normally, the received signals are very weak, and a high-gain low-noise RF front-end is expected with less distortion or highly linear performance. Meanwhile, low voltage levels and low-power consumption are design considerations because of for portability. Unfortunately, all those designs requirements are in contrast to each other as illustrated in Fig. 7.9, which increases the RF IC design difficulties. To obtain an optimised design of an RF IC for certain applications, one has to rely on experience and knowledge for achieving the goals, because electronic-design automation (EDA) tools for analysing and synthesizing RF IC are still in their infancy and inefficiently predict the performance. Simulation is just *models + simulator + experience*, because no EDA tools can generate circuits for the designs, as shown in Eq. (7.1).

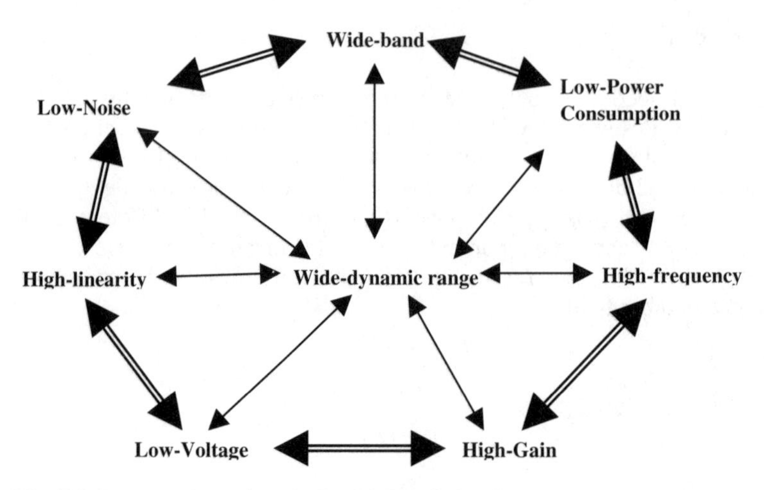

Fig. 7.9 Teamwork required in CMOS RF IC developments

$$\textbf{RF IC Designs = Models + Simulators + Experience} \quad (7.1)$$

Unfortunately, the models for RF components on silicon wafers such as RF on-wafer inductors, capacitors, resistors, varactors, diodes, and transistors are so new they generally cannot accurately predict the behaviours. One has to design and fabricate test chips for several times in order to get satisfactory performance. This increases the RF IC design costs dramatically.

7.3 Concepts of Distributed Circuits

It has been mentioned in Sect. 7.2 that RF circuits are analog circuits, but one has to consider so-called "distributed effects" that are not the considered in normal analog circuit designs, and analysis as well as synthesis, where discrete lumped elements are used. In general, analog circuit analysis and synthesis are very simple and systematic and can be described by four laws: *Kirchhoff's current law (KCL), Kirchhoff's voltage law (KVL), Ohm's law, and* the *superposition principle*. They contain three active components: *transistor, current source, and voltage source*; and three passive components: *resistor, capacitor, and inductor;* as well as *wire* for connecting all these physical components (Table 7.2). The analog circuits are designs following these four laws. Then the discrete lumped components with

Table 7.2 Basic laws and elements of analog circuits

Law	Active Element	Passive Element	Connection
KCL	Transistor	Resistor	
KVL	Current source	Inductor	Wire (perfect)
Ohm's	Voltage Source	Capacitor	
Superposition		Conductor	

the designed values are ordered to realize the circuits according to the designed connections. The connections using wires are assumed to be perfect connections without any extra effects.

For RF IC design on silicon, the principles are the same as those used in analog circuit designs, however, instead of the lumped elements, the distributed circuits must be taken into account. Let's investigate a normal inductor - capacitor (LC) tank with a coil and a two-parallel plate capacitor, for which the resonant angular frequency is $\omega_0 = 1/\sqrt{LC}$. There are only two ways to increase the resonant frequency: reduce the inductance and the capacitance. For a coil inductor one can reduce the turns and the total length to get the small inductance value as shown in Fig.7.10, until the minimum value when a unit-length perfectly conducting wire performs as an inductor!

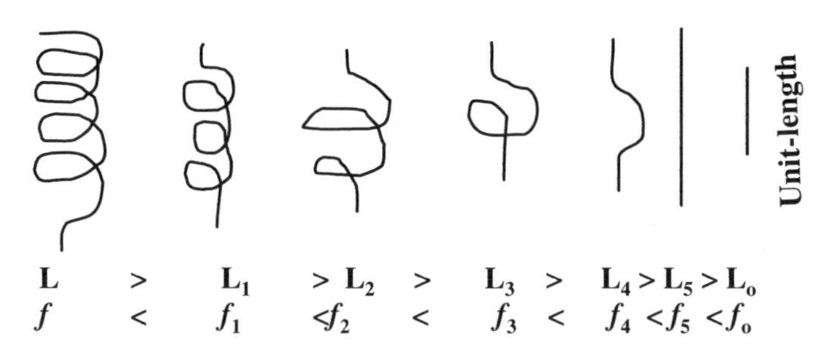

Fig. 7.10 Unit-length perfectly conducting wire has an inductive effect

The same procedure can be followed for a normal capacitor as illustrated in Fig. 7.11 for a two-plate parallel capacitor with the capacitance ($C = \varepsilon Wh / d$). If one reduces the width and the length of the capacitor to the unit lengths respectively, the capacitance value will become minimum as shown in Fig. 7.11. That means there is a capacitive effect between the two wires with unit-length.

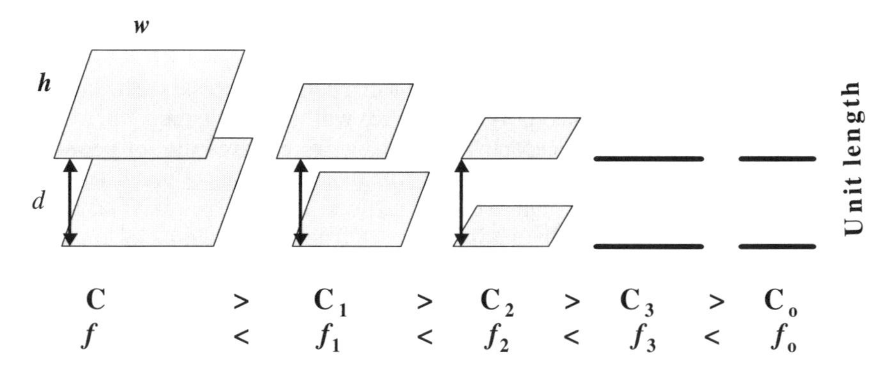

Fig. 7.11 Capacitance Between two parallel lines of unit length

The unit-length perfectly conducting wire has capacitive and inductive effects that are called the distributed effect. Therefore, a physical parallel wire with unit-length can be equivalently expressed as described in Fig. 7.12.

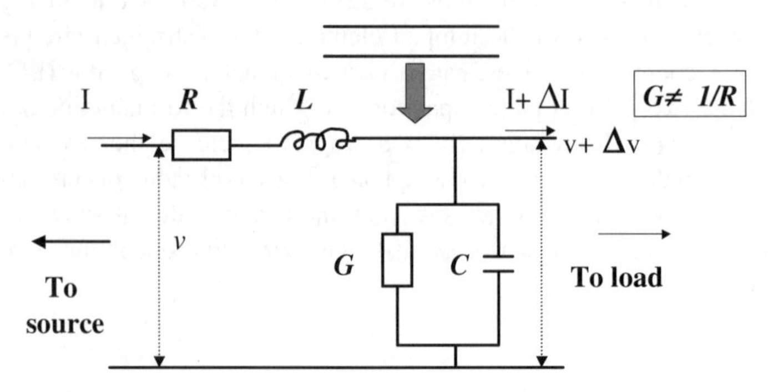

Fig. 7.12 Distributed circuit model for a unit-length parallel wire

Table 7.3 Impedance impedance for L=0.37 nH and C = 0.67 pF for different operating frequencies

Band	f (MHz)	$Z_L = j\omega L$ (Ω)	$Z_C = 1/j\omega C$ (Ω)
HF	3 ~ 30	0.00697 ~ 0.0697	80000~ 8000
VHF	30 ~ 300	0.0697 ~ 0.697	8000 ~ 800
UHF	300 ~ 1000	0.697 ~ 2.3248	800 ~ 238
GSM900	900	2.0943	264
GSM1800	1800	4.1846	132
3G Mobile	1800 ~ 2000	4.1846 ~ 4.650	132 ~ 115
Bluetooth	2450	5.6957	97
HiperLAN	5100 ~ 5900	11.8564 ~ 13.7162	47 ~ 40

The distributed effects are very crucial for the performance of a RF IC system, because the IC must be laid out and fabricated using certain processes. The layouts are the metal traces, metal vias, poly layers, etc. with certain lengths, therefore, they all exhibit the distributed capacitive and inductive effects even for perfectly conducting materials. The capacitance and the inductance values of the distributed effects might be very small, but we can neglect values. We only care the introduced impedance values about. It is expected that a drawn-metal trace to connect two points on a wafer be absolute open to the ground plane. However, because of the distributed capacitance between the trace and the ground plane, the impedance will not be infinity! Table 7.3 gives examples of the distributed effects for different frequencies.

Table 7.4 Basic laws and the elements for RF IC

Law	Active Element	Passive Element
KCL KVL Ohm's Superposition	Transistor Current source Voltage Source	Wire

It can be seen that only when the frequency is very low that the capacitor impedance can be approximately treated as an "open," and if the frequency is in the RF range, the impedance of the capacitance is very small and the effects must be included into the simulations. Therefore, the layouts will also determine the RF performance of a RF design. However, for conventional analog circuit design, these effects are not as crucial, and the introduced distributed effects can be neglected. Now for RF IC designs there is only one passive component – wire (Table 7.4). Therefore, RF IC designers need to design the shapes and the lengths of the used wire segments, which is not as familiar to the conventional IC designers.

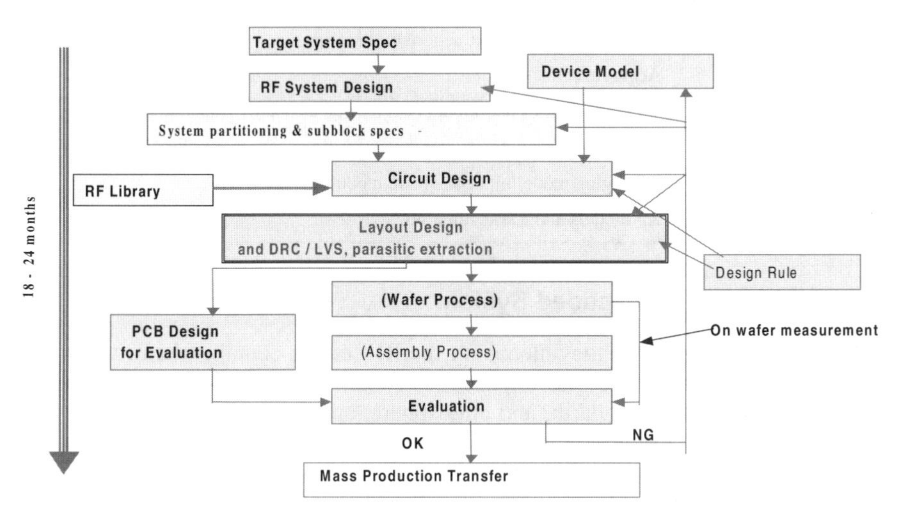

Fig. 7.13 RF IC development flow

7.4 General Design Considerations in RF IC Designs

The general design flow of RF ICs is illustrated in Fig. 7.13, which must be gone through entirely by an RF IC designer for several iterative routines before obtaining an acceptable design with satisfactory performance. The average development for a totally new RF IC is typically around 18 to 24 months, with several rounds of testing, chip fabrication, and measurements. The design difficulties and the final overall performance are mainly determined by the issues listed in Table 7.5. For practical applications, the sensitivity and the selectivity of a wireless terminal are defined by the wireless standards used and the environments. The output of an RF section drives the ADC, and the frequency step of the RF frequency synthesizer shown in Fig.7.6 must follow the given frequency channel spacing defined in the standards. Such necessary knowledge is to define the specifications for the sub-blocks of the RF section. This issue is discussed in detail.

The required design considerations are demonstrated in Fig. 7.9, and in the following section, then we discuss these considerations are explained in detail, then, we discuss the individual block at the circuit level.

Table 7.5 Issues to affect the final overall performance of RF ICs

Issue	Key factors affecting performance
Target system specs	Set by the standards required by the customers
RF system partitioning	EDA tools + good behaviour modelling
Device modelling	Passive and active device modelling including the temperature, biasing and frequency effects
Layouts	Parasitic parameter extractions and routines
Library	Accurate cells
Wafer processes	Technology files and model files as well as quality fluctuations
Interconnect	Extra distributed effects and parasitic effects, cross-coupling effects
Packaging	Packaging modelling, bond-wire models and the EMI
PCB design	RF models and test methodologies

Noise Figure of a Cascaded System

The signal obtained from the antenna by the RF section contains the useful signal and the noise introduced by the environment. The other signals around us include "natural" noise from sky, earth, and atmosphere, and man-made noise such as the signals generated by the power lines, electronic switches, vehicles, motors, machines, radio stations, TV stations, broadcasting systems, other wireless terminals, the signals from other wireless channels, radars. All these are undesired signals. These signals are useless with respect to the needed signal, thus are treated as noise to the needed useful signal.

In addition to the noise picked up by the antenna, the RF section itself generates further extra noise to the final output signal. The RF section must let the useful signal get through without or with less extra noise added to the useful signal. Normally, the detected useful signal is relative very weak for wireless applications, thus the noise added by the RF section becomes crucial. The quality of the useful signal is expressed in terms of its *signal-to-noise ratio* (SNR) [4],[5].

$$\mathrm{SNR} = \frac{\text{wanted signal power}}{\text{unwanted noise power}} \tag{7.2}$$

For example for a mobile radio-telephone system, SNR>15 dB is required from the receiver output [5]. The noise that occurs in an RF receiver masks weak signals and limits the ultimate sensitivity of the receiver.

In general, any sub-block in an RF section can be treated as a two-port network. The noise figure is a figure of merit which quantitatively specifies the noise of a general two-port system. The noise figure depends on several factors such as loss/gain in the block, the devices, the bias applied, and so on. The noise figure is simply the noise factor converted to decibel notation. The noise factor of a two-port network is defined as:

$$F = \frac{\text{SNR at the input}}{\text{SNR at the output}} = \frac{S_i / N_i}{S_o / N_o} = \frac{S_i}{GS_i}\frac{N_o}{N_i} = \frac{1}{G}\frac{N_o}{N_i}, \quad (7.3)$$

where S_i, N_i and S_o, N_o are the input signal power, input noise power and the output signal power as well the output noise power, respectively for a two-port network with a power gain G. Note $N_o \neq GN_i$, but $N_o = GN_i + N_a$, where N_a is the added noise power by the two-port network.

For a cascaded system with n sub-blocks having the power gain G_i and noise factor F_i, the total noise factor F can be found by the individual noise factor related gain as the Friis equation [6]:

$$F = F_1 + \frac{F_2 - 1}{G_1} + \frac{F_3 - 1}{G_1 G_2} + \frac{F_4 - 1}{G_1 G_2 G_3} + \cdots + \frac{F_n - 1}{G_1 G_2 G_3 \cdots G_{n-1}} \quad (7.4)$$

Equation (7.4) indicates that the overall noise factor (or noise figure) of a cascaded system is mainly determined by the first several stages, and particularly, the first stage is crucial in achieving a low overall noise figure. Thus, it is very desirable to have a low noise figure and high gain in the first stage. To use Eq. (7.4), all noise factor F and gain G are related in the ratio. It is clear that if one stage has a loss instead of gain, the noise factor will be "amplified" by this stage, for example, by a filter.

It is worth knowing how high the gain of the first stage would need to be in order to eliminate the noise figure contributions from the subsequent stages in a cascaded system. Suppose we have a five-stage system with following parameters: F_1=2 dB and G_1=0 dB to 20 dB, and all other stages have the parameters NF=5 dB and G=10 dB. The purpose of the example is to investigate how many stages will be required to ignor the noise figure contributions from the other stages.

It can be clearly seen from Fig. 7.14 that only the first two stages contribute much to the overall noise figure of the cascaded system for the example. Therefore, the noise figure contributions from the other stages can be ignored for calculating the overall noise figure of a multistage cascaded system, if there are more than three stages. Also it is known that if the gain of the first stage is higher than 16 dB, the overall noise figure of the system will be determined only by the first stage approximately.

Now we discuss an RF system having three stages with the parameters: F_1=2 dB, G_1=1, 2,⋯, 100; G_2=G_3=10 dB, while the second and the third stages have the same noise figures of F_2=F_3=5 dB, 10 dB, 15 dB, respectively. The purpose of this example is to find the required gain for the first stage, and at what value of the first stage gain that the overall noise figure is dominated by the first one. The results are shown in Fig. 7.15. It can be found that if the gain for the first stage is

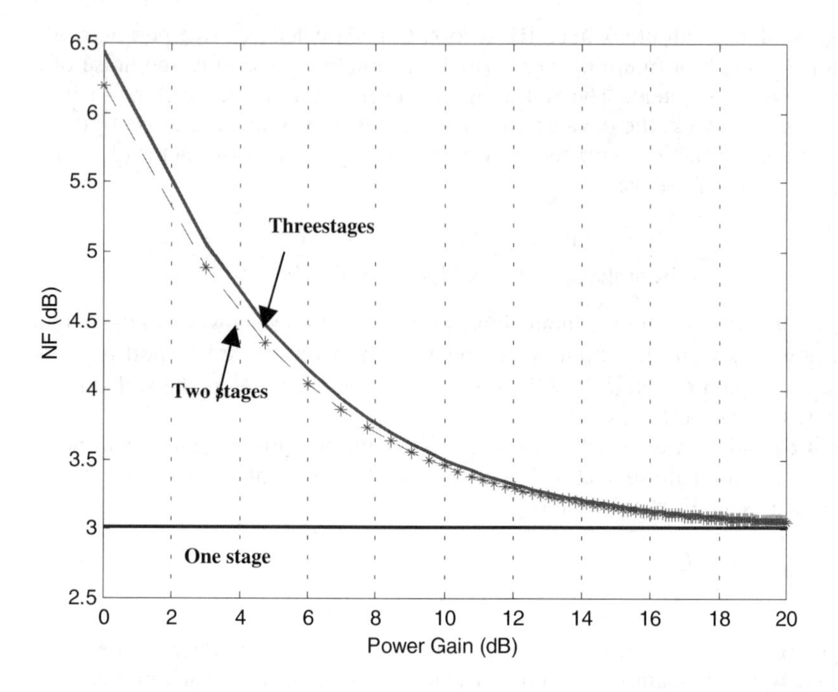

Fig. 7.14 System noise figure showing different numbers of stages versus the gain of the first stage

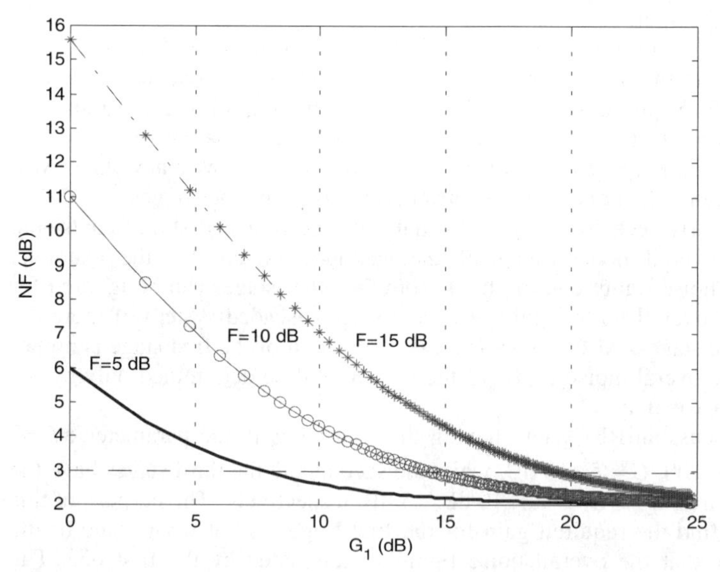

Fig. 7.15 Overall noise figure of a three-stage RF system versus different gain in the first stage

higher than 20 dB, the overall noise figure of the three-stage system will be below 3 dB, even if the noise figures of stages two and three are as high as 15 dB. Figures 7.14 and 7.15 are useful to determine the power gain of the low-noise amplifier (LNA) that is always the first stage in any wireless terminal and is usually followed by a mixer. The noise figure of a typical mixer is about 10~15 dB. If the gain of the LNA is about 15~20 dB, the overall noise figure will be below 3 dB if the LNA has noise figure of 2 dB.

7.5 Dynamic Range and Sensitivity

7.5.1 Sensitivity

For an RF system, operation is normally in a region where the output power is linearly proportional to the input power, while the coefficient is the desired power gain. This region is called the *dynamic range* (DR). Normally, an RF system is connected to the matched antenna, which is equivalent to a 50 Ω resistor, therefore, the input noise is only the noise floor due to the matched resistor as [5]:

$$N_i = kTB, \tag{7.5}$$

where k is the Boltzmann constant, B is the operating frequency bandwidth, and T is the temperature in Kelvin, which gives the noise floor of the input signal. The input signal power level should be at least 3 dB above the noise floor [5]. From Eq.(7.3) it is known that:

$$S_i = F\, SNR_{out}\, N_i \qquad \text{(W)} \tag{7.6}$$

Where S_i is the input signal power. For the given required noise factor F and the minimum required output signal-to-noise ratio SNR_{min}, Eq.(7.6) gives the required minimum input signal power, or sensitivity. Eq.(7.6) can be rewritten in dB more conveniently at room temperature (T=290 K):

$$P_{in,min}(\text{dBm}) = NF + SNR_{min}(\text{dB}) + 10\log(kT) + 10\log(B) + 3$$

$$= -171\ \text{dBm} + 10\log(B) + NF + SNR_{min}(\text{dB}) \tag{7.7}$$

Equation (7.7) is the minimum detectable signal (MDS), which increases with increasing width of the operating frequency bandwidth, or channel width. Generally, the operating frequency bandwidth and the MDS are given by the standards, and the minimum signal-to-noise ratio SNR_{min} is determined by the ADC following the RF section [7].

The power of the noise floor for a given bandwidth is $P_{N,\ floor} = 10\log(kT) + 10\log(B)$. Table 7.6 lists the noise floor powers for some popular wireless standards. Now Eq. (7.7) can be rewritten as:

$$P_{in,min}(\text{dBm}) = P_{N,\,floor} + 3\ \text{dBm} + NF + SNR_{min}(\text{dB}) \tag{7.8}$$

Table 7.6 Noise floor power of some wireless standards

	GSM900/1800	CDMA	PCS	3G	Bluetooth	HiperLAN
Channel Width	200 kHz	30 kHz	15 MHz	~2 MHz	1 MHz	20/23.5 MHz
Noise Floor Power (dBm)	–120.8	–129	–102	–110.8	–113.8	–100.8/–100.1

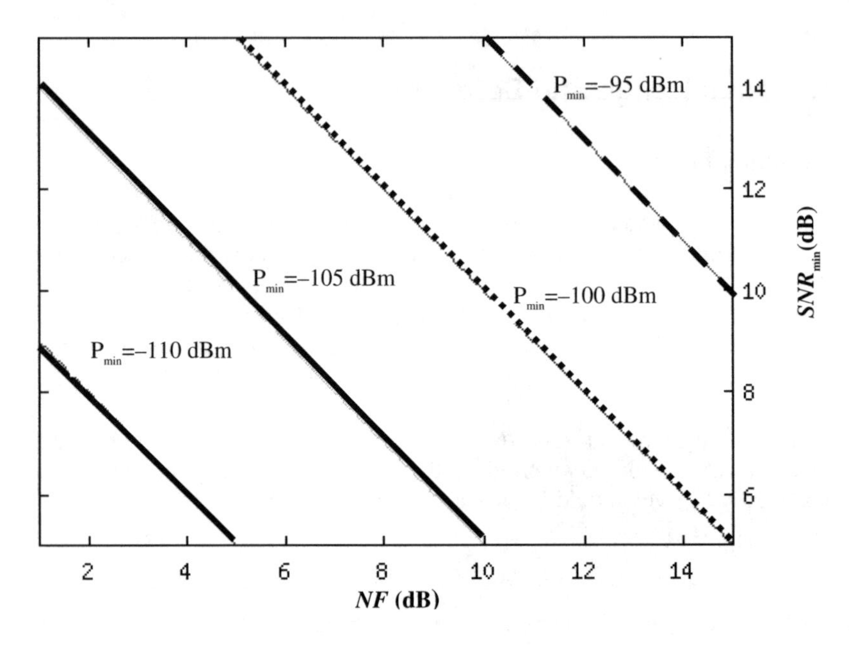

Fig. 7.16 The minimum detectable power of a GSM900 RF front-end for different *NF* and SNR_{min}

For given required minimum *SNR* by the ADC after the RF section, the overall allowed noise figure of the RF section can be calculated from Eq. (7.8) for different wireless standards using Table 7.6. For example, if an RF receiver has NF=9 dB, SNR_{min}=12 dB [4], then the achievable MDS for GSM900/1800, 3G or Bluetooth are –96.8 dBm, –86.8 dBm, or –89.8 dBm, respectively.

For different wireless standards, the minimum detectable input signal power P_{min} is given, which in turn will set the boundary for the choice of the maximum *NF* and the minimum signal-to-noise ratio. In Fig.7.16 the minimum detectable input signal power P_{min} is illustrated for different *NF* and SNR_{min} combinations. For example, if the required *SNR* is 10 dB, it is impossible to realize an RF section with the minimum detectable input signal power P_{min} of –110 dBm, because it can be seen from the figure that for this value of P_{min}, the achievable SNR is only less than 9 dB when NF=0 dB! Thus, Fig. 7.16 provides the trade-off relationship between *NF* and *SNR* for a given minimum detectable input signal power P_{min}.

7.5.2 Dynamic Range and Intermodulation

In an ideal RF system the output power remains a linearly amplified function of the input signal, regardless of the input signal level. Nonlinearities often exist in practical systems and lead to interesting phenomena, which limit the linear operating range. For simplicity, the outputinput relationship can be approximately modelled as:

$$y(t) \approx a_o + a_1 x(t) + a_2 x^2(t) + a_3 x^3(t) \tag{7.9}$$

Where $y(t)$ is the output and $x(t)$ is the input signal, a_o is the DC component, a_1 the gain, and a_2 and a_3 (less than zero) are the coefficients of the second and third nonlinear terms.

The input signal power for a RF system with gain G that produces a 1-dB gain in compression is shown in Fig. 7.17 and is given by:

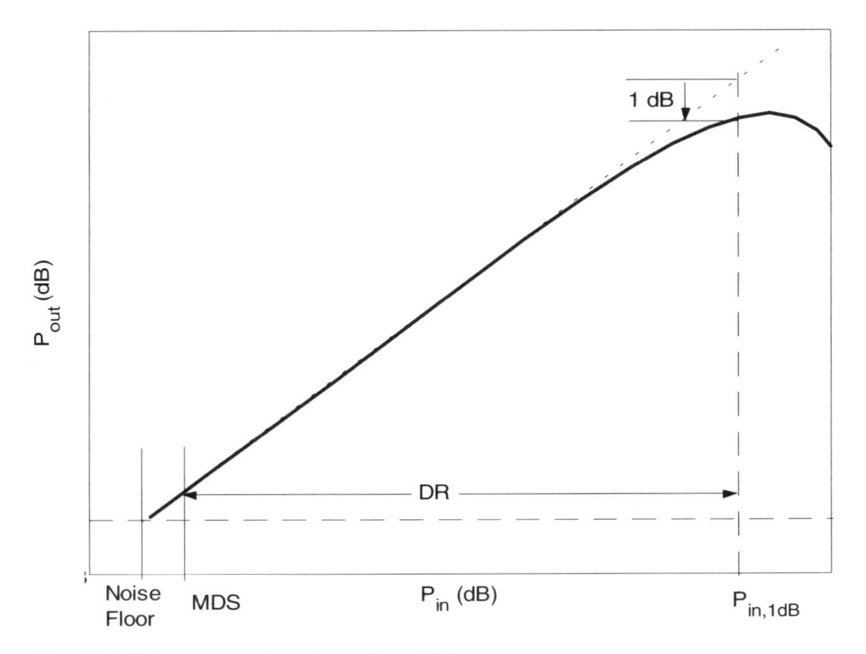

Fig. 7.17 Gain compression of a realistic RF system

$$P_{in,1\text{-}dB} = P_{out,1\text{-}dB} - G + 1\ \text{dB} \tag{7.10}$$

From the 1-dB compression point, the gain, the bandwidth, and the noise figure, the dynamic range (DR) can be calculated as:

$$DR = P_{\text{in,1dB}} - MDS, \qquad \text{(dB)} \tag{7.11}$$

where $P_{\text{in,1dB}}$ and MDS are in dBm.

Circuits with compressive characteristics resulting from the nonlinear relationship between the input and the output signals exhibit a "blocker" effect when a weak, desired signal is accompanied by a strong interferer with a very closed operating frequency to that of the desired signal. Before the desired weak signal is amplified sufficiently, the RF system is already in strong nonlinear region and the gain is reduced drastically because the input power of the unwanted interferer is already beyond the 1-dB point of the system. This leads to the weak signal experiencing a vanishingly small power gain, or even zero power gain. This effect can be investigated by assuming the input signal in Eq.(7.9) and is given by $x(t) = A_0 cos(\omega_0 t) + A_1 cos(\omega_1 t) + A_2 cos(\omega_2 t)$. Then, we have:

$$\begin{aligned} y(t) = a_o &+ a_1 (A_o \cos \omega_o t + A_1 \cos \omega_1 t + A_2 \cos \omega_2 t) \\ &+ a_2 (A_o \cos \omega_o t + A_1 \cos \omega_1 t + A_2 \cos \omega_2 t)^2 \\ &+ a_3 (A_o \cos \omega_o t + A_1 \cos \omega_1 t + A_2 \cos \omega_2 t)^3 \end{aligned} \tag{7.12}$$

We can rearrange Eq.(7.12) as:
DC term:

$$a_0 + a_2 \frac{A_0^2 + A_1^2 + A_2^2}{2} \tag{7.13}$$

Fundamental components:

$$\begin{aligned} &\left\{ a_1 + a_3 \left[\frac{3A_0^2}{4} + \frac{3}{2}(A_1^2 + A_2^2) \right] \right\} A_0 \cos \omega_0 t \\ &+ \left\{ a_1 + a_3 \left[\frac{3A_1^2}{4} + \frac{3}{2}(A_0^2 + A_2^2) \right] \right\} A_1 \cos \omega_1 t \\ &+ \left\{ a_1 + a_3 \left[\frac{3A_2^2}{4} + \frac{3}{2}(A_1^2 + A_0^2) \right] \right\} A_2 \cos \omega_2 t \end{aligned} \tag{7.14}$$

Harmonic components:

$$\begin{aligned} &\frac{a_2}{2}[A_0^2 \cos 2\omega_0 t + A_1^2 \cos 2\omega_1 t + A_2^2 \cos 2\omega_2 t] \\ &+ \frac{3a_3}{4}[A_0^3 \cos 3\omega_0 t + A_1^3 \cos 3\omega_1 t + A_2^3 \cos 3\omega_2 t] \end{aligned} \tag{7.15}$$

Mixing two-components:

$$\begin{aligned} a_2 \{ &A_0 A_1 [\cos(\omega_0 + \omega_1) + \cos(\omega_0 - \omega_1)] \\ &+ A_0 A_2 [\cos(\omega_0 + \omega_2) + \cos(\omega_0 - \omega_2)] \\ &+ A_2 A_1 [\cos(\omega_2 + \omega_1) + \cos(\omega_2 - \omega_1)] \} \end{aligned} \tag{7.16}$$

And the following intermodulation products:

$$
\begin{aligned}
& a_3\left\{\frac{3}{4}A_0^2\{A_1[\cos(2\omega_0+\omega_1)t+\cos(2\omega_0-\omega_1)t]\right. \\
& \left.+A_2[\cos(2\omega_0+\omega_2)t+\cos(2\omega_0-\omega_2)t]\}\right\} \\
& +a_3\left\{\frac{3}{4}A_1^2\{A_0[\cos(2\omega_1+\omega_0)t+\cos(2\omega_1-\omega_0)t]\right. \\
& \left.+A_2[\cos(2\omega_1+\omega_2)t+\cos(2\omega_1-\omega_2)t]\}\right\} \\
& +a_3\left\{\frac{3}{4}A_2^2\{A_1[\cos(2\omega_2+\omega_1)t+\cos(2\omega_2-\omega_1)t]\right. \\
& \left.+A_2[\cos(2\omega_2+\omega_0)t+\cos(2\omega_2-\omega_0)t]\}\right\} \\
& +\frac{3}{2}a_3A_0A_1A_2\left[\cos(\omega_0+\omega_1+\omega_2)t+\cos(-\omega_0+\omega_1+\omega_2)t\right. \\
& \left.+\cos(\omega_0+\omega_1-\omega_2)t+\cos(\omega_0-\omega_1+\omega_2)t+\right]
\end{aligned}
\tag{7.17}
$$

If ω_0 is the desired signal then the gain will be $a_1+1.5a_3(0.5A_0^2+A_1^2+A_2^2)$, which decreases because $a_3<0$. If the unwanted signal strengths A_1 and A_2 are very strong, the gain of the wanted signal drops to or below 1 when:

$$A_1^2+A_2^2\geq\frac{2}{3}\frac{a_1}{|a_3|}-\frac{1}{2}A_0^2 \tag{7.18}$$

Now the wanted signal is "blocked" by the unwanted strong signal, because the wanted signal cannot be amplified by the RF section. Many RF sections in wireless applications must able to withstand blocking signals 60–70 dB stronger than the wanted signal [4].

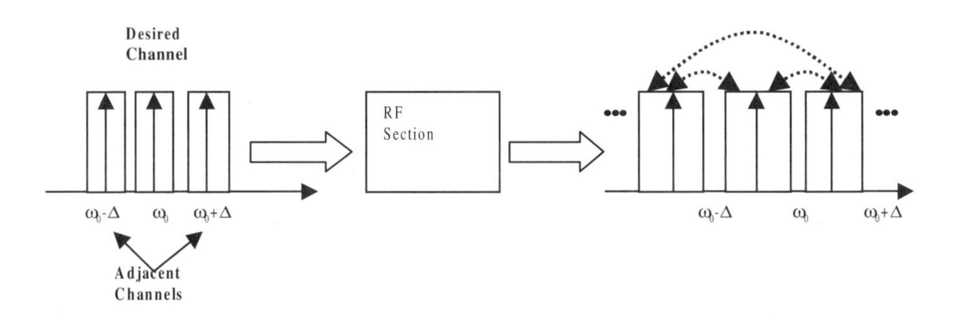

Fig. 7.18 Intermodulation in an RF section with nonlinear effects

Now we discuss the so-called "intermodulation" (IM) phenomena which arise from "mixing" of the wanted signal ω_0 with undesired signals of the adjacent channels $\omega_1 = \omega_0-\Delta$ and $\omega_1 = \omega_0+\Delta$. Using Eq.(7.17)we have the third-order IM products at:

$$2\omega_0-\omega_2=\omega_0-\varDelta;\qquad 2\omega_0-\omega_1=\omega_0+\varDelta;$$
$$2\omega_1-\omega_0=\omega_0-2\varDelta;\qquad 2\omega_2-\omega_0=\omega_0+2\varDelta;$$
$$2\omega_1-\omega_2=\omega_0-3\varDelta;\qquad 2\omega_2-\omega_1=\omega_0+3\varDelta;$$
$$\omega_1+\omega_2-\omega_0=\omega_0;\qquad \omega_1-\omega_2+\omega_0=\omega_0-2\varDelta;$$
$$\omega_2-\omega_1+\omega_0=\omega_0-2\varDelta;$$

then,

$$\begin{aligned}
y(t)=&\left\{1-\frac{|a_3|}{a_1}\left[\frac{3A_0^2}{4}+\frac{3}{2}(A_1^2+A_2^2+A_1A_2)\right]\right\}a_1A_0\cos\omega_0 t\\
&\left\{1-\frac{|a_3|}{a_1}\left[\frac{3(A_0^2\frac{A_2}{A_1}+A_1^2)}{4}+\frac{3}{2}(A_0^2+A_2^2)\right]\right\}a_1A_1\cos(\omega_0-\Delta)t\\
&+\left\{1-\frac{|a_3|}{a_1}\left[\frac{3(A_0^2\frac{A_1}{A_2}+A_2^2)}{4}+\frac{3}{2}(A_0^2+A_1^2)\right]\right\}a_1A_2\cos(\omega_0+\Delta)t\\
&+a_3\frac{3A_0}{2}\left\{\left[\frac{1}{2}A_1^2+A_1A_2\right]\cos(\omega_0-2\Delta)t+\left[\frac{1}{2}A_2^2+A_1A_2\right]\cos(\omega_0+2\Delta)t\right\}+\\
&+a_3\frac{3}{4}A_1A_2\{A_1\cos(\omega_0-3\Delta)t+A_2\cos(\omega_0+3\Delta)t\}+\text{others}
\end{aligned}\tag{7.19}$$

In the literature the third-order IM products are studied only for two-tone cases, that is, where there are only two input signals with the frequency ω_1 and ω_2, and in order to find the coefficients of $2\omega_1-\omega_2$ and $2\omega_2-\omega_1$ as third-order IM products, if the differences between the two frequencies are small [4, 5]. However there are actually many channels in a wireless communication system, which are very close to each other for the adjacent channels. For instance, in GSM900 there are 124 channels with channel spacing of 200 kHz, which becomes the total RF receiver bandwidth (from 935–960 MHz). That is, if a desired channel is at ω_0, its two adjacent channels are $\omega_o\pm\varDelta$ with $\varDelta$=200 kHz. Therefore, as the first-order approximation, the three-tone test should be closer to the practical cases for investigating third-order IM product effects. As can be seen from the above expression, besides the well-known the third-order IM products for the two-tone cases, $2\omega_1-\omega_0$ and $2\omega_0-\omega_1$ at $\omega_o-2\Delta$ and $\omega_o+\varDelta$, respectively, there are more terms in this range, even at ω_o. Thus, we obtain Eq. (7.19).

As shown in Fig. 7.18, there are IM effects between any two channels. It can be seen from Eq. (7.19) that the IM effects at the desired channel are different from those at the two adjacent channels. The IM products are located exactly in the desired channel, and it is impossible to remove them from the channel because their frequency is exactly the same as that of the desired channel.

Like the typical two-tone test [4, 5], $A_0 = A_1 = A_2 = A$, because statistically all channels should have the same magnitudes for their own signals. Traditionally, the third-order IM products are considered as components with frequencies at $2\omega_1-\omega_2$ and$-2\omega_2-\omega_1$ or $\omega_0-2\Delta$ and $\omega_0+\Delta$, respectively, and the coefficients of the conventional IM products are $3a_3A^3/4$ [4, 5]. However, from Eq. (7.19) it is obvious that the third-order IM products have the same frequency as the desired channel at ω_0 with the coefficient $9a_3\ A^3/2$, which is not the coefficient of the components at frequency $\omega_0-2\Delta$ and $\omega_0+\Delta$, and the gain compression is due to the term $3a_3A^3/4$.

The linear gain is a_1 and the output is proportional to the input signal level A, whereas the third-order IM products increase in proportion to A^3. Now Eq. (7.19) can be rewritten as:

$$\begin{aligned} y(t) &= \left\{1-\frac{|a_3|}{a_1}\left[\frac{3A_0^2}{4}+\frac{9}{2}A^2\right]\right\}a_1A_0\cos\omega_0 t \\ &+\left\{1-\frac{|a_3|}{a_1}\left[\frac{3A_1^2}{4}+3\frac{3}{4}A^2\right]\right\}a_1A_1\cos(\omega_0-\Delta)t \\ &+\left\{1-\frac{|a_3|}{a_1}\left[[\frac{3A_2^2}{4}+3\frac{3}{4}A^2\right]\right\}a_1A_2\cos(\omega_0+\Delta)t+\text{others} \end{aligned} \tag{7.20}$$

It is clear from Eq. (7.20) that the third-order IM product effects are different for the desired channel and the adjacent channels. Additionally, the three-tone output faces the blocker effect earlier than that of the two-tone test, as shown in Fig. 7.19.

The 1-dB compression point can be obtained from Eq. (7.20) for the wanted channel as:

$$20\log(a_1A_{1dB})-20\log(a_1A_{1dB}-|a_3|\frac{21}{4}A_{1dB}^3)=1 \quad \text{(dB)} \tag{7.21}$$

or

$$|a_3|=0.0207\frac{a_1}{A_{1-dB}^2} \quad \text{or} \quad A_{1-dB}=\sqrt{0.0207\frac{a_1}{|a_3|}} \tag{7.22}$$

For comparison, the 1-dB compression point for the traditional two-tone test is [4]:

$$A_{1dB}=\sqrt{0.145\frac{a_1}{|a_3|}} \tag{7.23}$$

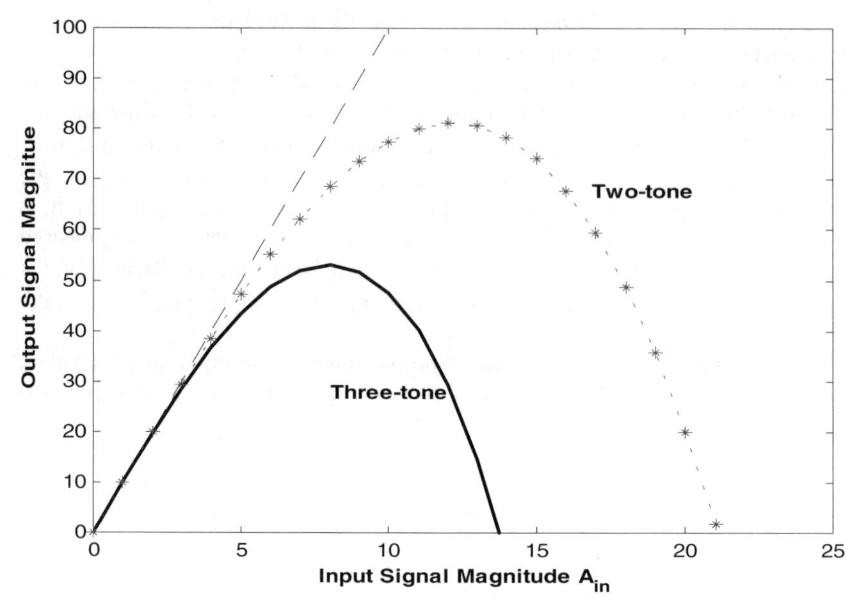

Fig. 7.19 The output signal versus the input signal amplitude for the three-tone and the two-tone tests, respectively

Thus, from the measured linear gain a_1 and the input level at the 1-dB compression point, one can calculate the nonlinear coefficient $|a_3|$ using Eq.(7.22). Obviously, the 1-dB compression point for the three-tone test is lower than that of the two-tone test as:

$$20\log(A_{1dB}[\text{two tone}]) - 20\log(A_{1dB}[\text{three tone}]) = 8.45\,(\text{dB})$$

Similarly, the third-order IM products occur at the point when the output component of the desired channel signal at ω_0 has the same magnitude as the effects from the adjacent channels' signals (the term $9A^2/2$) as:

$$a_1 A_{IP3} = \frac{9}{2}|a_3|A_{IP3}^3, \quad \text{or} \quad A_{IP3} = \sqrt{\frac{2}{9}\frac{a_1}{|a_3|}} \tag{7.24}$$

$$\text{or} \qquad \frac{A_{IP3}}{A_{1dB}} = \sqrt{\frac{2}{9\times 0.0207}} = 3.276 \qquad (\sim 5.154\,\text{dB})$$

Equations (7.22) and (7.24) are much different from those in the literature [4–5]. We show the output magnitude and the newly defined IM product as well as the traditionally defined IM product in Fig. 7.20. The figure shows that the traditional intersection point (IP3) prediction is much higher than the value defined here. It also indicates that before the output reaches the output intersection point (OIP3), the blocker effect has already been occurred. The comparison of the IP3 for the three-tone and the two-tone tests are depicted in Fig. 7.20 and

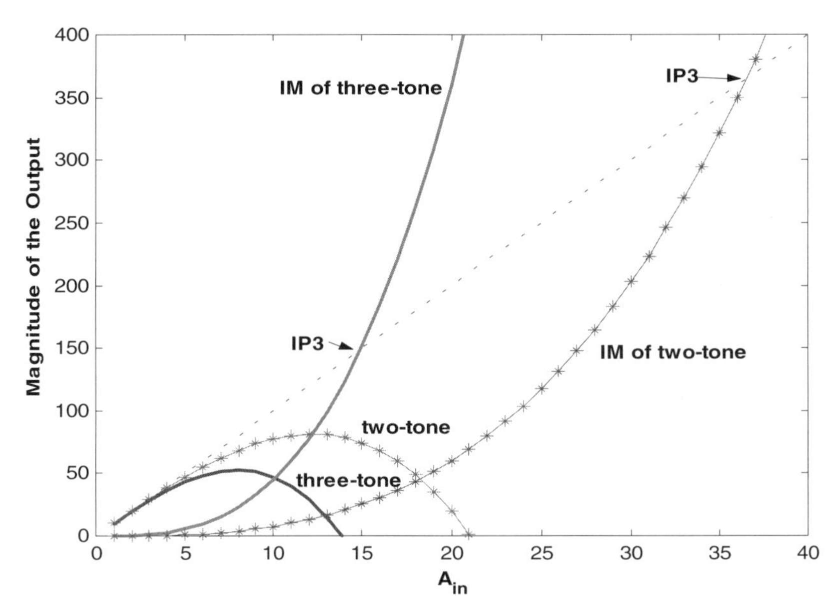

Fig. 7.20 Output of the desired channel signal with IM distortion and compression effects and the growth of the intermodulation for the new definition and the definition in the literatures

7.21, which indicate that the IP3 as well as the input intersection point (IIP3) and OIP3 will be lower because of the adjacent channel effects than that of the two-tone test.

The quality of the wanted channel signal can be estimated using the ratio of the intermodulation signal strength to the total output signal strength at the desired frequency in percentages as:

$$\text{IM Effect} = \frac{\text{IM output at } \omega_0}{\text{Total output}} \frac{18\frac{|a_3|}{a_1}A_0^2}{4-21\frac{|a_3|}{a_1}A_0^2} \tag{7.25}$$

The IM effect on the desired channel signal is displayed in Fig. 7.22. Here it is assumed that the signal magnitudes of all the channels are the same. The IM effect of the adjacent channel on the desired channel signal is equivalent to the add noise contribution on the desired output signal. Note that now the IM products have the same frequency as the wanted signal. The noise performance discussed in this section has not taken this noise contribution into account. Practically, the noise contributions added by the IM of the adjacent channels must be considered.

From Fig. 7.22 it can be seen that if a_1/a_3 is great than 10^4 for the input signal magnitude less than 20, the IM contribution to the desired signal output is negligible. However, the noise contributed by the adjacent channels is much stronger than that from the circuit devices.

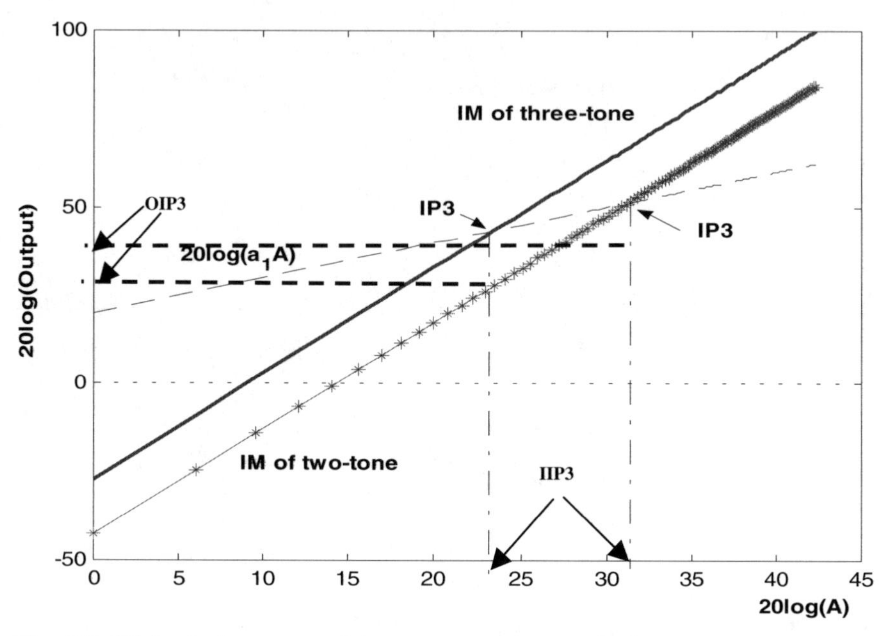

Fig. 7.21 IP3 for the three-tone and the two-tone tests, respectively

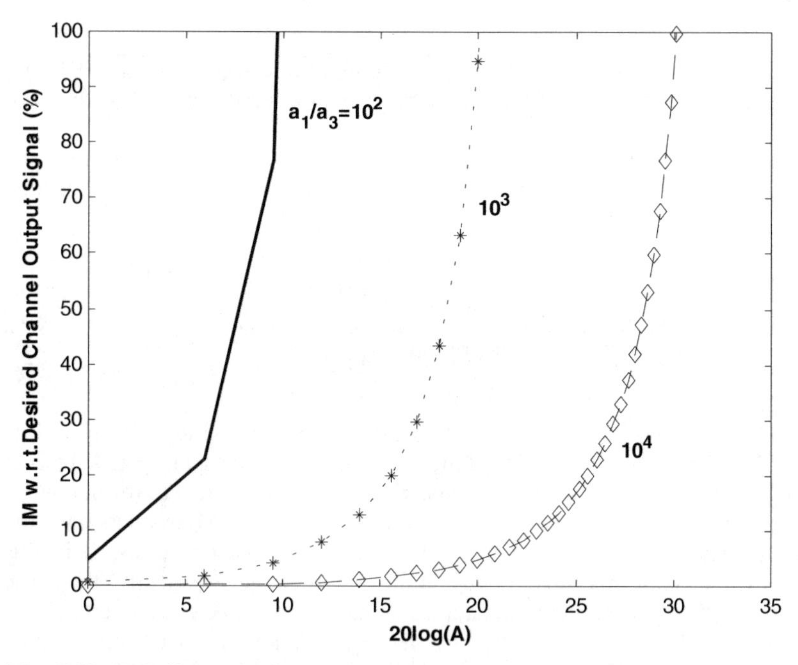

Fig. 7.22 IM effects of adjacent channels on desired channel signal in percentage with respect to (w.r.t) the desired channel output signal for different ratio of a_1/a_3

7.5.3 Nonlinear Effects of the Cascaded RF Systems

Since RF systems consist of many stages to process the signals, each stage also faces the linearity problem because of the overall dynamic range requirements for wireless applications. In Sect 7.5.2 the nonlinear effects such as the 1-dB compression point and the intermodulation products for different tones were discussed. It is also desirable to know how the entire 1-dB compression point and the IM products are determined in terms of the 1-dB points and the IP3 of each stage.

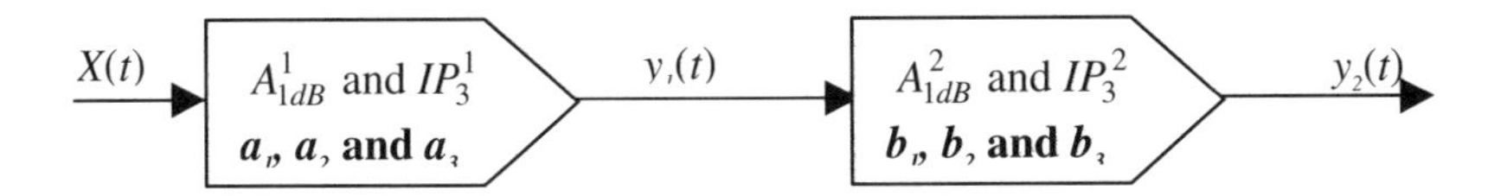

Fig. 7.23 Two RF stages in cascade

Consider two RF stages with nonlinearities shown in Fig. 7.23:

$$y_1 = a_1 x + a_2 x^2 + a_3 x^3, \qquad y_2 = b_1 y_1 + a_2 y_1^2 + a_3 y_1^3 \tag{7.26}$$

Or

$$y_2 = a_1 b_1 x + [b_1 a_3 + 2a_1 a_2 b_2 + a_1^3 b_3] x^3 + \cdots \tag{7.27}$$

The input signal is:

$$x = A_0 \cos \omega_0 t + A_1 \cos \omega_1 t + A_2 \cos \omega_2 t \tag{7.28}$$

With $\omega_1 = \omega_0 - \Delta$ and $\omega_2 = \omega_0 + \Delta$, where Δ is the channel spacing.

Substituting Eq.(7.28) into (7.29) and using (7.19), we have:

$$y_2 = a_1 b_1 A_0 \cos \omega_0 t + [b_1 a_3 + 2a_1 a_2 b_2 + a_1^3 b_3] \cdot \left[\frac{3}{4} A_0^2 + \frac{3}{2}(A_1^2 + A_2^2 + A_1 A_2) \right] A_0 \cos \omega_0 t + \cdots \tag{7.29}$$

Normally, $a_2 b_2 << a_1$ and b_1, thus the term $2a_2 b_2 / b_1$ can be ignored.

1-dB Compression Point

Assume that the magnitudes of the channels' signals are the same, i.e. $A = A_0 = A_1 = A_2$, then we have:

$$\frac{1}{(A_{1dB})^2} = \frac{1}{(A_{1dB}^1)^2} + \frac{a_1^2}{(A_{1dB}^2)^2} \tag{7.30}$$

where A_{1dB} is the overall 1-dB compression point (voltage quantities) of the two-stage system, and the terms in the right side are the 1-dB points for each stage. Then a_l is the voltage gain of the first stage, and Eq. (7.30) can be written in power quantities as:

$$\frac{1}{P_{1dB}} = \frac{1}{P_{1dB,1}} + \frac{G_1}{P_{1dB,2}} \tag{7.31}$$

Equation (7.31) can be extended to multistage systems:

$$\frac{1}{P_{1dB}} = \frac{1}{P_{1dB,1}} + \frac{G_1}{P_{1dB,2}} + \frac{G_1 G_2}{P_{1dB,3}} + \frac{G_1 G_2 G_3}{P_{1dB,4}} + \cdots \tag{7.32}$$

Therefore, the later stages have more effect on the overall 1-dB compression point of the system. By increasing the 1-dB compression point of the multistage system, the gains of the each stage should be lower, particularly for the first stage, which is in contrast to the noise figure requirement as shown in Fig. 7.15. This is another trade-off for any RF IC design.

IP3

The intermodulation effect of the adjacent channels on the desired channel should be the term contributed by the adjacent channels in Eq.(7.29):

$$[b_1 a_3 + 2a_1 a_2 b_2 + a_1^3 b_3]\frac{3}{2}(A_1^2 + A_2^2 + A_{\cdot}A_1)A_0 \cos\omega_0 t$$

Then

$$\frac{1}{A_{IP3}^2} = \frac{1}{A_{IP3,1}^2} + \frac{a_1^2}{A_{IP3,2}^2} - 9a_2\frac{b_2}{b_1} \approx \frac{1}{A_{IP3,1}^2} + \frac{a_1^2}{A_{IP3,2}^2} \tag{7.33}$$

Note here A denotes the magnitude of the signals, and (7.33) can be expressed in terms of power with respect to the same impedance as:

$$\frac{1}{P_{IP3}} = \frac{1}{P_{IP3,1}} + \frac{G_1}{P_{IP3,2}} - 9a_2\frac{b_2}{b_1} \approx \frac{1}{P_{IP3,1}} + \frac{G_1}{P_{IP3,2}} \tag{7.34}$$

This equation readily gives a general form for a multistage system:

$$\frac{1}{P_{IP3}} = \frac{1}{P_{IP3,1}} + \frac{G_1}{P_{IP3,2}} - 9a_2\frac{b_2}{b_1} \approx \frac{1}{P_{IP3,1}} + \frac{G_1}{P_{IP3,2}} + \frac{G_1 G_2}{P_{IP3,3}} + \cdots \tag{7.35}$$

Like the case for the 1-dB compression point of a multistage system, the overall IP3 depends mainly on the later stages divided by the power gain of the proceeding stages. That is because the IP3 and 1-dB point of each stage are equivalently scaled down by the gains before that stage. Equation (7.35) can be rewritten as:

$$P_{IP3}\mid_{\text{dBm}} = P_{IP3,N}\mid_{\text{dBm}} - \sum_{i=1}^{N-1} G_i \mid_{\text{dBm}} - 10\log\left[1 + P_{IP3,N}\sum_{i=1}^{N-1}\frac{1}{P_{IP3,i}}\frac{1}{\prod_{j=1}^{N-i} G_j}\right] \tag{7.36}$$

The above equation indicates that the overall IP3 is less than the first two terms. Reducing the gains in the proceeding stages can improve the overall IP3. However, that will degrade the NF performance.

As an example, we consider a two-stage RF front-end consisting of a low-noise amplifier (LNA) with NF=2 dB, $IIP3$=−10 dBm and a mixer with NF = 12 dB and $IIP3$ = 5 dBm, 10 dBm, and 15 dBm, respectively. The plots are given in Fig. 7.24 for the gain of the LNA from 0 dBm to 45 dBm. It can be seen that higher values of $IIP3$ of the second stage do not necessarily lead to increase overall values of $IIP3$. The overall $IIP3$ is poorer than that of the first stage because of the gain of the first stage. It is seen from Fig. 7.24 that if the gain of the LNA is around 15 dBm and the IP3 of the mixer is about 10 dBm, the overall NF and the IP3 of the two-stage RF front-end can perform optimum.

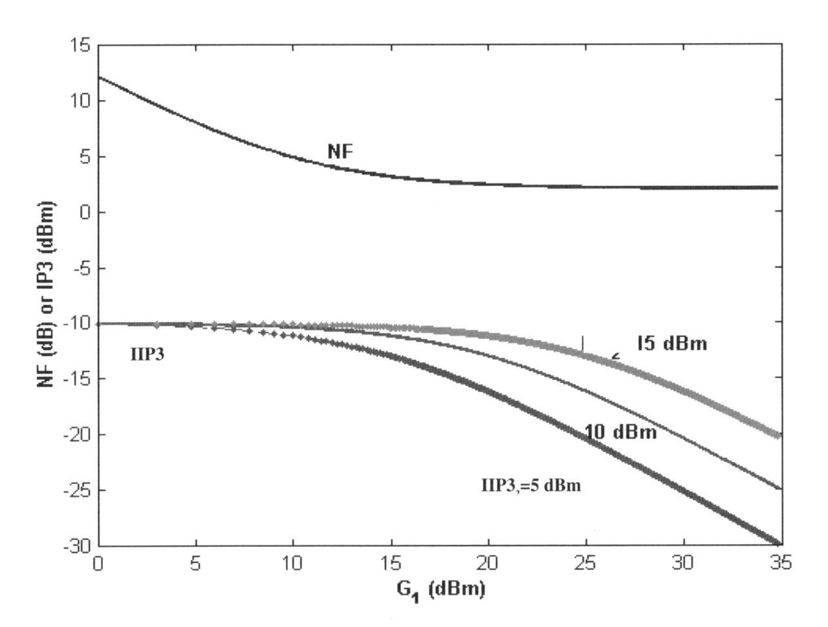

Fig. 7.24 The overall NF and $IIP3$ of a two-stage RF system with NF_1 = 2 dB, NF_2 = 12 dB; $IIP3_1$ = −10 dBm, and IIP_2 = 5 dBm, 10 dBm, and 15 dBm, respectively, where G_1 is the power gain of the first stage.

Spurious-Free Dynamic Range (SFDR)

Another often used definition of dynamic range is the "spurious-free dynamic range (SDR)" that characterizes a receiver with more than one signal applied to the input. For the case of input signals at equal levels like the multichannel mobile systems, the SFDR is illustrated in Fig. 7.25. It is obvious that the adjacent channel signals greatly reduce the SFDR.

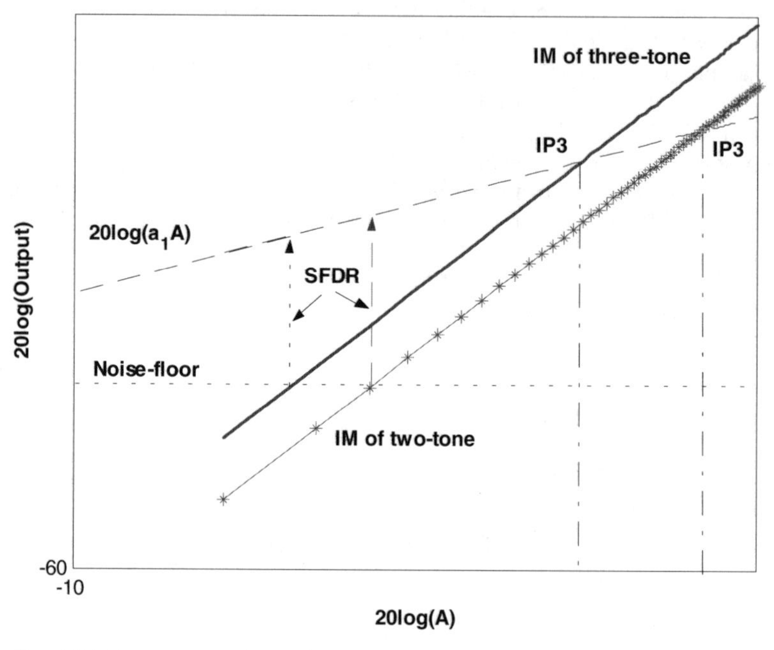

Fig. 7.25 Spurious-free dynamic range

All the above-mentioned discussions indicate that the effects of the adjacent channels on the main channel signal must be taken into account when a RF system is designed.

7.6 RF Front-End Receiver Architectures

An RF front-end receiver picks up the modulated RF carrier signal from its antenna with a very high carrier frequency, thus, the signal cannot be processed directly in the baseband processors. The carrier signal must be downconverted into a low-frequency signal and meanwhile provide sufficient gain to the signal, because the received RF signal from the antenna is normally too weak to drive the baseband processors. After downconversion, the wanted signal is recovered by the baseband processors. The performance of the RF receiver depends strongly on the system design (receiver architectures), circuit design (circuit configurations), and the working environment (operating frequency/bandwidth, EDA tools, models of the used devices as well elements, and fabrication processes). The acceptable distortion and noise vary with the applications. Noise and interference determine a lower limit on the wanted signal level at the output of the receiver. The conceptual RF receiver is illustrated in Fig. 7.26, where a local oscillator (LO) generates a signal with frequency f_{LO}, and the output is the difference between the two frequencies.

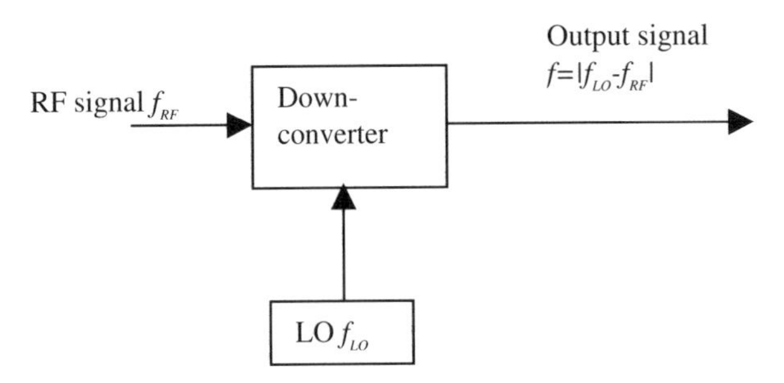

Fig. 7.26 Conceptual RF receiver

To design an RF front-end receiver, the general technical considerations are: *sensitivity, selectivity, IM rejection, spurious rejection, frequency stability, circuit quality factor (Q), and EMI suppression.* On the other hand, the commercial market requires more and more miniaturization, which leads RF IC designers to develop innovative architectures with higher and higher integration as well as fewer and fewer external components. There is always a trade-off between the technical considerations and the market trends to decide which architectures should be used. An important transition from voice-oriented wireless (like 2G cellular systems) to multimedia-oriented systems (such as 3G mobile systems), is occurring, which will bring a different focus to the technical considerations. This emphasizes the new challenges caused by the required feature integration.

The main task of a front-end RF receiver is the frequency translation from RF to intermediate-frequency (IF) as shown in Fig. 7.27. In the frequency domain it is equivalent to shifting the reference frequency point from the LO frequency to the intemediate frequency point.

The sensitivity refers to the absolute magnitude of the desired signal, and the selectivity refers to separating the desired signal from other in-band and out-band channel signals. Architectures in the RF front-end receiver IC [4–25] mainly determine how to choose the LO and the IF frequencies so that the required specifications can be met. Assuming the bandwidth of the RF carrier is BW, the IF frequency ($=|f_{LO}-f_{RF}|$) selection is key compared to the BW. There are only three cases, namely, IF equals zero, IF is less than BW/2, or IF is greater than BW/2. Therefore, there are only three possible architectures:

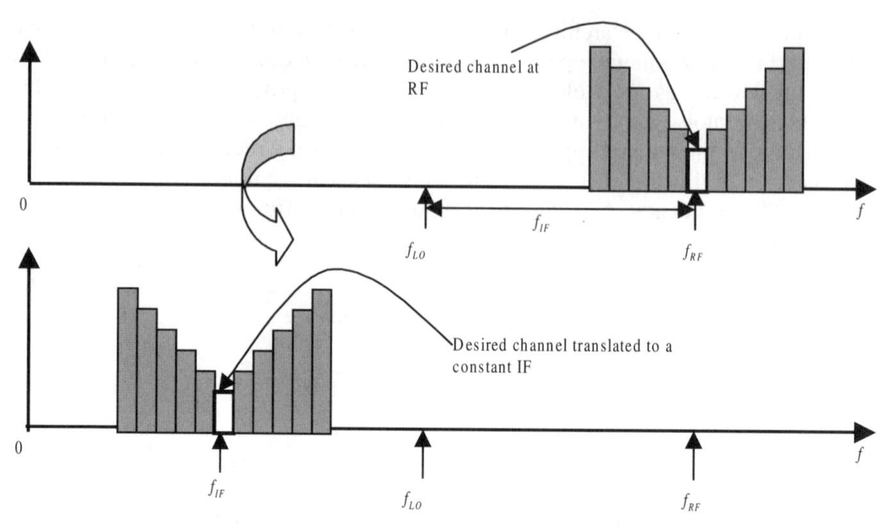

Fig. 7.27 Conceptual frequency translation of a general RF receiver

- IF=0 ($f_{LO} = f_{RF}$): This architecture is called the zero-IF receiver (ZIF) [12]. Since the output frequency of the receiver is in the baseband, the architecture is also named as direct-conversion receiver (DCR) [4, 14, 16–18].
- IF<BW/2 (IF<<f_{LO}): This architecture is called the near-zero-IF (NZIF) receiver [12], because the IF is close to zero compared to the LO and RF frequencies. Most often, it is called the low-IF receiver [4, 21, 22, 24].
- IF>BW/2: In this case the used IF is much higher than the baseband, and the architecture is called a superheterodyne receiver (SHD) [4, 9, 13–15, 19, 20, 23, 25],

7.6.1 Direct Conversion Receiver

In 1924 the first direct-conversion receiver (DCR) was introduced by Colebrook [26]. At that time he used the term of *homodyne* in order to distinguish from the *heterodyne* invented by Amstrong in 1918. The DCR was further developed by Tucker [27, 28]. Over the last decade the concept has been revived because DCR has the potential to reach the "single-chip RF receiver" goal (14, 17, 18). The concept is shown in Fig. 7.28.

The architecture is very simple and fewer external components are needed. In addition, it is very suitable for integration as well as multiband, wideband, and multistandard applications. Even though this architecture is very simple, many issues have to be solved. The most serious problem could be the DC offset problem, because the RF carrier frequency and the local oscillator frequency are the same. DC offset is generated mainly by the "self-mixing" phenomenon (Fig. 7.29).

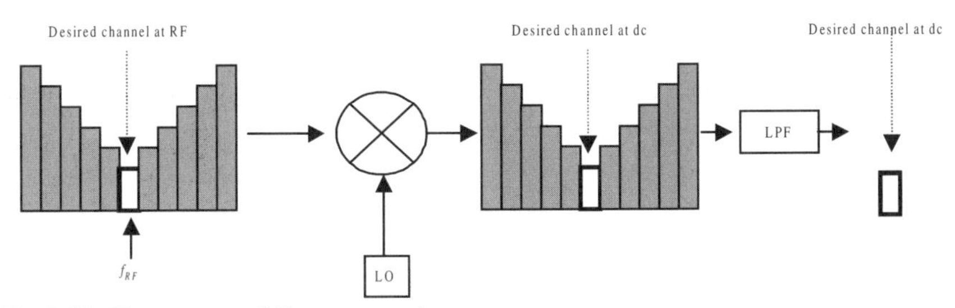

Fig. 7.28 The concept of direct-conversion

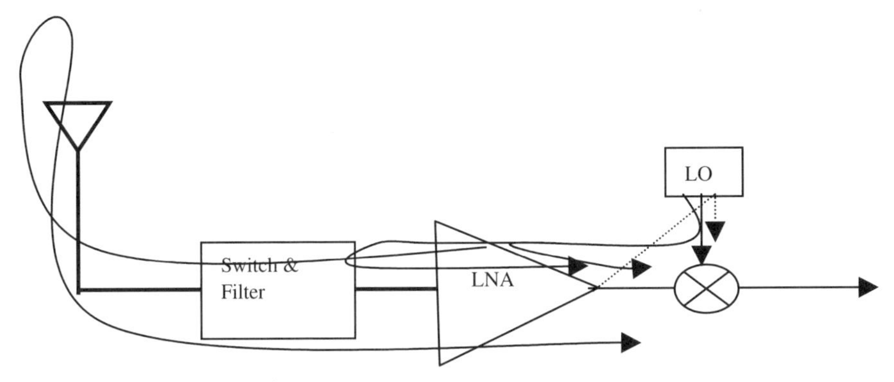

Fig. 7.29 DC offset mechanism

The "self-mixing" effect has two main causes: LO leakage (solid lines in Fig. 7.29), and the strong interferer received by the receiver (dashed line in Fig. 7.29). The Si-CMOS substrate is a high-loss substrate ,and therefore the LO signal can be coupled to the input port of the receiver. Then, the coupled LO signal is reflected by the amplifier to the mixer because of the mismatch between these two elements; or LO signal provides feedback to the input of the amplifier and comes to the mixer with the wanted signal, or the LO leaks to the antenna and from the antenna receives the LO signal again. The LO frequency is the same as the wanted channel signal; therefore, it is impossible to distinguish the LO leakage signal from the wanted signal and the unwanted leakages will mix with the direct LO signal in the mixer. If a very strong interferer is received by the receiver, even if the frequency is different from that of the wanted signal, it can go through the amplifier and leak to the LO and be futher reflected by the LO to the mixer. This reflected interferer will mix with itself in the mixer. This self-mixing produces a DC output, because the signal have the same frequencies, i.e.

$$V_{out} = V_{leakage} V_{LO} \cos^2 \omega_{LO} t = \frac{1}{2} V_{leakage} V_{LO} (1 + \cos 2\omega_{LO} t) \tag{7.37}$$

The first term in Eq. (7.37) contributes a DC component, which gives the DC offset. Why is the DC offset harmful in a direct-conversion receiver? Now the input signals to the input port (the output of the RF amplifer) of the mixer for the received signal path, including the self-mixing effects will be:

$$V_{in} = V_{RF} \cos \omega t + \sum_i V_{RF}^i \cos(\omega + \Delta_i)t + V_{inter} \cos \omega_{inter} t + V_{leakage} \cos \omega t \tag{7.38}$$

And the signal from the LO is

$$V_L = V_{LO} \cos \omega t + \gamma V_{inter} \cos \omega_{inter} t \tag{7.39}$$

where ω_{inter} is the angular frequency of the received strong interferer, while factor γ is the degree of the reflection of the VCO. Mixing the two signals, which can be expressed mathematicallv as:

$$\begin{aligned} V_{out} &= V_{in} V_L \\ &= \frac{1}{2}[V_{LO}(V_{RF} + V_{leakage}) + \gamma V_{inter}^2] \\ &+ \frac{1}{2}\sum_i V_{LO} V_{RF}^i \cos \Delta_i t + \text{others} \end{aligned} \tag{7.40}$$

where Δ_i are the angular frequencies of the signalsin the desired band with the related amplitudes. The DC offset is the first term in Eq. (7.40). Statistically, the values of V_{RF} and V_{RF}^i are comparable, and the DC offset affects the digital processing in the baseband.

Assume the isolation of the substrate is β, the coefficients of the LNA and the mixer are α, and

$$V_{LO} = g_{LO} V_{RF}, \text{ and } V_{inter} = g_{int} V_{RF}$$

We can define a DC-offset ratio (DCOR) as the ratio of the DC offset to the wanted signal magnitude:

$$DCOR = \frac{\frac{1}{2}V_{LO}[V_{RF} + V_{leakage}] + \frac{1}{2}\gamma V_{inter}^2}{\frac{1}{2}V_{LO}V_{RF}} = 1 + \alpha\beta\left[g_{LO} + \frac{g_{int}^2}{g_{LO}}\right] \tag{7.41}$$

If the signal itself contains a DC component from the coding systems, the DCOR will be

$$DCOR = \frac{\frac{1}{2}V_{LO}[V_{leakage}] + \frac{1}{2}\gamma V_{inter}^2}{\frac{1}{2}V_{LO}V_{RF}} = \alpha\beta\left[g_{LO} + \frac{g_{int}^2}{g_{LO}}\right] \tag{7.42}$$

Assume that the parameters g_{LO} and g_{int} are equal, Eq. (7.42) reads:

$$DCOR(\text{dB}) = 6 + (g + \alpha + \beta)\,|_{\text{dB}} \tag{7.43}$$

Figure 7.30 shows the DC offset ratio compared to the wanted carrier in dBc for g=0 dB, namely, $V_{LO}=V_{RF}$ and $V_{inter}=V_{RF}$. If g is not zero, the value of g simply add into the figure, e.g. if g=40 dBc typically, we have Fig. 7.31. Figure 7.31 indicates that if we expect –60 dBc DC offset ratio, the substrate isolation must better than –76 dB and the reflection must be better than –26 dB. Normally, the isolation and the reflection could be as poor as –40 dB and –10 dB, then the DCOR is only around –3 dBc!

Whether or not the DC offset will desensitize the receiver depends on the modulation system. It is clear that AC coupling after the mixer is preferable in order to eliminate the DC offset. Modulations such as frequency shift-keying (FSK) used in paging applications typically have signal energy away from the DC; therefore DC offset can be removed by using a capacitor just after the mixer. Some of other modulation schemes, such as Gaussian-filtered minimum-shift keying (GMSK), quadrature phase-shift keying (QPSK), and so on, have DC peaks. That means the DC component is the major part of the modulation signal, while the DC offset directly adds to the DC component of the spectrum. Therefore, one cannot use capacitice coupling to remove the DC offset in the baseband signal paths. However, digital DC compensation could be employed. From Eqs. (7.42) and (7.43) it is known that the major contributions of DC offset are from the LO leakage and the strong interferer. Because the interfer signal is

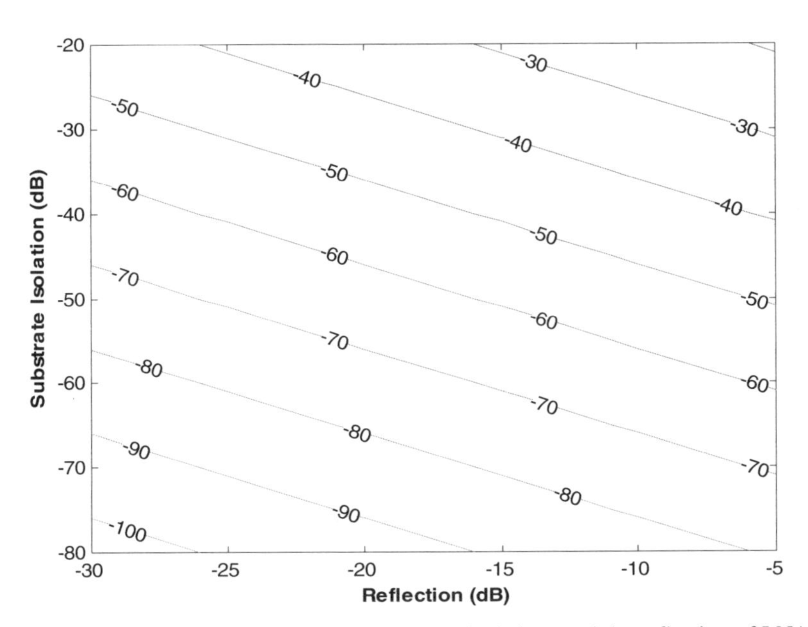

Fig. 7.30 DCOR in dBc for different substrate isolations and the reflection of LNA/LO for g=0 dBc

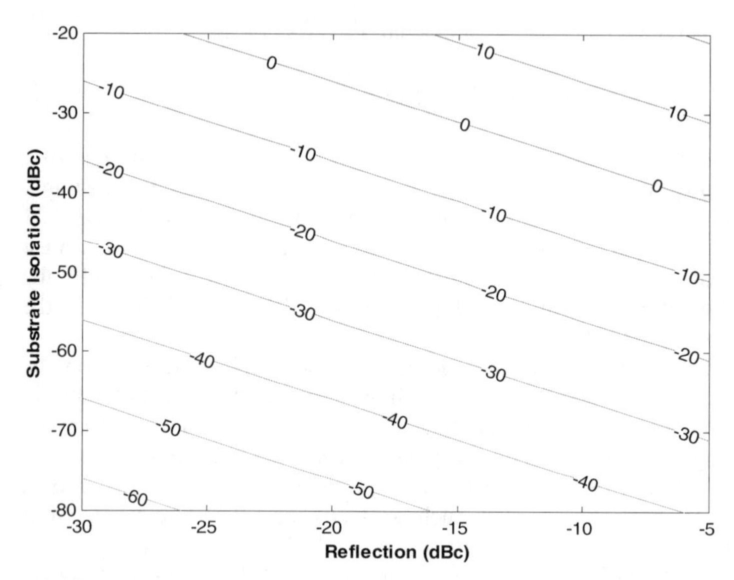

Fig. 7.31 DCOR in dBc for different substrate isolations and the reflection of LNA/LO for g=40 dBc

dymanic it cannot be controlled in the design. The self-mixing by the LO leakage can be compensated through so-called "pre-measurement," because it is independent of the received signal and depends only on the LO signal and the isolation as well as the reflection of the LNA/oscillator output port. If there is no received signal in the input port of the receiver, after direct conversion the DC output of the mixer will be the DC offset contributed by the LO leakage self-mixing. This DC offset is measured and stored in a memory digitally, then in normal operation the actual DC output of the mixer minus the memorized DC offset value removes the DC offset created by the LO leakage. However, the DC offset of the strong interferer cannot be removed.

The noise of the architecture is dominated by the flicker noise rather than the thermal noise, which is the dominant noise for other architectures, because the output of the architecture is baseband. Therefore, the SNR will be lower. Particular attention must be paid to suppressing the flicker noise.

If only one path is employed as given in Fig. 7.28 in a direct-conversion receiver, the mixing process is just equivalent to multiplying the RF signal with the LO signal as shown in Fig. 7.32. Where Δ is the channel width defined in the used wireless systems, e.g. see Table 7.6. After direct downconversion, the desired channel and the adjacent channel are entirely overlapped. This adjacent channel is also called the image -channel. Effort must be made to suppress the imagininary-channel with respect to the LO/RF frequency. The quadrature structure can be employed to suppress the adjacent channel interference effect Fig. 7.33, and detailed discussion can be found in [4, 17]. Of course, the mismatch between the I and Q paths is another problem for I/Q direct-conversion [4, 14, 17, 18], but it is not crucial as the DC offset.

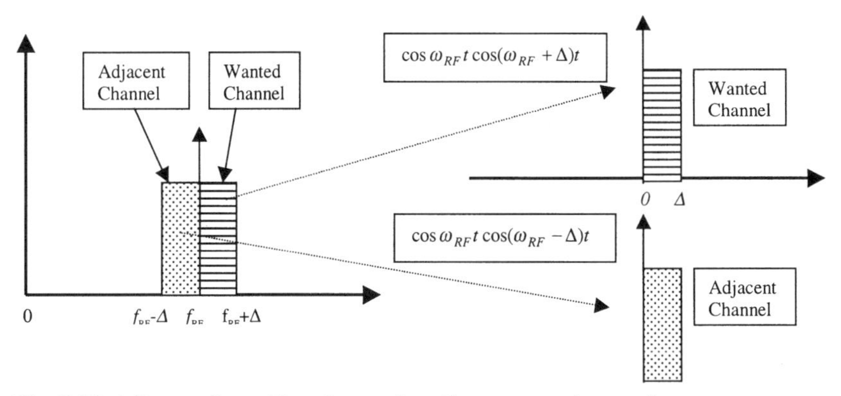

Fig. 7.32 Adjacent channel interference in a direct-conversion receiver

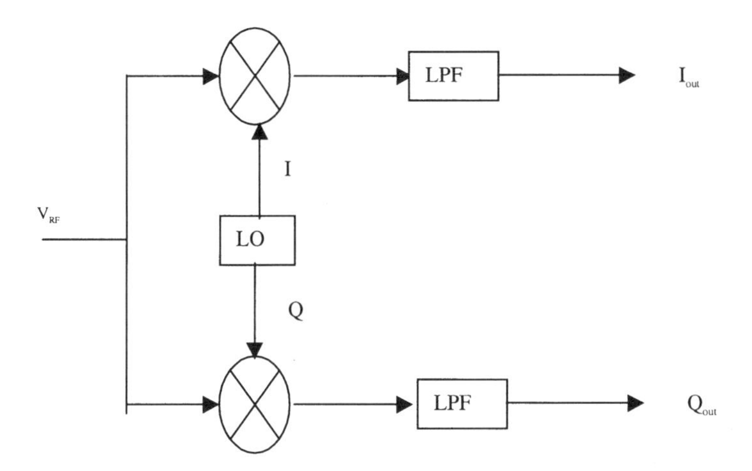

Fig. 7.33 Quadrature direct conversion

7.6.2 Low-IF Architecture and Superheterodyne Architecture

It is obvious that direct conversion is very simple, but its fatal disadvantage is the DC offset because LO=RF. If LO is not the same as RF, clearly the DC offset problem will no longer exist. However, the big problem for this architecture is the image-frequency rejection [4, 5, 12], because the image frequency also falls in the bandwidth. A high-speed wideband AD converter is required after the receiver. Using I/Q paths similar to those in Fig. 7.33 can suppress the image-frequency problem [4, 13, 21, 22, 24].

If the difference between the LO and the RF increases further, the image-frequency problem can be eliminate totally, which is the standard superheterodyne architecture. More downconversion stages must be used in this architecture because the IF is far from the baseband frequency. Also more nonintegratable external components have to be used and the power consumption will be significant.

The detailed discussion regarding to these two architectures can be found in any text books (e.g [4, 5]) and is summerized in Table 7.7.

It can be concluded that the selection of the architecture depends strongly on the required performance and the degree of integration. There is a trade-off among the simplicity, the number of external components, and the power-consumption. As a compromise, the low-IF architecture would be a good choice if a high-speed, high-dynamic range and higher sampling frequency AD converter is available [12].

Table 7.7 Comparisons of the three architectures

Architecture	Pros	Contras
Direct-conversion	• No IF filtering required • Eliminates many off-chip components • On-chip filtering at baseband • No image frequency problem • On-chip filtering facilitates multimode, multiband and multistandard operation • Easier to be SoC • Low power consumption	• LO leakage to mixer or antenna (radiation) • Self-mixing • Time-varying DC offset from the above processing • DC offset from baseband I/Q-circuit offsets • Reduced dynamic range from second-order IM and flicker noise • High-Q synthesizer is still needed
Low-IF	• No DC-offset problems • Eliminates many high-Q off-chip components as in DCR • Possible on-chip filtering at baseband • Lower-power consumption	• Image frequency problem • Lower image-suppression ratio on silicon • Require high performance ADC with respect to bandwidth, resolution and the speed • Difficult in multistandards operation
Super heterodyne	• High selectivity • High sensitivity • Good SNR	• High-Q (discrete component like SAW) filters needed • Full integration impossible • High power consumption

7.7 Conclusion

In this article the system-level considerations of RF IC design for wireless applcations are investigated systematically. The design considerations and the pros and cons of the RF receivers are given.

References

[1] Sene G. (2001) Riding the 3G wireless revolution. International High-Tech Forum 2001, 6-8 June, 2001, Zhuhai, China

[2] Larson L. L. (1998) Integrated circuit technology options for RFIC's – Present status and future directions. IEEE J Solid-State Circuits 33:387–399

[3] Do M.A., Zhao R., Yeo K.S., Ma J.-G. (2001). Fully integrated 10 GHz CMOS VCO. Electronics Letters 37:1021–1023

[4] Razavi B. (1998) RF Microelectronics. Prentice-Hall, Upper Saddle River, New Jersey

[5] Chang K. (2000) RF and Microwave Wireless Systems. John Wiley, New York

[6] Friis H. T. (1944) Noise figure of radio receivers. Proc IRE 32:419–422

[7] Brannon B., Cloninger, C. (2001) Redefining the role of ADCs in wireless. Applied Microwave & Wireless 13 (3):94–109

[8] Domino W., Agahi D. (2001) Polynomial model of blocker effects on LNA/mixer devices. Applied Microwave & Wireless 13:30–44

[9] Abou-Allam E., Nisbet J.J., Maliepaard M. C. (2001) Low-voltage 1.9-GHz front-end receiver in 0.5-µm CMOS technology. IEEE J Solid-State Circuits 36:1434–1443

[10] Ryynänen J., Kivekäs K., Jussila J., Pärssinen A. Halonen A.I. (2001) A dual-band RF front-end for WCDMA and GSM applications. IEEE J Solid-State Circuits 36:1198–1204

[11] Yamamoto K., Heima T., Furukawa A., Ono M., Hashizume Y., Komurasaki H., Maeda S., Sato H. Kato N. (2001) A 2.4-GHz-Band 1.8-V operation single-chip Si-CMOS T/R_MMIC front-end with a low insertion loss switch. IEEE J Solid-State Circcits 36:1186– 1197

[12] Droinet Y. (2001) Advanced RF technologies for the wireless market. Microwave Journal 44:148–159

[13] Rudell J., Ou J., Cho T. B., Chien G., Brianti F., Weldon J.A. Gray P.R. (1997) A 1.9-GHz wide-band IF double conversion CMOS receiver for cordless telephone applications. IEEE J Solid-State Circuits 32:2071–2068

[14] Abidi A.A. (1995) Direct-conversion radio transceivers for digital communications. IEEE J Solid-State Circuits 30:1399–1410

[15] Piazza P., Huang Q. (1998) A 1.57-GHz RF front-end for triple conversion GPS receiver. IEEE J Solid-State Circcits 33:202–209

[16] Rofougaran A., Chang J. Y.-C., Rofougaran M. Abidi A.A. (1996) A 1 GHz CMOS RF front-end IC for a direct-conversion wireless receiver. IEEE J Solid-State Circuits 31:880–889

[17] Mashhour A., Domino W. Beamish N. (2001) On the direct conversion receiver – A tutorial. Microwave J. 44:114–128

[18] Razavi B. (1997) Design considerations for direct-conversion receivers. IEEE Trans. On Circuits and Systems-II: Analog and Digital Signal Processing 44:428–435

[19] Wu S. Razavi B. (1998) A 900-MHz/1.8-GHz CMOS receiver for dual-band applications. IEEE J Solid-State Circuits 33:2178–2185

[20] Stetzler T.D., Post I.G., Havens J.H. Koyama M. (1995) A 2.7-4.5 V single chip GSM transceiver RF integrated circuit. IEEE J Solid-State Circuits 30:1421–1429

[21] Crols J. Steyaert M.S. (1995) A single-chip 900 MHz CMOS receiver front-end with a high performance low-IF topology. IEEE J Solid-State Circuits 30:1483–1492

[22] Razavi B. A 5.2-GHz CMOS receiver with 62-dB image rejection. IEEE J Solid-State Circuits 36:810–815

[23] Maligeorgos J.P. Long J.R. (2000) A low-voltage 5.1–5.8-GHz image-rejection receiver with wide dynamic range. IEEE J Solid-State Circuits 35:1917–1926

[24] Behbahani F., Leete J.C., Kishigami Y., Roithmeier A., Hoshino K. Abidi A.A. (2000) A 2.4-GHz low-IF receiver for wideband WLAN in 0.6-μm CMOS – Architecture and front-end. IEEE J Solid-State Circuits 35:1908–1916
[25] Steyaert M.S., Janssens J., Muer B.D., Borremans M. Itoh N. (2000) A 2-V CMOS cellular transceiver front-end. IEEE J Solid-State Circuits 35:1895–1907
[26] Colebrook F.M. (1924) Homodyne. Wireless World and Radio Rev. 13:774
[27] Tucker D.G. (1947) The synchrodyne. Electronic Eng. 19:75–76
[28] Tucker D.G. (1954) The history of the homodyne and the synchrodyne. J British Institute of Radio Eng. 14:143–154